STARBURSTS

ASTROPHYSICS AND SPACE SCIENCE LIBRARY

VOLUME 329

EDITORIAL BOARD

Chairman

W.B. BURTON, National Radio Astronomy Observatory, Charlottesville, Virginia, U.S.A.
(bburton@nrao.edu); University of Leiden, The Netherlands (burton@strw.leidenuniv.nl)

Executive Committee

J. M. E. KUIJPERS, *University of Nijmegen, The Netherlands*
E. P. J. VAN DEN HEUVEL, *University of Amsterdam, The Netherlands*
H. VAN DER LAAN, *University of Utrecht, The Netherlands*

MEMBERS

J. N. BAHCALL, *The Institute for Advanced Study, Princeton, U.S.A.*
F. BERTOLA, *University of Padua, Italy*
J. P. CASSINELLI, *University of Wisconsin, Madison, U.S.A.*
C. J. CESARSKY, *European Southern Observatory, Garching bei München, Germany*
O. ENGVOLD, *University of Oslo, Norway*
A. HECK, *Strasbourg Astronomical Observatory, France*
R. McCRAY, *University of Colorado, Boulder, U.S.A.*
P. G. MURDIN, *Institute of Astronomy, Cambridge, U.K.*
F. PACINI, *Istituto Astronomia Arcetri, Firenze, Italy*
V. RADHAKRISHNAN, *Raman Research Institute, Bangalore, India*
K. SATO, *School of Science, The University of Tokyo, Japan*
F. H. SHU, *University of California, Berkeley, U.S.A.*
B. V. SOMOV, *Astronomical Institute, Moscow State University, Russia*
R. A. SUNYAEV, *Space Research Institute, Moscow, Russia*
Y. TANAKA, *Institute of Space & Astronautical Science, Kanagawa, Japan*
S. TREMAINE, *Princeton University, U.S.A.*
N. O. WEISS, *University of Cambridge, U.K.*

STARBURSTS

From 30 Doradus to Lyman Break Galaxies

Edited by

RICHARD DE GRIJS

The University of Sheffield, UK

and

ROSA M. GONZÁLEZ DELGADO

Instituto de Astrofísica de Andalucía (CSIC),
Granada, Spain

A C.I.P. Catalogue record for this book is available from the Library of Congress.

ISBN-10 1-4020-3538-1 (HB) Springer Dordrecht, Berlin, Heidelberg, New York
ISBN-10 1-4020-3539-X (e-book) Springer Dordrecht, Berlin, Heidelberg, New York
ISBN-13 978-1-4020-3538-8 (HB) Springer Dordrecht, Berlin, Heidelberg, New York
ISBN-13 978-1-4020-3539-5 (e-book) Springer Dordrecht, Berlin, Heidelberg, New York

Published by Springer,
P.O. Box 17, 3300 AA Dordrecht, The Netherlands.

Cover illustration:
NASA/ESA Hubble Space Telescope panoramic portrait of the 30 Doradus Nebula, the most intense star-forming region that can be resolved into individual stars. 30 Doradus, located in the Large Magellanic Cloud, contains in its centre R136, the most spectacular starburst cluster of massive stars in the Local Group of galaxies. The mosaic picture shows that ultraviolet radiation and high-speed material unleashed by the stars in the cluster, are weaving a tapestry of creation and destruction, triggering the collapse of looming gas and dust clouds and forming pillar-like structures that are incubators for nascent stars.

Credit:
NASA/ESA, N.Walborn and J. Maíz-Apellániz (Space Telescope Science Institute, USA), R. Barba (La Plata Observatory, Argentina).

Printed on acid-free paper

springeronline.com

All Rights Reserved
© 2005 Springer
No part of this work may be reproduced, stored in a retrieval system, or transmitted in any form or by any means, electronic, mechanical, photocopying, microfilming, recording or otherwise, without written permission from the Publisher, with the exception of any material supplied specifically for the purpose of being entered and executed on a computer system, for exclusive use by the purchaser of the work.

Printed in the Netherlands.

Contents

Conference Participants   xxi
Preface   xxvii

Session I: Local Starbursts as Benchmarks for Galaxy Evolution   1

Local Starbursts in a Cosmological Context   3
    Timothy M. Heckman

Starbursts in the Evolving Universe: A Local Perspective   11
    John S. Gallagher, III

Are there local analogs of Lyman break galaxies?   17
    James D. Lowenthal, R. Nick Durham, Brian J. Lyons, Matthew A. Bershady, Jesus Gallego, Rafael Guzmán, and David C. Koo

The disk Wolf-Rayet population of the nuclear starburst galaxy M83   21
    Paul A. Crowther, Lucy J. Hadfield, Hans Schild, and Werner Schmutz

Laser Illuminates Compact Galaxies   27
    J. Melbourne

First Spectroscopic Results from the Spitzer Infrared Nearby Galaxies Survey   31
    John-David T. Smith, Robert C. Kennicutt, and the SINGS Team

Near-Infrared Super Star Clusters in Starburst and Luminous Infrared Galaxies   35
    Almudena Alonso-Herrero

High-resolution imaging of the SSCs in NGC 1569 and NGC 1705   41
    M. Sirianni, G. Meurer, N. Homeier, M. Clampin, R. Kimble, and the ACS Science Team

Massive Star Clusters, Feedback, and Superwinds   45
    Andrea M. Gilbert and James R. Graham

30 Doradus – a Template for "Real Starbursts"? 49
Bernhard R. Brandl

Session II: The Initial Mass Function in Starburst Regions: Environmental Dependences? 55

The Initial Mass Function in Starbursts 57
Bruce G. Elmegreen

Dynamical models of star formation and the initial mass function 65
Ian A. Bonnell

Red supergiants, mass segregation and M/L ratios in young star clusters 71
Ariane Lançon and Christian Boily

IMF Variation in M82 Super Star Clusters 75
Nate McCrady, James R. Graham, and William D. Vacca

Session III: Starbursts as a Function of Wavelength 79

Colourful starbursts 81
Jürgen Knödlseder

A Far-Ultraviolet View of Starburst Galaxies 89
Claus Leitherer

Local Starbursts: Perspectives from the Optical 97
Daniela Calzetti

The starburst phenomenon from the optical/near-IR perspective 103
Nils Bergvall, Thomas Marquart, Göran Östlin, and Erik Zackrisson

Dissecting starburst galaxies with infrared observations 109
Paul P. van der Werf and Leonie Snijders

What fraction of stars formed in infrared galaxies at high redshift? 115
Neil Trentham

SHADES: The Scuba HAlf Degree Extragalactic Survey 121
James S. Dunlop

Compact Extragalactic Star Formation: Peering Through the Dust at Centimeter Wavelengths 127
James S. Ulvestad

Contents vii

Dust attenuation and star formation in the nearby Universe 133
    Véronique Buat and Jorge Iglesias-Páramo

Theoretical Pan-Spectral Energy Distributions of Starburst Galaxies 137
    Michael A. Dopita

Session IV: Triggering and Quenching of Starbursts and the Effects of Galactic Interactions   141

Merger-Induced Starbursts 143
    François Schweizer

Galaxy Collisions: Modeling Star Formation in Different Environments 153
    Chris Mihos

Star and cluster formation in extreme environments 157
    Richard de Grijs

The Recurrent Nature of Central Starbursts 163
    Curtis Struck

Efficiency of the dynamical mechanism 167
    F. Combes

2D kinematics and mass derivations in ULIRGs 173
    M. García-Marín, L. Colina, S. Arribas, and A. Monreal

Internal Kinematics of LCBGs 177
    Matthew A. Bershady, M. Vils, C. Hoyos, R. Guzmán, and D.C. Koo

Fuelling starbursts and nuclear rings 181
    Johan H. Knapen

Session V: Star-Formation Rates in Relation to the Host Galaxy Properties   185

Starburst Galaxy Demographics 187
    Robert C. Kennicutt, Jr., Janice C. Lee, Jose G. Funes, S.J., Shoko Sakai, and Sanae Akiyama

Star-forming, recently star-forming, and "red and dead" galaxies at $1 < z < 2$ 195
    Roberto G. Abraham, Karl Glazebrook, Patrick J. McCarthy, David Crampton, Sandra Savaglio, Stephanie Juneau, Damien Le Borgne, Hsiao-Wen Chen, Raymond G. Carlberg, Richard Murowinski, Inger Jørgensen, Kathy Roth, and Ron Marzke

Global Star-Formation Rates 201
Joseph Silk

Star Formation Efficiencies and Star Cluster Formation 209
Uta Fritze – v. Alvensleben

Star Cluster Populations in Nearby Starburst Galaxies 215
Jason Harris, Daniela Calzetti, Denise A. Smith, John S. Gallagher, III, and Christopher J. Conselice

Young Massive Clusters in Non-interacting Galaxies 219
Søren S. Larsen, Jean P. Brodie, Deidre A. Hunter, and Tom Richtler

Nascent starbursts in synchrotron-deficient galaxies 223
Hélène Roussel, George Helou, James Condon, and Rainer Beck

HST/STIS Spectroscopy of the Starburst Core of M82 227
L.J. Smith, M.S. Westmoquette, J.S. Gallagher, R.W. O'Connell, and R. de Grijs

Session VI: Starburst Tracers: Gas, Dust and Star Formation 231

Starburst Galaxies: an Infrared Perspective 233
Natascha M. Förster Schreiber

Dusty starbursts as a standard phase in galaxy evolution 241
David Elbaz

Is the interstellar gas of starburst galaxies well mixed? 247
Vianney Lebouteiller and Daniel Kunth

Star Clusters in M51: Connection between Molecular Gas, Stars and Dust 251
Eva Schinnerer, Axel Weiß, Susanne Aalto, Nicolas Z. Scoville, Michael P. Rupen, Robert C. Kennicutt, Rainer Beck, and Andrew Fletcher

Session VII: Starbursts at Intermediate Redshifts and the Starburst versus AGN Paradigm 255

Starbursts in nearby radio galaxies 257
Clive Tadhunter

Starbursts in Low Luminosity Active Galactic Nuclei 263
Rosa M. González Delgado and Roberto Cid Fernandes

GALEX UltraViolet Spectroscopy of Luminous InfraRed Galaxies 269
Denis Burgarella, Véronique Buat, Todd Small, Tom A. Barlow, and the GALEX Team

A recent rebuilding of most spirals? 273
François Hammer, Hector Flores, Xianzhong Zheng, and Yanchun Liang

Evolution of the IR energy density and SFH up to $z \sim 1$: first results from MIPS 279
E. Le Floc'h, C. Papovich, H. Dole, E. Egami, P. Pérez-González, G. Rieke, M. Rieke, E. Bell, and the Spitzer/MIPS GTO team

Session VIII: Violent Star Formation and the Properties of Star-Forming Galaxies at High Redshift 283

Starbursts from redshift $z \sim 3$ to 7–10   CD-ROM: T1
Daniel Schaerer

Understanding Infrared–Luminous Starbursts in Distant Galaxies 285
C. Papovich, E. Le Floc'h, H. Dole, E. Egami, P. Pérez-González, G. Rieke, and M. Rieke (for the Spitzer/MIPS GTO Team)

Starbursts in the Ultra Deep Field 289
Rodger I. Thompson

Properties of Lyα and Gamma Ray Burst-selected starbursts at high redshifts 293
Johan P.U. Fynbo, Brian Krog, Kim Nilsson, Gunnlaugur Björnsson, Jens Hjorth, Páll Jakobsson, Cédric Ledoux, Palle Møller, and Bjarne Thomsen

High-z Galaxies in the FORS Deep Field 299
D. Mehlert, C. Tapken, I. Appenzeller, S. Noll, D. de Mello, and T.M. Heckman

Star-Forming Galaxies at $z \sim 2$: Stellar and Dynamical Masses 303
Dawn K. Erb, Charles C. Steidel, Alice E. Shapley, Max Pettini, Naveen A. Reddy, and Kurt L. Adelberger

Metallicity of Star-Forming Galaxies 307
Lisa Kewley and Henry A. Kobulnicky

New Metallicity Diagnostics for High-Redshift Star-Forming Galaxies 311
Samantha A. Rix, Max Pettini, Claus Leitherer, Fabio Bresolin, Rolf-Peter Kudritzki, and Chuck C. Steidel

UV Luminosity Function at $z \sim 4, 3$, and 2 315
Marcin Sawicki and David Thompson

Massive galaxies at $z = 2$ 319
  Kentaro Nagamine, Renyue Cen, Lars Hernquist, Jeremiah P. Ostriker,
  and Volker Springel

K-luminous galaxies at $z \sim 2$ 323
  Duília de Mello and the K20 Team

Resolved Molecular Gas Emission in a QSO host galaxy at $z = 6.4$ 327
  Fabian Walter

Session IX: Conference Summary 331

Conference Summary: Starbursts and Galaxy Evolution 333
  Robert W. O'Connell

Poster Contributions CD-ROM

Stellar and gas kinematics in the core and bar regions of M100 P1
  E.L. Allard, J.H. Knapen, and R.F. Peletier

Metals in the neutral interstellar medium of starburst galaxies P2
  A. Aloisi, T. Heckman, C.G. Hoopes, C. Leitherer, S. Savaglio, and
  K.R. Sembach

Measuring sizes and compactnesses of young star clusters P3
  P. Anders, M. Gieles, R. de Grijs, and U. Fritze–v. Alvensleben

The complex star-formation history of NGC 1569 P4
  L. Angeretti, M. Tosi, L. Greggio, E. Sabbi, A. Aloisi, and C. Leitherer

Recent imaging results from SINGS P5
  G.J. Bendo, R.C. Kennicutt, L. Armus, D. Calzetti, D.A. Dale, B.T.
  Draine, C.W. Engelbracht, K.D. Gordon, A.D. Grauer, G. Helou, D.J.
  Hollenbach, T.H. Jarret, L.J. Kewley, C. Leitherer, A. Li, S. Malhotra,
  M. Meyer, E. Murphy, M.W. Regan, G.H Rieke, M.J. Rieke, H. Roussel,
  K. Sheth, J.-D.T. Smith, M.D. Thornley, and F. Walter

Galactic winds and transport of metals into the IGM in semi-analytic simulations P6
  S. Bertone, F. Stoehr, and S.D.M. White

High-redshift galaxy evolution P7
  R. Bouwens, G. Illingworth, R. Thompson, the UDF NICMOS Team,
  and the ACS GTO Science Team

Spitzer imagery of embedded ultra-young star clusters in M33 P8
  B. Buckalew, H. Kobulnicky, R. Gehrz, C.E. Woodward, M. Ashby, P.

Barmby, B. Brandl, N. Devereux, C. Engelbracht, G. Fazio, K. Gordon, J. Hinz, R. Humphreys, K. Misselt, M. Pahre, P. Pérez-González, E. Polomski, G. Rieke, T. Roellig, J. van Loon, and S. Willner

Starbursts in dwarf galaxies: A multi-wavelength case study of NGC 625   P9
J.M. Cannon

A multi-wavelength study of the starburst galaxy NGC 7673   P10
P. Castangia, A. Pasquali, and P. Benvenuti

Gamma-ray bursts in starburst galaxies   P11
L. Christensen, J. Hjorth, and J. Gorosabel

Magnetic fields and starbursts: from irregulars to mergers   P12
K.T. Chyży

Westerlund 1: A super star cluster in the Milky Way   P13
J.S. Clark, I. Negueruela, P.A. Crowther, S. Goodwin, and L.J. Hadfield

Star formation in close pairs selected from the Sloan Digital Sky Survey   P14
H. Cullen, B. Nikolic, and P. Alexander

Evolution of the chemical properties of galactic systems in hierarchical universes   P15
M.E. De Rossi, P.B. Tissera, and C. Scannapieco

Search for $z \sim 5$ galaxies   P16
L. Douglas, M. Bremer, and M. Lehnert

An ultraviolet spectral library of metal-poor OB stars   P17
C.J. Evans, D.J. Lennon, N.R. Walborn, C. Trundle, and S.A. Rix

A possible formation scenario for the heavy-weight young cluster W3 in NGC 7252   P18
M. Fellhauer and P. Kroupa

Integral-field spectroscopy and *HST* imaging of ULIRGs at low and high redshift: IRAS 16007+3743   P19
M. García Marín, L. Colina, and S. Arribas

A kinematic study of the nuclear stellar population in Seyfert galaxies   P20
A. García-Rissmann, R. Cid Fernandes, N. do Vale Asari, L.R. Vega, H. Schmitt, and R. González Delgado

Tracing star formation in the X-ray: From ULIRGs to dwarf starbursts   P21
J. Grimes, T. Heckman, D. Strickland, and A. Ptak

Clustering of simulated galaxies at redshift 4   P22
A.G. Harford and N.Y. Gnedin

Does size matter (in the SFRs)? P23
  A. Hidalgo-Gámez

Star formation in intermediate-redshift cluster galaxies P24
  N.L. Homeier, R. Demarco, P. Rosati, M. Postman, J.P. Blakeslee, R.J.
  Bouwens, L.D. Bradley, H.C. Ford, T. Goto, C. Gronwall, B. Holden,
  G.D. Illingworth, M.J. Jee, A.R. Martel, S. Mei, F. Menanteau, A. Zirm,
  M. Clampin, G. Hartig, and the ACS Science Team

New frontiers opened by Chandra in cosmological studies of galaxies P25
  A.E. Hornschemeier, T. Heckman, C. Tremonti, B. Mobasher, B.
  Poggianti, F. Bauer, and D. Alexander

On the nature of the intermediate-redshift population of luminous compact blue
  galaxies P26
  C. Hoyos, R. Guzmán, A.I. Díaz, D. Koo, and M. Bershady

Constraints on Lyman continuum flux escaping from galaxies at $z \sim 3$ using VLT
  narrow-band photometry P27
  A.K. Inoue, I. Iwata, J.-M. Deharveng, V. Buat, and D. Burgarella

Near-infrared spectral properties of metal-poor red supergiants P28
  V.D. Ivanov, M.J. Rieke, A. Alonso-Herrero, and D. Alloin

Wide and deep survey of Lyman break galaxies at $z \sim 5$ P29
  I. Iwata, K. Ohta, N. Tamura, M. Ando, M. Akiyama, and K. Aoki

On the importance of the few most massive stars: The ionizing cluster of NGC
  588 P30
  L. Jamet, E. Pérez, C. Cerviño, G. Stasińska, R.M. González Delgado,
  and J.M. Vílchez

Non-isothermal gravoturbulent fragmentation: Effects on the IMF P31
  A.-K. Jappsen, R.S. Klessen, R.B. Larson, Y. Li, and M.-M. Mac Low

A study of the near-IR [SIII] lines in HII galaxies P32
  C. Kehrig, J.M. Vílchez, E. Pérez, E. Telles, and F. Cuisinier

Star (cluster) formation in 3D: Integral-field spectroscopy at the VLT P33
  M. Kissler-Patig

A Lyα emitter at $z = 6.5$ found with slitless spectroscopy P34
  J. Kurk, A. Cimatti, S. di Serego Alighieri, J. Vernet, A. Ferrara,
  E. Daddi, and B. Ciardi

Neutral ISM surrounding starburst regions P35
  V. Lebouteiller, D. Kunth, J. Lequeux, A. Aloisi, J.-D. Désert, A.
  Lecavelier des Etangs, G. Hébrard, A. Vidal-Madjar

Contents

The dwarf starburst galaxy NGC 1705: New HII region element abundances and reddening variations near the center P36
H. Lee and E. Skillman

The dwarf galaxy duty cycle: Measurements from a complete sample of the local volume P37
J.C. Lee, R.C. Kennicutt, S. Akiyama, J.G. Fuentes, and S. Sakai

Optical and near-IR luminosity-metallicity relations of star-forming emission-line galaxies P38
J.C. Lee, J. Salzer, and J. Melbourne

Physical properties of low-luminosity high-redshift lensed galaxies P39
M. Lemoine-Busserolle, T. Contini, R. Pelló, J.-F. Le Borgne, J.-P. Kneib, J. Richard, and C. Lidman

The luminosity-metallicity relation of distant luminous infrared galaxies P40
Y. Liang, F. Hammer, H. Flores, D. Elbaz, D. Marcillac, L. Deng, and C. Cesarsky

New tools for the tracing of ancient starbursts: Analysing globular cluster systems using Lick indices P41
T. Lilly, U. Fritze–v. Alvensleben, and R. de Grijs

The rest-frame FUV morphologies of star-forming galaxies at $z \sim 4$ and $z \sim 1.5$ P42
J. Lotz, P. Madau, J. Primack, and M. Giavalisco

Oxygen abundances, SFRs and dust of CFRS galaxies at intermediate redshift P43
C. Maier, S. Lilly, and M. Carollo

NGC 604: The scaled OB association (SOBA) prototype. Spatial distribution of the different gas phases and attenuation by dust P44
J. Maíz-Apellániz, E. Pérez, and J.M. Mas-Hesse

Star-formation history of distant luminous infrared galaxies P45
D. Marcillac, D. Elbaz, S. Charlot, Y. Liang, F. Hammer, H. Flores, and C. Cesarsky

Evolution of thermally conducting clouds embedded in a galactic wind P46
A. Marcolini, A. D'Ercole, D. Strickland, and T. Heckman

The dynamics of emission-line galaxies from new Fabry-Perot observations P47
T. Marquart, N. Bergvall, G. Östlin, P. Amram, J. Masegosa, J. Boulesteix, E. Zackrisson, J.-L. Gach, and P. Balard

Secular evolution of stellar bars, vertical instabilities and starbursts P48
I. Martínez-Valpuesta and I. Shlosman

Enhanced tidally-induced starbursts in cluster galaxies: Evidence for transformation of spirals to S0s by gravitational interactions associated with cluster virialisation P49
C. Moss

NGC 1569: A dwarf galaxy with a giant starburst P50
S. Mühle, U. Klein, S. Hüttemeister, and E.M. Wilcots

The unusual tidal dwarf galaxy in the merger system NGC 3227/6. Star formation in a tidal shock? P51
C.G. Mundell, P. James, N. Loiseau, E. Schinnerer, and D. Forbes

Blue compact galaxies in the Ultra Deep Field P52
K. Noeske, D. Koo, P. Papaderos, J. Melbourne, A. Gil de Paz, A. Phillips, C. Willmer, and the Deep Team

Exploring galaxy evolution at high redshift P53
S. Noll, D. Mehlert, I. Appenzeller, and the FDF Team

On the X-ray contribution from young supernovae in starbursts P54
L. Norci and E.J.A. Meurs

ISO observations of Markarian 297 P55
B. O'Halloran

Chemical abundances in starburst galaxies: M82 and NGC 253 P56
L. Origlia, P. Ranalli, R. Maiolino, and A. Comastri

The temperature distribution of dense gas in starburst cores P57
J. Ott, A. Weiß, C. Henkel, and F. Walter

The photometric structure of young blue compact dwarf (BCD) candidates P58
P. Papaderos, Y.I. Izotov, K.G. Noeske, N.G. Guseva, T.X. Thuan, and K.J. Fricke

Luminous compact blue galaxies in the local Universe: A key reference for high-redshift studies P59
J. Pérez Gallego, R. Guzmán, F.J. Castander, C.A. Garland, and D.J. Pisano

Ionising stellar populations in circumnuclear star-forming regions P60
E. Pérez-Montero, A.I. Díaz, and M. Castellanos

Mid-ultraviolet spectral templates for single-age, single-metallicity systems P61
R.C. Peterson, B.W. Carney, B. Dorman, E.M. Green, W. Landsman, J. Liebert, R.W. O'Connell, R.T. Rood, and R.P. Schiavon

Compact galaxies in the GOODS-N field P62
 A. Phillips, J. Melbourne, D. Koo, and K. Noeske

Optical and near-IR properties of submillimetre galaxies in the GOODS-N field P63
 A. Pope, D. Scott, C. Borys, C. Conselice, M. Dickinson, and B. Mobasher

The X-ray number counts and luminosity function of galaxies P64
 P. Ranalli, A. Comastri, and G. Setti

X-ray inferred bolometric star-formation rates of UV-selected galaxies at $z \sim 2$ P65
 N. Reddy, C. Steidel, and D. Erb

Star-forming galaxies in a nearby group: Abell 634 P66
 D. Reverte-Payá, J.M. Vílchez, and J. Iglesias-Páramo

Chemical evolution of late-type dwarf galaxies P67
 D. Romano, M. Tosi, and F. Matteucci

The 2–850 μm SED of star-forming galaxies P68
 A. Sajina, D. Scott, M. Dennefeld, H. Dole, M. Lacy, and G. Lagache

Star-formation histories of $z \leq 0.25$ galaxies from GALEX P69
 S. Salim and the GALEX Science Team

Multi-wavelength star-formation indicators P70
 H.R. Schmitt, D. Calzetti, L. Armus, K.D. Gordon, T.M. Heckman, R.C. Kennicutt, C. Leitherer, and G.R. Meurer

High-resolution near-UV imaging of Seyfert galaxies P71
 H.R. Schmitt, R.M. González Delgado, R. Cid Fernandes, T. Storchi-Bergmann, T.M. Heckman, and C. Leitherer

Integral-field spectroscopy in the IR: Gemini-CIRPASS observations and star-formation history in the nucleus of M83 P72
 R. Sharp, S. Ryder, J. Knapen, L. Mazzuca, and I. Parry

Extended tidal structure in two Lyα-emitting starburst galaxies P73
 E.D. Skillman, J.M. Cannon, D. Kunth, C. Leitherer, M. Mas-Hesse, G. Östlin, and A. Petrosian

Near-infrared properties of starbursts at $z = 6$ P74
 E. Stanway, R. McMahon, and A. Bunker

Are gamma-ray bursts good star-formation indicators? P75
 N. Tanvir

Tracing the inner gas disks in starbursts and radio galaxies G. Taylor	P76
Parametric relations of local HII galaxies E. Telles and V. Bordalo	P77
Star formation in three nearby galaxy systems S. Temporin, S. Ciroi, A. Iovino, E. Pompei, M. Radovich, and P. Rafanelli	P78
Far-Ultraviolet imaging of the Hubble Deep Field North H.I. Teplitz, T.M. Brown, C. Conselice, D.F. de Mello, M.E. Dickinson, H.C. Ferguson, J.P. Gardner, M. Giavalisco, and F. Menanteau	P79
What shuts off star formation? Post-burst galaxies from $z = 0$ to 1 C. Tremonti, R. Kennicutt, and T. Heckman	P80
The $z \sim 1.5$ Tully-Fisher relation L. van Starkenburg, P. van der Werf, L. Yan, and A. Moorwood	P81
The dust in Lyman break galaxies U.P. Vijh, A.N. Witt, and K.D. Gordon	P82
Implications for the formation of star clusters from extragalactic star-formation rates C. Weidner and P. Kroupa	P83
Atomic carbon and CO at redshift 2.5 A. Weiß, D. Downes, C. Henkel, and F. Walter	P84
Revealing the complex structure of the M82 superwind M.S. Westmoquette, J.S. Gallagher, III, and L.J. Smith	P85
Modelling the Red Halos of Blue Compact Galaxies E. Zackrisson, N. Bergvall, T. Marquart, L. Mattsson, and G. Östlin	P86
The enigmatic Local Group starburst galaxy IC 10 D. Zucker	P87
Memorable Quotes	341
Author Index	343
Object Index	349

xviii   *Starbursts – From 30 Doradus to Lyman Break Galaxies*

Conference Photograph

Front row –

1, N. Trentham; 2, R. Terlevich; 3, R.M. González Delgado; 4, A. Alonso-Herrero; 5, L. Norci; 6, J. Harris; 7, J. Kurk; 8, G. Kauffmann; 9, I. Bonnell; 10, C. Clarke; 11, M. McCaughrean; 12, E. Le Floc'h; 13, D. Elbaz; 14, E. Skillman; 15, C. Leitherer; 16, M. Pettini; 17, R. de Grijs; 18, S. Howard; 19, A. Smith; 20, G. Gilmore; 21, P. Aslin; 22, A. Batey; 23, N. Orr; 24, T. Heckman; 25, A. Aloisi; 26, D. de Mello; 27, E. Terlevich; 28, L. Origlia; 29, S. Rix; 30, Y.C. Liang; 31, V. Ivanov; 32, G.J. Bendo.

Second row –

33, M. García-Marín; 34, L. Colina; 35, F. Combes; 36, D. Reverte-Payá; 37, C. Maier; 38, J.P. Gallego; 39, N.A. Reddy; 40, Y. Li; 41, A.-K. Jappsen; 42, A. Sajina; 43, H. Teplitz; 44, F. Fynbo; 45, A. Marcolini; 46, L. Yan; 47, D. Erb; 48, P. Castangia; 49, L. Kewley; 50, D. Romano; 51, E.L Allard; 52, I. Martínez-Valpuesta; 53, C. Moss; 54, A.G. Harford; 55, L. Angeretti; 56, P. Ranali; 57, S. Larsen; 58, J. Lotz; 59, J. Dunlop; 60, R. Abraham; 61, J. Knapen.

Third row –

62, U. Vijh; 63, E.R. Stanway; 64, D. Burgarella; 65, J. Lowenthal; 66, D. Koo; 67, C. Mihos; 68, M. Bershady; 69, D. Calzetti; 70, J. Grimes; 71, M. Dopita; 72, P. Appleton; 73, J. Knödlseder; 74, M. Lemoine-Busserolle; 75, S.F. Beaulieu; 76, S.G. Temporin; 77, F. Schweizer; 78, E. Schinnerer; 79, S. Higdon; 80, J. Higdon; 81, R.W. O'Connell; 82, A. Bunker.

Fourth row –

83, L.J. Storrie-Lombardi; 84, E. Telles; 85, D. Marcillac; 86, C. Struck; 87, R. Kennicutt; 88, C. Hoyos; 89, P. van der Werf; 90, S. Mengel; 91, R.C. Peterson; 92, A. Phillips; 93, K.G. Noeske; 94, J. Melbourne; 95, L. Christensen; 96, V. Lebouteiller; 97, L. Snijders; 98, C. Tremonti; 99, A. Hornschemeier; 100, N. Homeier; 101, K. Nagamine; 102, R. Bouwens; 103, D. Mehlert; 104, S. Noll; 105, A. Inoue; 106, M. Kissler-Patig; 107, T. Nikola; 108, G. Taylor; 109, H. Schmitt; 110, S. Heap; 111, A. Lançon.

Fifth row –

112, L. Sampson; 113, C. Kehrig; 114, A. Garcia-Rissmann; 115, C. Papovich; 116, A. Weiss; 117, S. Bertone; 118, L. van Starkenburg; 119, C. Scannapieco; 120, T. Lilly; 121, E. Zackrisson; 122, U. Fritze; 123, V. Buat; 124, P. Anders; 125, A. Pope; 126, M. Sawicki; 127, F. Walter; 128, S. Salim; 129, C. Akerman; 130, J. Brinchmann; 131, J.-D. Smith; 132, M. Sirianni; 133, I. Iwata; 134, B. Brandl; 135, L.J. Smith; 136, J. Gallagher; 137, B. Elmegreen; 138, N. Tanvir.

Sixth row –
139, N. Förster Schreiber; 140, J. Huang; 141, P. Papaderos; 142, T. Marquart; 143, J. Ott; 144, H. Roussel; 145, A. Gilbert; 146, B. O'Halloran; 147, C. Weidner; 148, D. Schaerer; 149, R. Thompson; 150, L. Hadfield; 151, M. Westmoquette; 152, F. Sidoli; 153, L. Douglas; 154, S. Mhle; 155, K. Chyży; 156, M. Fellhauer; 157, G. Silvestro; 158, P. Crowther; 159, C. Mundell; 160, N. Bergvall; 161, J. Cannon; 162, H. Lee; 163, D. Zucker.
Seventh row –
164, E. Pérez-Montero; 165, L. Jamet; 166, C. Evans; 167, B. Buckalew; 168, C. Tadhunter; 169, J. Ulvestad.

Scientific Organising Committee:

Richard de Grijs (Chair)	University of Sheffield, UK
Cathie J. Clarke	Institute of Astronomy, Cambridge, UK
Françoise Combes	Observatoire de Paris, France
Uta Fritze–v. Alvensleben	Universitäts-Sternwarte Göttingen, Germany
Rosa M. González Delgado	Instituto de Astrofísica de Andalucía, Spain
Claus Leitherer	Space Telescope Science Institute, USA
Max Pettini	Institute of Astronomy, Cambridge, UK
Evan Skillman	University of Minnesota, USA
Roberto J. Terlevich	Instituto Nacional de Astrofísica, Óptica y Electrónica, Mexico
Paul P. van der Werf	Leiden Observatory, The Netherlands

Local Organising Committee:

Richard de Grijs	(Chair)
Andrew J. Bunker	
Max Pettini	
Gerry F. Gilmore	
Paul Aslin	(Senior departmental administrator)
Suzanne Howard	(Conference secretary)
Andrew Batey	(Computer officer)
Richard Sword	(Graphics officer)

Conference Participants

Abraham	Bob	abraham@astro.utoronto.ca
Akerman	Chris	cja@ast.cam.ac.uk
Alexander	David	dma@ast.cam.ac.uk
Allard	Emma L.	allard@star.herts.ac.uk
Aloisi	Alessandra	aloisi@stsci.edu
Alonso-Herrero	Almudena	aalonso@damir.iem.csic.es
Anders	Peter	panders@uni-sw.gwdg.de
Angeretti	Luca	luca.angeretti2@studio.unibo.it
Appleton	Philip N.	apple@ipac.caltech.edu
Beaulieu	Sylvie F.	sbeaulieu@phy.ulaval.ca
Bendo	George J.	gbendo@as.arizona.edu
Bergvall	Nils	nils.bergvall@astro.uu.se
Bershady	Matthew A.	mab@astro.wisc.edu
Bertone	Serena	s.bertone@sussex.ac.uk
Bonnell	Ian A.	iab1@st-and.ac.uk
Bouwens	Rychard J.	bouwens@ucolick.org
Brandl	Bernhard R.	brandl@strw.leidenuniv.nl
Bremer	Malcolm	m.bremer@bristol.ac.uk
Brinchmann	Jarle	jarle@astro.up.pt
Buat	Veronique	veronique.buat@oamp.fr
Buckalew	Brent A.	mrk1236@uwyo.edu
Bunker	Andrew	bunker@astro.ex.ac.uk
Burbidge	Margaret	mburbidge@ucsd.edu
Burgarella	Denis	denis.burgarella@oamp.fr
Calzetti	Daniela	calzetti@stsci.edu
Cannon	John M.	cannon@mpia.de
Carswell	Bob	rfc@ast.cam.ac.uk
Castangia	Paola	pcastang@ca.astro.it
Christensen	Lise	lchristensen@aip.de
Chyży	Krzysztof T.	chris@oa.uj.edu.pl
Clarke	Cathie	cclarke@ast.cam.ac.uk
Colina	Luis	colina@damir.iem.csic.es
Combes	Françoise	francoise.combes@obspm.fr
Crowther	Paul A.	Paul.Crowther@sheffield.ac.uk
de Grijs	Richard	R.deGrijs@sheffield.ac.uk
de Mello	Duília	duilia@ipanema.gsfc.nasa.gov
Doherty	Michelle	md@ast.cam.ac.uk
Dopita	Michael A.	Michael.Dopita@anu.edu.au
Douglas	Laura	L.Douglas@bristol.ac.uk
Dunlop	James	jsd@roe.ac.uk
Elbaz	David	delbaz@cea.fr
Elmegreen	Bruce	bge@watson.ibm.com

Erb	Dawn K.	dke@astro.caltech.edu
Evans	Chris	cje@ing.iac.es
Fellhauer	Michael A.D.	mike@astro.uni-bonn.de
Ferguson	Annette	ferguson@mpa-garching.mpg.de
Ford	Dominic	dcf21@mrao.cam.ac.uk
Förster Schreiber	Natascha M.	forster@mpe.mpg.de
Fritze-v. Alvensleben	Uta	ufritze@uni-sw.gwdg.de
Fynbo	Johan P.U.	jfynbo@astro.ku.dk
Gallagher	John S.	jsg@astro.wisc.edu
García-Marín	Macarena	maca@damir.iem.csic.es
Garcia-Rissmann	Aurea	arissmann@cristina.lna.br
Gilbert	Andrea M.	agilbert@mpe.mpg.de
Gilmore	Gerard F.	gil@ast.cam.ac.uk
González Delgado	Rosa M.	rosa@iaa.es
Grimes	John P.	john@jhu.edu
Hadfield	Lucy J.	l.hadfield@sheffield.ac.uk
Hammer	François	francois.hammer@obspm.fr
Harford	A. Gayler	gharford@CASA.colorado.edu
Harris	Jason	jharris@as.arizona.edu
Heap	Sara R.	Sally.Heap@nasa.gov
Heckman	Timothy M.	heckman@pha.jhu.edu
Hidalgo-Gámez	Ana M.	anamaria@astroscu.unam.mx
Higdon	James L.	jhigdon@astro.cornell.edu
Higdon	Sarah J.U.	sjuh@astro.cornell.edu
Homeier	Nicole L.	nhomeier@pha.jhu.edu
Hornschemeier	Ann E.	annh@milkyway.gsfc.nasa.gov
Hoyos	Carlos C.H	carlos.hoyos@uam.es
Huang	Jiasheng	jhuang@cfa.harvard.edu
Inoue	Akio K.	akio.inoue@oamp.fr
Ivanov	Valentin D.	vivanov@eso.org
Iwata	Ikuru	iwata@optik.mtk.nao.ac.jp
Jamet	Luc	Luc.Jamet@obspm.fr
Jappsen	Anne-Katharina	akjappsen@aip.de
Kauffmann	Guinevere	gamk@mpa-garching.mpg.de
Kehrig	Carolina	kehrig@iaa.es
Kennicutt	Robert C.	rkennicutt@as.arizona.edu
Kewley	Lisa J.	kewley@ifa.hawaii.edu
Kissler-Patig	Markus	mkissler@eso.org
Knapen	Johan H.	j.knapen@star.herts.ac.uk
Knödlseder	Jürgen	knodlseder@cesr.fr
Koo	David C.	koo@ucolick.org
Kunth	Daniel	kunth@iap.fr

Kurk	Jaron	kurk@arcetri.astro.it
Lançon	Ariane	lancon@astro.u-strasbg.fr
Larsen	Søren S.	slarsen@eso.org
Lebouteiller	Vianney	leboutei@iap.fr
Lee	Henry	hlee@astro.umn.edu
Lee	Janice C.	jlee@as.arizona.edu
Le Floc'h	Emeric M.	elefloch@as.arizona.edu
Leitherer	Claus H.	leitherer@stsci.edu
Lemoine-Busserolle	Marie	lemoine@ast.cam.ac.uk
Li	Yuexing	yuexing@amnh.org
Liang	Yanchun	ycliang@bao.ac.cn
Lilly	Thomas	tlilly@uni-sw.gwdg.de
Lotz	Jennifer M.	jlotz@scipp.ucsc.edu
Lowenthal	James D.	james@ast.smith.edu
Maier	Christian	chmaier@phys.ethz.ch
Marcillac	Delphine	marcilla@discovery.saclay.cea.fr
Marcolini	Andrea	andrea.marcolini@bo.astro.it
Marquart	Thomas S.	thomas.marquart@astro.uu.se
Martínez-Valpuesta	Inma	martinez@star.herts.ac.uk
McCaughrean	Mark J.	mjm@astro.ex.ac.uk
McCrady	Nate	nate@astro.berkeley.edu
McMahon	Richard	rgm@ast.cam.ac.uk
Mehlert	Dörte	d.mehlert@lsw.uni-heidelberg.de
Melbourne	Jason L.	jmel@ucolick.org
Mengel	Sabine	smengel@eso.org
Mihos	Chris	mihos@case.edu
Moss	Chris	cmm@astro.livjm.ac.uk
Mühle	Stefanie	muehle@astro.utoronto.ca
Mundell	Carole C.G.	cgm@astro.livjm.ac.uk
Munshi	Dipak	munshi@ast.cam.ac.uk
Nagamine	Kentaro	knagamine@ucsd.edu
Nikola	Thomas J.	tn46@cornell.edu
Noeske	Kai Gerhard	kai@ucolick.org
Noll	Stefan	snoll@mpe.mpg.de
Norci	Laura	ln@dunsink.dias.ie
O'Connell	Robert W.	rwo@virginia.edu
O'Halloran	Brian O.	boh@physics.gmu.edu
Origlia	Livia	livia.origlia@bo.astro.it
Orr	Nancy	scienceinstitute@earthlink.net
Ostriker	Jeremiah P.	jpo@astro.princeton.edu
Ott	Jürgen	Juergen.Ott@csiro.au
Papaderos	Polychronis	papade@uni-sw.gwdg.de

Papovich	Casey	papovich@as.arizona.edu
Pérez	Enrique	eperez@iaa.es
Pérez Gallego	Jorge	jgallego@astro.ufl.edu
Pérez-Montero	Enrique	Enrique.Perez@Uam.es
Peterson	Ruth C.	peterson@ucolick.org
Pettini	Max	pettini@ast.cam.ac.uk
Phillips	Andrew C.	phillips@ucolick.org
Pope	Alexandra	pope@physics.ubc.ca
Ranalli	Piero	piero.ranalli2@studio.unibo.it
Reddy	Naveen A.	nar@astro.caltech.edu
Reverte-Payá	Daniel	drp@iaa.es
Rix	Samantha A.	srix@ing.iac.es
Romano	Donatella	donatella.romano@bo.astro.it
Roussel	Hélène	hroussel@irastro.caltech.edu
Sajina	Anna	sajina@astro.ubc.ca
Salim	Samir	samir@astro.ucla.edu
Sampson	Leda	lsampson@ast.cam.ac.uk
Santos	Michael	mrs@ast.cam.ac.uk
Sawicki	Marcin	sawicki@physics.ucsb.edu
Scannapieco	Cecilia	cecilia@iafe.uba.ar
Schaerer	Daniel	daniel.schaerer@obs.unige.ch
Schinnerer	Eva	eschinne@nrao.edu
Schmitt	Henrique	hschmitt@ccs.nrl.navy.mil
Schweizer	François	schweizer@ociw.edu
Sharp	Rob	rgs@aaoepp.aao.gov.au
Sidoli	Fabrizio	fs@star.ucl.ac.uk
Silk	Joe	silk@astro.ox.ac.uk
Silvestro	Giovanni	silvestro@ph.unito.it
Sirianni	Marco	sirianni@stsci.edu
Skillman	Evan	skillman@astro.umn.edu
Smith	J.-D.T.	jdsmith@as.arizona.edu
Smith	Linda J.	ljs@star.ucl.ac.uk
Snijders	Leonie	snijders@strw.leidenuniv.nl
Stanway	Elizabeth R.	ers@ast.cam.ac.uk
Storrie-Lombardi	Lisa J.	lisa@ipac.caltech.edu
Struck	Curtis	curt@iastate.edu
Tadhunter	Clive N.	c.tadhunter@sheffield.ac.uk
Tanvir	Nial R.	nrt@star.herts.ac.uk
Tasker	Elizabeth J.	ejt@astro.ox.ac.uk
Taylor	Greg B.	gtaylor@nrao.edu
Telles	Eduardo	etelles@on.br
Temporin	Sonia G.	giovanna.temporin@uibk.ac.at

Teplitz	Harry I.	hit@ipac.caltech.edu
Terlevich	Elena	et@ast.cam.ac.uk
Terlevich	Roberto J.	rjt@ast.cam.ac.uk
Thompson	Rodger I.	rthompson@as.arizona.edu
Tremonti	Christy A.	tremonti@as.arizona.edu
Trentham	Neil	trentham@ast.cam.ac.uk
Ulvestad	Jim	julvesta@nrao.edu
Vacca	William	wvacca@mail.arc.nasa.gov
van der Werf	Paul P.	pvdwerf@strw.leidenuniv.nl
van Starkenburg	Lottie	vstarken@strw.leidenuniv.nl
Vijh	Uma	uvijh@astro.utoledo.edu
Walter	Fabian	fwalter@nrao.edu
Weidner	Carsten	cweidner@astro.uni-bonn.de
Weiss	Axel	aweiss@iram.es
Westmoquette	Mark S.	msw@star.ucl.ac.uk
Yan	Lin	lyan@ipac.caltech.edu
Zackrisson	Erik	ez@astro.uu.se
Zucker	Daniel B.	zucker@mpia.de

Preface

It all started with a thought along the lines of *"Wouldn't it be nice to get a few collaborators together for a small, focused meeting on violent star formation?"* In the ensuing discussions, Max Pettini's enthusiasm for the idea proved infectious, and as a consequence we set our aims higher, at organising a medium-sized conference on the premises of the Institute of Astronomy in Cambridge. Little did we know what we had unleashed upon ourselves...

With swaying palm trees, sweltering summer temperatures, azure seas and white sand beaches well and truly out of reach (except for the two small palm trees in front of the old Observatory building in Cambridge, and a full week of sunshine and blue skies – unexpected? Well, perhaps, but not unwelcome!), we set our initial goals at attracting around 80–100 participants. In view of the non-tropical location, the generally high living expenses in the UK, and the decidedly notorious reputation of the British cuisine in mind, any larger number of participants would be a small wonder. Little did we know...

How wrong our predictions really were was evidenced by the uptake of our conference announcement... The breadth of the conference topic generated immense interest. The conference proved far more popular than space availability allowed for, once again showing that the starburst community is very much alive and kicking! With more than 190 participants, 71 oral presentations and some 100 poster papers, most of the networking and surely some of the most interesting scientific discussions happened off-line in the corridors, during the lunch breaks or even at the sumptuous conference dinner hosted by the University of Cambridge's ancient Queens' College.

Five days of a densely packed programme proved to be like drinking from a fire hose; yet it also generated a highly satisfactory and stimulating environment. We hope that new collaborations have found roots at this meeting, and that it will be remembered as a watershed conference, bringing together experts from a wide variety of backgrounds. Despite the unabated flow of information, Bob O'Connell managed to synthesize it all very much to the point in his conference summary – no mean feat under the circumstances! We would like to thank, specifically, Max Pettini, Linda Smith, Jay Gallagher, Bruce Elmegreen, Cathie Clarke, Daniela Calzetti, Rob Kennicutt, Clive Tadhunter, Guinevere

Kauffmann, and Evan Skillman for keeping the speakers to a very tight schedule during the sessions they chaired.

We were indeed very fortunate, in that the conference took place at an opportune time. Data on star-forming galaxies are now pouring in from large telescopes across the globe, and especially from the deep extragalactic surveys being made at all UV through infrared wavelengths. This, in combination with a concerted theoretical and modeling effort, will help resolve many of the issues raised at the conference, and which are covered in great detail in these proceedings. We are already very much looking forward to the rapid progress that will be delivered by GALEX, the Spitzer Space Telescope, ALMA, and the large number of dedicated ground-based surveys by the time of the next major starburst conference!

At this point, we would like to express our heartfelt thanks to the excellent support staff at the Institute of Astronomy, who made this conference possible and – above all – successful. Thank you, both to the support staff on the Local Organising Committee (in particular our conference secretary Suzanne Howard), and to the Institute's secretarial, catering and reception staff at large!

We would like to thank Richard Sword, in particular, for his great help – despite a general overload of work – with all graphical questions raised in relation to preparing these proceedings, and Enrique Pérez for technical and catering support. In addition, we thank Amanda Smith for taking time out from her job at the Institute's reception to capture the conference's atmosphere on camera, as well as Michael Fellhauer and Kai Gerhard Noeske for making their photographs available for inclusion in the printed and electronic proceedings. The latter can be found on the accompanying CD-ROM, containing a large number of high-quality poster presentations, with significantly more in-depth coverage than would have been allowed in, necessarily short, printed proceedings.

Finally, we acknowledge a generous grant from the Royal Astronomical Society, which allowed us to waive the conference fee for twelve young scientists.

Richard de Grijs and Rosa M. González Delgado
Granada, January 2005

Session I

Local Starbursts as Benchmarks for Galaxy Evolution

LOCAL STARBURSTS IN A COSMOLOGICAL CONTEXT

Timothy M. Heckman
Center for Astrophysical Sciences, Johns Hopkins University, USA
heckman@pha.jhu.edu

Abstract In this contribution I introduce some of the major issues that motivate the conference, with an emphasis on how starbursts fit into the "big picture". I begin by defining starbursts in several different ways, and discuss the merits and limitations of these definitions. I will argue that the most physically useful definition of a starburst is its "intensity" (star-formation rate per unit area). This is the most natural parameter to compare local starbursts with physically similar galaxies at high redshifts, and indeed I will argue that local starbursts are unique laboratories to study the processes at work in the early Universe. I will describe how NASA's GALEX mission has uncovered a rare population of close analogs to Lyman Break Galaxies in the local Universe. I will then compare local starbursts to the Lyman Break and sub-mm galaxies' high-redshift populations, and speculate that the multi-dimensional "manifold" of starbursts near and far can be understood largely in terms of the Schmidt/Kennicutt law and galaxy mass-metallicity relation. I will briefly summarize the properties of starburst-driven galactic superwinds and their possible implications for the evolution of galaxies and the intergalactic medium. These complex multiphase flows are best studied in nearby starbursts, where we can study the hot X-ray gas that contains the bulk of the energy as well as newly produced metals.

1. Introduction: What is a Starburst?

Why are local ($z \ll 1$) starbursts important? First of all, they are a very significant component of our present-day Universe, and as such deserve to be understood in their own right. They provide roughly 10% of the radiant energy production and about 20% of all the high mass star formation in the local Universe (e.g., Heckman 1998, Brinchmann et al. 2004). Their cosmological relevance has been highlighted by their many similarities to star-forming galaxies at high redshift. In particular, local UV-bright starbursts appear to be good analogs to the Lyman Break Galaxies (Meurer et al. 1997, Shapley et al. 2003, Heckman et al. 2005). Local starbursts provide a laboratory in which to study the complex ecosystem of stars, gas, black holes, galaxies, and the intergalactic medium up close and in detail. Finally, starbursts can contain millions of OB stars, and hence they also offer a unique opportunity to test theories of the evolution of massive stars.

Perhaps the most fundamental definition of a starburst would be that it is a galaxy in which the star-formation rate approaches the upper limit set by causality. For a self-gravitating system this upper limit is reached if the entire gas reservoir is converted into stars in one dynamical time. For a total mass $M_{\rm tot}$, a gas-mass fraction $f_{\rm gas}$, and a velocity dispersion σ, this upper bound can be written as

$$\text{SFR} \leq M_{\rm tot} f_{\rm gas}/t_{\rm dyn} \sim f_{\rm gas}\sigma^3/G \sim 115(\sigma/100{\rm km\,s}^{-1})^3 f_{\rm gas} {\rm M}_\odot\,{\rm yr}^{-1} \quad (1)$$

If the star formation occurs with a standard IMF, the implied bolometric luminosity is:

$$L_{\rm max} \sim 10^{12}(\sigma/100{\rm km\,s}^{-1})^3 f_{\rm gas} {\rm L}_\odot \quad (2)$$

Figure 1 compares this upper limit to observations of starburst galaxies in both the local Universe and at high redshift. Starbursts approach the limit, and in extreme cases consistency requires gas-mass fractions approaching unity.

The classic definition of a starburst is in terms of its duration, and has two variations. First, a starburst is commonly defined as a galaxy in which the time it would take at the current star-formation rate to consume the remaining reservoir of interstellar gas is much less than the age of the Universe. This is commonly called the gas consumption time. The inverse of the gas consumption time is sometimes called the "efficiency". The related definition is that a starburst is a galaxy in which the time it would take to produce the current stellar mass at the current star-formation rate is much less than a Hubble time. The inverse of this time can be recast as the "birth-rate parameter" (b, i.e., the ratio of the current to past average star-formation rate).

These are sensible definitions and can be measured relatively easily. However, it is important to note that the mass ratio of gas and stars varies significantly and systematically as a function of galaxy properties (e.g., Boselli et al. 2001). Thus, (for example) using the birth-rate parameter as the definition, leads to a steep decline in the fraction of starbursts with increasing galaxy mass (e.g., Brinchmann et al. 2004). On the other hand, since the gas-mass fraction is so much smaller in massive galaxies, using the gas consumption time to define a starburst leads to very little mass-dependence of the starburst phenomenon. An additional point is that either of these timescale definitions of a starburst build in a strong redshift dependence: since the age of the Universe at $z \sim 5$ is only 10% of its present value, a galaxy with $b = 10$ (strong starburst) today would have $b = 1$ (not a starburst) at $z \sim 5$.

An alternative definition is that a starburst has a high intensity: the star-formation rate per unit area ($I_{\rm SF}$) is very large compared to normal galaxies. As shown by Kennicutt (1998), this definition is functionally equivalent to the

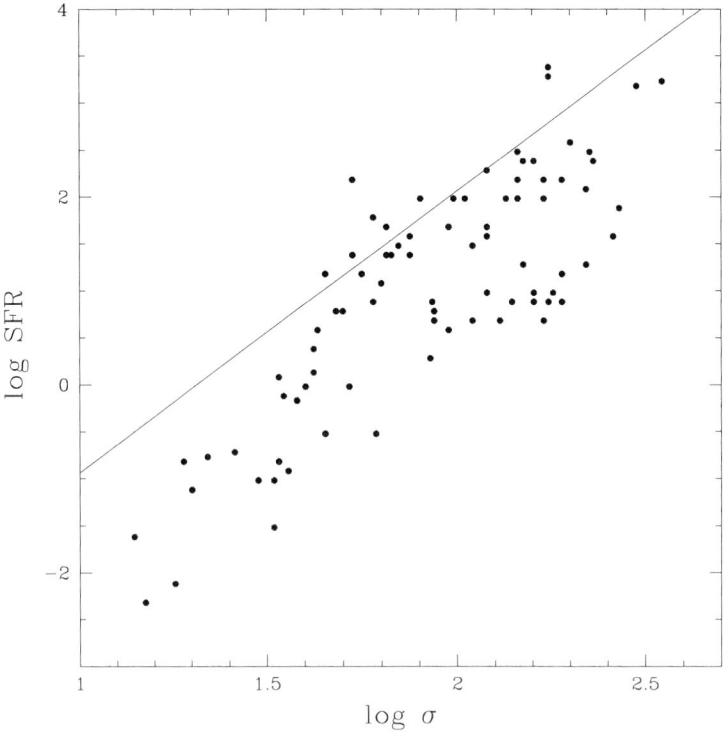

Figure 1. The logarithm of the star-formation rate ($M_\odot$ yr^{-1}) is plotted vs. the logarithm of the galaxy velocity dispersion (km s^{-1}) for the sample of local and high-redshift starbursts described by Murray et al. (2005). The diagonal line corresponds to the upper bound on the star-formation rate allowed by causality for a gas-mass fraction of 100% (it corresponds to the conversion of the entire mass of the system into stars in a single dynamical time). See Heckman (1994) for an older version of this figure. Note that Murray et al. (2005) interpret this in terms of an upper limit on star formation set by the Eddington limit for radiation pressure acting on dust.

classic definition in terms of gas depletion time. Since he found that $I_{SF} \propto N_{gas}^{3/2}$, this means that the gas depletion time $\propto I_{SF}^{-1/3}$. Extreme starbursts have star-formation rates per unit area thousands of times larger than in the disk of the Milky Way, and gas consumption times of only $\sim 10^8$ yr. I will argue below that defining starbursts in terms of their intensity is the most physically useful way to proceed.

2. The Consequences of High Intensity

The very high star-formation rate per unit area in starbursts has immediate consequences for the basic physical properties of the galaxy. It immediately implies a high interstellar gas surface mass density (Σ_g; Kennicutt 1998) and also a high stellar surface mass density (Σ_*). A typical case would have $I_{SF} \sim 10$ M$_\odot$ yr^{-1} kpc^{-2} and $\Sigma_g \sim \Sigma_* \sim 10^9$ M$_\odot$ kpc^{-2}. These are roughly 10^3 (Σ_{SFR}), 10^2 (Σ_g) and 10^1 (Σ_*) times larger than the corresponding values in the disks of normal galaxies.

The basic physical and dynamical properties of starbursts follow directly from the above. A gas surface mass density of 10^9 M$_\odot$ kpc^{-2} corresponds to an extinction of $A_B \sim 10^2$ for a Milky Way dust-to-gas ratio. The characteristic dynamical time in the star-forming region will be short: $t_{dyn} \sim (G\rho)^{-1/2} \sim (G\Sigma_{tot}H)^{-1/2} \sim$ few Myr, where $H \sim 10^2$ pc is the thickness of the disk. A surface brightness of a few $\times 10^{10}$ L$_\odot$ kpc^{-2} corresponds to a radiant energy density inside the star-forming region that is roughly 10^3 times the value in the ISM of the Milky Way. Finally, simple considerations of hydrostatic equilibrium imply correspondingly high total pressures in the ISM: $P \sim G\Sigma_g\Sigma_{tot} \sim$ few $\times 10^{-9}$ dyne cm^{-2} ($P/k \sim$ few $\times 10^7$ K cm^{-3}, or several thousand times the value in the local ISM in the Milky Way). The rate of mechanical energy deposition (supernova heating) per unit volume is also 10^3 or 10^4 times higher than in the ISM of our Galaxy.

As shown by Meurer et al. (1997), local starbursts and Lyman Break Galaxies have very similar values for I_{SF} (1 to 100 M$_\odot$ yr^{-1} kpc^{-2}). Thus, this immediately implies that there are also strong similarities in the basic physical properties of local starbursts and Lyman Break Galaxies.

3. Lyman Break Galaxy Analogs at Low Redshift

While Meurer et al. (1997) showed that local starbursts and Lyman Break Galaxies have similar UV surface brightnesses, the former are generally smaller and less luminous than the latter. Local starbursts are either dwarf galaxies or small regions (usually nuclear) inside big galaxies. Are there true local analogs of the Lyman Break Galaxies in terms of size and ultraviolet luminosity? The success of NASA's Galaxy Evolution Explorer (GALEX; Martin et al. 2004) mission now makes it possible to find out.

We (Heckman et al. 2005) have used the first matched set of GALEX and Sloan Digital Sky Survey (SDSS) data to investigate the properties of a sample of 74 nearby ($z < 0.3$) galaxies with far-ultraviolet luminosities greater than 2×10^{10} L$_\odot$ (with no extinction correction). This was chosen to overlap the luminosity range of typical Lyman Break Galaxies. GALEX deep surveys have shown that ultraviolet-luminous galaxies similar to these are the fastest evolving component of the UV galaxy population (Arnouts et al. 2005,

Schiminovich et al. 2005). Model fits to the combined GALEX and SDSS photometry yield typical FUV extinctions in these galaxies of 0.5 to 2 magnitudes (similar to Lyman Break Galaxies). The implied star-formation rates are SFR $\sim$ 3 to 30 M$_\odot$ yr^{-1}. This overlaps the range for Lyman Break Galaxies. We found a strong inverse correlation between galaxy mass and far-ultraviolet surface brightness, and on this basis divided our sample into "large" and "compact" systems. The large ultraviolet-luminous galaxies are relatively massive ($M_* \sim 10^{11}$M$_\odot$) late-type disk galaxies forming stars at a rate similar to their past average (M_*/SFR $\sim t_{\rm Hubble}$). They are metal rich ($\sim$ solar), and have intermediate optical-UV colors (FUV $- r \sim 2$ to 3). In contrast, the compact ultraviolet-luminous galaxies have half-light radii of a few kpc or less (similar to Lyman Break Galaxies). They are relatively low-mass galaxies ($M_* \sim 10^{10}$M$_\odot$) with typical velocity dispersions of 60 to 150 km s^{-1}. They span a range in metallicity from $\sim$ 0.3 to 1$\times$ solar, have blue optical-UV colors (FUV $- r \sim 0.5$ to 2), and are forming stars at a rate sufficient to build the present galaxy in of order a Gigayear. In all these respects they appear similar to the Lyman Break Galaxies.

The GALEX mission will ultimately find over a thousand such "living fossils" This will provide an opportunity for detailed local investigation of the physical processes occurring in typical star-forming galaxies in the early Universe.

4. Understanding the Starburst Manifold

Let us adopt the idea that a starburst by definition has a high star-formation rate per unit area and a short gas depletion time. The most fundamental properties of a starburst would then be its star-formation rate, metallicity, and dust opacity, and the mass of its host galaxy. In principle, starbursts could uniformly populate the multi-dimensional manifold defined by these parameters. Instead, the parameters show very strong systematic relations. More powerful local starbursts (higher SFR) are more metal rich, more dust obscured, and occur in more massive galaxies (e.g., Heckman et al. 1998). At high redshifts we know that the highest SFRs also occur in the most dust-obscured galaxies (the sub-mm sources compared to the Lyman Break Galaxies).

This systematic behavior can be understood in simple terms as the consequences of three effects. First, as discussed in the introduction, causality implies that the maximum possible SFR is higher in more massive galaxies because they have higher velocity dispersions. This effect is mitigated somewhat by the systematically lower gas mass fraction ($f_{\rm gas}$) in more massive galaxies. Second, more massive star-forming galaxies have systematically higher metallicity (Tremonti et al. 2004). Assuming a constant dust-to-metals ratio, this implies that more massive galaxies have higher dust-to-gas ratios in the ISM.

Third, the Schmidt-Kennicutt law implies that a high SFR requires a high ISM column density.

The natural assumption is that the amount of extinction in a starburst will be strongly related to the dust column density in the ISM (gas column density times dust-to-gas ratio). These three effects then naturally explain why the more powerful starbursts are the more dust obscured (Wang & Heckman 1996, Martin et al. 2004b). A higher SFR requires both a higher gas column and a more massive galaxy with its associated higher dust-to-gas ratio.

The argument can be easily extended to explain why galaxies with a given SFR are less dust-obscured at high redshift than in the local Universe (Adelberger & Steidel 2000). Consider a galaxy with a specific mass (stars plus gas plus dark matter). The Schmidt-Kennicutt law implies that the specific star-formation rate is $\mathrm{SFR}/M \propto f_{\mathrm{gas}} N_{\mathrm{gas}}^{1/2}$. As long as f_{gas} is systematically higher at higher redshifts, this then implies that a correspondingly smaller N_{gas} is required at higher redshifts to sustain the same SFR/M (e.g., if f_{gas} increases by three, N_{gas} would decrease by nine). If the mass-metallicity relation is displaced to lower metallicity (dust-to-gas ratio) at higher redshifts this would only reinforce the effect.

The arguments above imply that both rest-frame UV and far-IR data are required to get a complete picture of the population of star-forming galaxies, both locally at at high redshifts. A UV(FIR)-selected sample will preferentially sample the population with lower (higher) metallicity, SFR, and mass (Martin et al. 2004b, Buat et al. 2005). In light of the above, it is natural to speculate that the main physical difference between the population of star-forming galaxies at high z selected as Lyman Break Galaxies or sub-mm sources is the galaxy mass. The latter may well be the progenitors of giant elliptical galaxies.

5. Starburst-Driven Galactic Winds

By now, it is well-established that galactic-scale outflows of gas are a ubiquitous phenomenon in the most actively star-forming galaxies in the local Universe (see Heckman 2002 for a recent review). These outflows are potentially very important in the evolution of galaxies and the intergalactic medium. For example, by selectively blowing metals out of shallow galactic potential wells, they may explain the tight relation between the mass and metallicity in galaxies (Larson 1974, Tremonti et al. 2004). This same process would have enriched and heated the intergalactic medium in metals at early times (e.g., Adelberger et al. 2003), and would explain why the majority of metals in galaxy clusters are in the intracluster medium (e.g., Loewenstein 2004).

We know that galactic winds are ubiquitous in the population of Lyman Break Galaxies (e.g., Shapley et al. 2003), and have also been observed in sub-mm-selected galaxies (Smail et al. 2003). The big advantage in studying them in local starbursts is that we can investigate their physics in considerably more detail, and in so-doing better understand the form and magnitude of the mass, energy, and metals being carried out in the wind.

Observations of local starburst winds show that they are complex multiphase phenomena. The hot ($\sim 10^7$ K) gas traced by X-ray emission appears to arise in shocks in the wind fluid as it impacts cooler, denser material in the galaxy halo (e.g., Strickland et al. 2004, Lehnert, Heckman & Weaver 1999). As this ambient material encounters the wind, it is heated and accelerated, giving rise to regions of optical line emission and the blueshifted interstellar absorption lines that are characteristic of Lyman Break Galaxies (Shapley et al. 2003) and local starbursts (e.g., Heckman et al. 2000). Dust contained in these clouds is revealed as it reddens the background star light (Heckman et al 2000) and scatters the starburst's UV radiation (Hoopes et al. 2005).

The picture emerging from these panchromatic investigations of local starburst winds is that the fate of the outflow depends strongly on the phase of the outflow and the mass of the galaxy. The hot gas (which contains most of the energy and metals) has nearly the same temperature, independent of the escape velocity from the galaxy blowing the wind (Martin 1999). This hot gas is thus more likely to escape from low-mass galaxies (with their shallower potential wells). This could naturally account for the galaxy mass-metallicity relation. In contrast, the outflow velocity in the cooler gas traced by the interstellar absorption lines is lower in the much less powerful starbursts in dwarf galaxies (Martin 2004). This is most likely because the low-power winds in the dwarfs have insufficient thrust to accelerate interstellar clouds up to the velocity of the hot wind.

The combination of insights like these from the local Universe and systematic investigations of the redshift dependence of the outflow rate derived through rest-frame UV spectroscopy will prove quite powerful in terms of addressing the cosmological significance of starburst-driven winds.

6. Conclusions

- Starbursts are an important component of the local Universe and worth understanding in their own right.

- The key astrophysical property of a starburst is its "intensity" (SFR/area). This property has far-reaching consequences for the physical and dynamical properties of the ISM.

- Local starbursts provide excellent laboratories for the study of the astrophysics of high-z star-forming galaxies; they have very similar SFR/area.

- GALEX has begun to provide a large sample of low-z analogs to Lyman Break Galaxies.

- The systematic properties of starbursts are largely a consequence of the Schmidt/Kennicutt law plus the mass-metallicity relation. This explains why more massive galaxies host more powerful, more metal-rich, and more highly obscured starbursts. It also explains why high-z starbursts are less obscured, on average for a given SFR, than low-z starbursts.

- Only in local starbursts can the multi-phase physics of galactic winds be fully investigated. The hot metal-rich phase traced by X-rays is the key to understanding how the IGM was chemically enriched by outflows from low-mass galaxies.

References

Adelberger K., Steidel C., 2000, ApJ, 544, 218
Adelberger K., Steidel C., Shapley A., Pettini M., 2003, ApJ, 584, 45
Arnouts S., et al., 2005, ApJ, 619, L43
Boselli A., Gavazzi G., Donas J., Scodeggio M., 2001, AJ, 121, 753
Brinchmann J., Charlot S., White S., Tremonti C., Kauffmann G., Heckman T., Brinkmann J., 2004, MNRAS, 351, 1151
Buat V., et al., 2004, ApJ, 619, L51
Larson R., 1974, MNRAS, 169, 229
Heckman T., 1994, in: Mass-Transfer Induced Activity in Galaxies, Shlosman I., ed., (CUP: Cambridge), p. 234
Heckman T., 1998, in: Origins, Woodward C., Shull J.M., Thronson Jr. H., eds., ASP Conf. Ser., (ASP: San Francisco), vol. 148, p. 127
Heckman T., Robert C., Leitherer C., Garnett D., van der Rydt F., 1998, ApJ, 503, 646
Heckman T., 2002, in: Extragalactic Gas at Low Redshift, Mulchaey J., Stocke J., eds., ASP Conf. Ser., (ASP: San Francisco), vol. 254, p. 292
Heckman T., Lehnert M., Strickland D., Armus L., 2000, ApJS, 129, 493
Heckman T., et al., 2005, ApJ, 619, L35
Hoopes C., et al., 2005, ApJ, 619, L99
Lehnert M., Heckman T., Weaver K., 1999, ApJ, 523, 575
Loewenstein M., 2004, in: Origin and Evolution of the Elements, McWilliam A., Rauch M., eds., (CUP: Cambridge), p. 425
Martin C.L., 1999, ApJ, 513, 156
Martin C.L., 2004, ApJ, subm. (astro-ph/0410247)
Martin C.D., et al., 2004a, ApJ, in press
Martin C.D., et al., 2004b, ApJ, in press
Meurer G., Heckman T., Leitherer C., Lowenthal J., Lehnert M., 1997, AJ, 114, 54
Murray N., Quataert E., Thompson T., 2005, ApJ, 618, 569
Kennicutt R., 1998, ApJ, 498, 541
Schiminovich D., et al., 2005, ApJ, 619, L47
Shapley A., Steidel C., Pettini M., Adelberger K., 2003, ApJ, 588, 65
Smail I., Chapman S., Ivison R., Blain A., Takata T., Heckman T., Dunlop J., Sekiguchi K., 2003, MNRAS, 342, 1185
Strickland D., Heckman T., Colbert E., Hoopes C., Weaver K., 2004, ApJS, 151, 193
Tremonti C., et al., 2004, ApJ, 613, 898
Wang B., Heckman T., 1996, ApJ, 457, 645

STARBURSTS IN THE EVOLVING UNIVERSE: A LOCAL PERSPECTIVE
Nearby Starbursts

John S. Gallagher, III
Department of Astronomy, University of Wisconsin, 475 N. Charter St., Madison, WI 53706, USA
jsg@astro.wisc.edu

Abstract   Nearby starburst galaxies provide fascinating astrophysical laboratories for studies of intense modes of star formation. This paper briefly considers some key features of starbursts as revealed by nearby examples, and discusses their implications for galaxy evolution.

1.   Introduction

Nearby galaxies offer opportunities to study the detailed astrophysics of the starburst phenomenon, which then can be more widely applied to cosmic starburst samples. Local starbursts display a range in star-formation rates (SFRs), and care is needed in defining which galaxies are starbursts. One approach distinguishes starbursts relative to the behavior of normal galaxies. Considerable evidence shows that most galactic disks have SFRs that are constant over time to factors of 2–3. For example, the Large Magellanic Cloud experienced an increase in SFR by a factor of a few in the last few Gyr, but is not a starburst system. Thus, one definition is that a starburst involves an increase in SFR by a factor of > 3 during a short time interval (e.g., ~ 0.1 Gyr). Starbursts can be distinguished astrophysically by star formation that is out of balance with the available resources; e.g., a SFR that cannot be sustained for a significant fraction of a Hubble time with the available interstellar gas.

The fraction of nearby galaxies that are experiencing starbursts is relatively small, $< 10\%$, and is even less if one considers only starbursts with global impact. However, since starbursts have short durations, this implies many galaxies are likely to have experienced multiple starbursts in the last few Gyr. This mode of star formation therefore remains important in the present-day Universe.

Nearby starbursts offer a variety of observational advantages. Spatial resolution can be achieved to the few-pc level, permitting connections to be made between the products of star formation, such as compact star clusters, and the structure of the host galaxy. Proximity allows multi-wavelength observations with high sensitivity. For example, we can map the multi-phase ISM in a starburst using molecular lines, thermal infrared radiation from dust and ionized gas, the H I 21cm line, optical–ultraviolet emission and absorption lines, radio continuum from bremsstrahlung and synchrotron emission, and X-rays from hot gas. This access allows us to address key issues, such as the operation of feedback mechanisms, properties of the products of intense star formation, high-energy particle acceleration processes and sites, and production and distribution of newly synthesized chemical elements.

2.   Classes of Starbursts

Starbursts range in spatial extent within their host galaxies. The most compact are nuclear starbursts that are confined within the inner few hundred pc of their hosts, and yet provide a substantial fraction of the total SFR. This type of starburst is common, and is found in systems ranging from giant spirals (e.g., the SBc galaxy M83; Harris et al. 2001) to dwarfs, such as NGC 1705 (Tosi et al 2001). Starbursts that cover more than a disk scale length are rare, but include nearby compact luminous emission-line galaxies like NGC 7673 (Homeier et al. 2002) or NGC3310. An extensive starburst may also be present in the extremely peculiar central galaxy of the Perseus cluster, NGC 1275. Most of the other well-known starbursts are intermediate cases, with the burst being somewhat larger than the nuclear region of the host but not extending very far into the disk (e.g., M82). Starbursts associated with merging galaxies probably also fall in the intermediate category, with the burst occurring in the inner regions of the galaxy, but not being strictly confined to the nuclear zone (e.g., NGC 6240; Pasquali et al. 2004).

A second parameter is the intensity of the starburst as given by its SFR, which – combined with a size – leads to the measure of the SFR per area. However, this factor alone does not describe a starburst. We still need to consider the spatial scale of the event relative to the normal mode of star formation in that galaxy. Thus, from this perspective, a large but normal star-forming region, such as Orion in the Milky Way, is not a starburst, even though within that region some conditions may resemble those found in starburst galaxies. Figure 1 sketches the properties of some nearby starburst galaxies.

Another dimension in the world of starbursts comes from efforts to characterize triggering events. In some cases, such as mergers, these are obvious. However, caution is in order as not all mergers lead to starbursts, but the most intense bursts, the ULIRGs, occur in gas-rich mergers between two or more

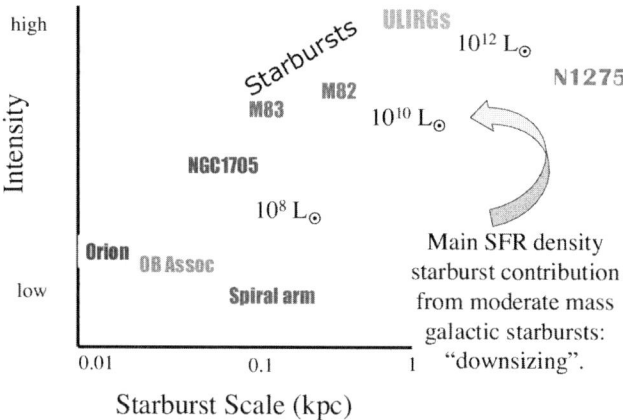

Figure 1. Sketch showing the approximate scales and intensities (ranking by intensity or SFR per area) of selected nearby starbursts. Rough scales of the starbursts are included in terms of luminosities from young stars; 10^{10} $L_\odot$ corresponds to a SFR of ≈ 1 $M_\odot$ yr^{-1} (Kennicutt 1998). SFR intensities of normal star-forming regions are shown for comparison.

galaxies. However, mergers can also be associated with less energetic starbursts, e.g., that in NGC 3310 (Wehner & Gallagher 2005).

However, in other cases the relationship between the trigger and the starburst is less clear. M82 is in orbit around the M81 spiral, and close passages apparently suffice to make giant starbursts (de Grijs et al. 2003). Thus, in M82 we have an example of a galaxy where episodic starbursts are a major evolutionary factor, even though the triggering perturbation is relatively weak (Förster Schreiber et al. 2003). Even milder events, such as gas infall, may be the causes of bursts in blue compact dwarf galaxies, including their Local Group relative, IC10 (Wilcots & Miller 1998).

3. Dissipation, Winds, and Star Formation

It is not surprising that starbursts occur in regions with high gas densities. The trick is how the gas was collected without first experiencing disruptive star formation. To make a starburst, the high-density gas reservoir must form over a time span that is short as compared with its internal star-formation time scale. Circumstances that increase the star-forming time scales, such as shearing ISM flows in bars, and assemble gas on dynamical time scales, as in galaxy mergers, thus favor starbursts. We also require that the resulting starburst-driven galactic wind not remove gas faster than the SFR or the burst would fizzle. Understanding the conditions within the ISM of starbursts is essential (e.g., Melioli & de Gouveia Dal Pino 2004).

We think the ISM in starbursts is not simply a scaled-up version of the solar neighborhood for several reasons: (i) Young massive stars frequently are born in dense star clusters which themselves are clustered to yield extremely high stellar densities. This produces concentrated inputs of mechanical and ionizing luminosity into the surrounding ISM. It also allows interactions to readily occur between adjacent star-forming sites, e.g., leading to locally collimated gas outflows that can feed galactic winds (Tenorio-Tagle et al. 2003). (ii) As a result of the large numbers and densities of OB stars, photo-ionization can be widespread over time and space. Starbursts, and especially those in dwarf galaxies, have socialized photo-ionization: multiple star-forming sites contribute to photo-ionization rather than the star-forming location–HII region connection seen in spirals. As a result, photo-ionization is more extensive and more durable than in cases where it depends on a single event. (iii) The pressure in the ISM of starbursts is 10–100 times or more higher than in typical spiral disks. This probably leads to a rather different ISM phase structure. Furthermore, the presence of an extensive and dense hot ISM phase, prominently seen as diffuse X-ray emission in starburst zones, offers the possibility of rapid communication throughout the region, due to the high sound speed. (iv) In extreme cases, such as the nuclear region starbursts in ULIRGS, the *mean* ISM densities over hundreds of pc scales equal those in typical molecular clouds. In these systems we are in a unique ISM regime.

"Starburst clumps", kpc-scale regions that support intense star formation, are frequently seen in starbursts (e.g., M82; O'Connell & Mangano 1978), and were noted as general features of galaxies with "hyperactive star formation" by J. Heidemann and his collaborators. Numerical simulations demonstrate that massive gas clumps form in galactic disks when the kinematically cool ISM is dynamically important (Noguchi 1998, Immeli et al. 2004). This seems to be the case in the unusual giant irregular starbursts, such as NGC 7673, where a clumpy structure is evident even in NIR images (Fig. 2). This starburst mode may be important in gas-rich young galaxies. On the other hand, starburst rings associated with gas build-ups exterior to bars may be more common in relatively mature galaxies.

The compact young massive star clusters and super star clusters (SSCs) in starburst clumps are signatures of the high levels of dissipation within starbursts. The SSCs appear to define the end of the dissipation sequence for single-generation stellar systems outside of galactic nuclei. For example, an SSC with $M \approx 10^5$ $M_\odot$ has a mean density of $n = 10^5$ cm^{-3} within a half-mass radius of 3 pc. High star-formation efficiencies, $\sim 30\%$, are required to keep a compact cluster bound after formation. These are several times higher than the *mean* star-formation efficiencies found in molecular clouds, and suggests the star formation in starbursts might be more violent and more efficient than in normal galaxies. We then expect the SFR per unit molecular gas to

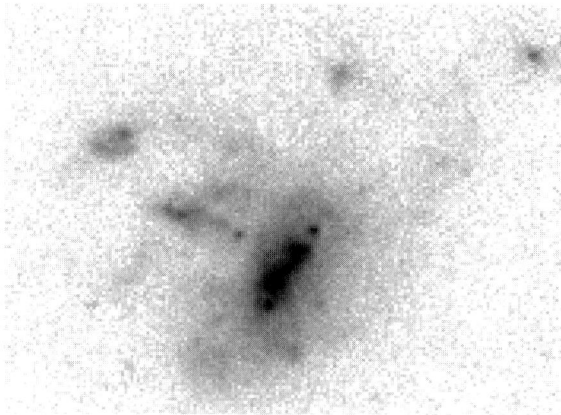

Figure 2. Image of the clumpy irregular galaxy NGC 7673 taken in the K_s filter with the NIRIM camera on the WIYN 3.5-m telescope. The similar structure of this galaxy in the optical through K band is typical of clumpy starbursts, and suggests that the disk is highly unstable in the starburst region.

be enhanced in starbursts, but this effect is not seen in the study of Gao & Solomon (2004). While feedback may change the nature of star formation and structure of the ISM, it does not seem to enhance the efficiency of the conversion of dense molecular gas into stars.

Since starbursts tend to be centrally concentrated, they can produce rapid changes in the structures of their galactic hosts. Post-burst galaxies should be denser and more luminous than the initial systems. If a galaxy experiences several starbursts during its lifetime, then its velocity field also must change in order for it to stay on the Tully-Fisher relationship. Writing the Tully-Fisher correlation as $L = C v_{\rm rot}^{\alpha}$ where C is a constant, we then have $\Delta L/L = \alpha (\Delta v_{\rm rot}/v_{\rm rot})$. Since for normal galaxies $\alpha \approx 3 - 4$, the increases in $v_{\rm rot}$ required for typical starbursts are modest, but go in the direction expected for increasing galactic densities. As galaxies evolve via bursts, they can walk their way up the Tully-Fisher relationship, increasing in both $v_{\rm rot}$ and L.

4. Summary

⋆ Starbursts occur in a range of galactic hosts, where they produce star formation that is out of equilibrium with the available ISM supplies. As a result, starbursts should be and usually are short-lived. The bursts are also out of balance with the gravitational fields in that they frequently power substantial galactic winds, as discussed elsewhere in this conference.

⋆ Triggering of starbursts requires the collection of a dense ISM with a significant amount of mass. This can occur in mergers, but other less violent events also lead to starbursts. No single mechanism fits all of the observed cases.

★ The intense starburst mode of star formation favors the production of dense young stellar systems. These range from starburst clumps and rings, which mainly reflect the distribution of the gas supporting the starbursts, to dense star clusters, including SSCs. Despite mass loss due to winds, the overall impact of a starburst is to increase the mean density of the host galaxy, an effect that supports the Tully-Fisher relationship.

★ Starbursts in nearby systems may differ from those in young galaxies seen at high redshifts. Factors to consider include the lower metallicities and higher gas reserves of younger galaxies, as well as the higher rate of mergers and/or interactions. It also seems possible that in some high-redshift galaxies star formation is taking place within a 3-dimensional structure, while in the well-studied nearby starbursts, it is confined to two dimensions in disks.

Acknowledgments

Studies of starburst galaxies at Wisconsin involve students, recently Ph.D. students Chris Conselice, Nicole Homeier, Elizabeth Wehner and Mark Westmoquette (at UCL with L.J. Smith), and undergraduates Phil Cigan and Gwen Rudie (from Dartmouth College). I also thank my external collaborators, especially D. Calzetti, R. de Grijs, J. Harris, A. Lançon, R.W. O'Connell, A. Pasquali, L.J. Smith, and A.M. Watson. This research has been supported by NASA through *HST* grants, the Wisconsin Alumni Research Foundation via funds awarded by the Graduate School, and by the National Science Foundation through grant AST98-03018.

References

de Grijs R., Bastian N., Lamers H.J.G.L.M., 2003, MNRAS, 340, 197
Förster Schreiber N.M., Genzel R., Lutz D., Sternberg A., 2003, ApJ, 599, 193
Gao Y., Solomon P.M., 2004, ApJ, 606, 271
Harris J., Calzetti D., Gallagher J.S., Conselice C.J., Smith D.A., 2001, AJ, 122, 3046
Homeier N., Gallagher J., Pasquali A., 2002, A&A, 391, 857
Immeli A., Samland M., Gerhard O., Westera P., 2004, A&A, 413, 547
Kennicutt R.C. Jr., 1998, ARA&A, 36, 1
Melioli C., de Gouveia Dal Pino E.M., 2004, A&A, 424, 817
Noguchi M., 1998, Nature, 392, 253
O'Connell R.W., Mangano J.J., 1978, ApJ, 221, 62
Pasquali A., Gallagher J., de Grijs R., 2004, A&A, 415, 103
Tenorio-Tagle G., Silich S., Muñoz-Tuñón C., 2003, ApJ, 597, 279
Tosi M., Sabbi E., Bellazzini M., Aloisi A., Greggio L., Leitherer C., Montegriffo P., 2001, AJ, 122, 1271
Wehner E., Gallagher J., 2005, ApJ, 618, L21
Wilcots E., Miller B.W., 1998, AJ, 116, 2363

ARE THERE LOCAL ANALOGS OF LYMAN BREAK GALAXIES?

James D. Lowenthal[1], R. Nick Durham[1], Brian J. Lyons[2], Matthew A. Bershady[3], Jesus Gallego[4], Rafael Guzmán[5], and David C. Koo[6]

[1]*Smith College, Northampton, MA 01060, USA;* [2]*Amherst College, USA;* [3]*University of Wisconsin, USA;* [4]*Universidad Complutense de Madrid, Spain;* [5]*University of Florida, USA;* [6]*UCO/Lick Observatory, USA*

Abstract To make direct comparisons in the rest-frame far-ultraviolet (FUV) between LBGs at $z \sim 3$ and more local star-forming galaxies, we use *HST*/STIS to image a set of 12 nearby ($z < 0.05$) H II galaxies in the FUV, and a set of 14 luminous compact blue galaxies (LCBGs) at moderate redshift ($z \sim 0.5$) in the NUV, corresponding to the rest-frame FUV. We then subject both sets of galaxy images and those of LBGs at $z \sim 3$ to the same morphological and structural analysis. We find many qualitative and quantitative similarities between the rest-frame FUV characteristics of distant LBGs and of the more nearby starburst samples, including general morphologies, sizes, asymmetries, and concentrations. Along with some kinematic similarities, this implies that nearby H II galaxies and LCBGs may be reasonable local analogs of distant Lyman break galaxies.

1. Lyman break galaxies

Lyman break galaxies (LBGs) at redshifts $z \sim 3$ are the current gold standard for star-forming galaxies in the early Universe, at least rest-frame UV and optically selected ones. Many of their properties are revealed by deep multiwavelength imaging and spectroscopic surveys (see, e.g., presentations at this conference by Erb, Mehlert, Papovich, Sawicki, and others). These include small sizes $r_{1/2} < 4$ kpc, high luminosities $L \sim L^*$, significant clustering, and diverse morphologies. LBGs are also copiously star forming, easily qualifying as starbursts according to the star-formation intensity (SFR per unit mass or gas mass) definition advocated by Tim Heckman at this conference (e.g., Meurer et al. 1997).

Significant questions remaining about the nature of LBGs include their masses, their mass assembly histories, their fate, and their environments, including any dark matter. Comparison to local galaxies with similar properties may help illuminate some or all of those issues, since nearby systems can generally be studied in much greater detail.

2. Compact starbursts at $z < 1$ in the Rest-frame UV

We face two problems in attempting to draw parallels between distant LBGs and nearby starbursts: (i) LBGs are best seen in the optical, corresponding at $z = 3$ to the rest-frame UV; and (ii) it is not obvious which kind or kinds of local systems are the best proxies.

Two classes of galaxy at $z < 1$ seem especially promising as nearby cousins of LBGs: H II galaxies at $z \sim 0$, and luminous compact blue galaxies (LCBGs) at $0.4 < z < 1$. Both classes show, in the optical, the small sizes, high luminosities, diverse morphologies, and copious star formation that also characterize LBGs. To push the comparison further, we have obtained rest-frame UV images with the Space Telescope Imaging Spectrograph onboard the *Hubble Space Telescope* (*HST*/STIS) of 12 H II galaxies and 14 LCBGs. The samples are drawn from the UCM survey (Pérez-González et al. 2001) and the Kitt Peak Galaxy Redshift Survey (Munn 1997), respectively. Many have also been imaged with *HST*/WFPC2 and/or NICMOS and/or studied spectroscopically at Keck and Arecibo (e.g., Pisano et al. 2001).

3. Rest-frame UV Morphologies

The *HST*/STIS UV images of the 26 low and intermediate-redshift starburst galaxies in our sample are shown in Fig. 1. It is immediately obvious that the rest-frame UV morphologies represent a diverse panoply, rather than a uniform class. Multiple knots, tails, and extended emission – almost all invisible at ground-based resolution in the optical – are the rule rather than the exception.

We attempted to quantify the UV morphologies of our nearby and intermediate-z compact starbursts using the CAS (compactness, asymmetry, and clumpiness) methodology of Conselice et al. (2003). The measured asymmetry is very sensitive to the exact radius at which it is measured; we tried both half-light radii and Petrosian radii. Fig. 2 shows the distribution of asymmetries $A(r_\mathrm{P})$. We find that the mean $A(r_\mathrm{P})$ of our two samples is roughly consistent with that of LBGs measured in the HDF by Conselice et al. (2003), although the formal measurement uncertainties are very large.

4. Simulating LBGs at $z = 3$

What would the H II galaxies and LCBGs look like if placed at redshift $z = 3$? We simulated that view by resampling our *HST*/STIS images and adding noise to reproduce the Hubble Deep Field sensitivity. The H II galaxies are too faint to detect, but the LCBGs, which are more luminous, are all easily detected

Are there local analogs of Lyman break galaxies? 19

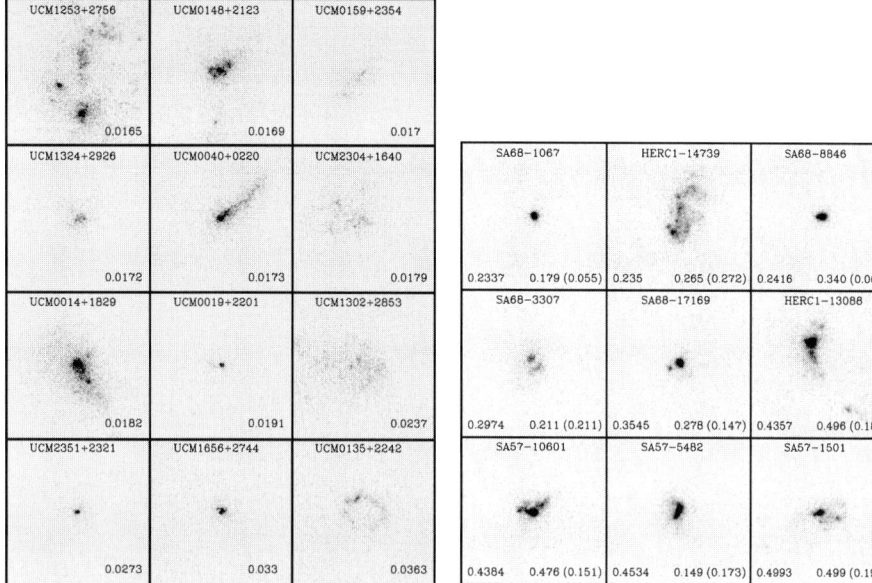

Figure 1. *Left:* Mosaic of *HST*/STIS FUV images ($\lambda_c = 1590\text{Å}$) of 12 H II galaxies from the UCM survey. Redshifts are labelled for each galaxy. Each image is 6.5 arcsec on a side, corresponding to 2.5 kpc at $z = 0.02$. *Right:* Mosaic of *HST*/STIS NUV images ($\lambda_c = 2320\text{Å}$) of 9 of the 14 LCBGs. Each image is 3.75 arcsec on a side, corresponding to 22.5 kpc at $z = 0.5$. Note the small sizes and disturbed, varied rest-frame UV morphologies of both samples, reminiscent of LBGs at higher redshift.

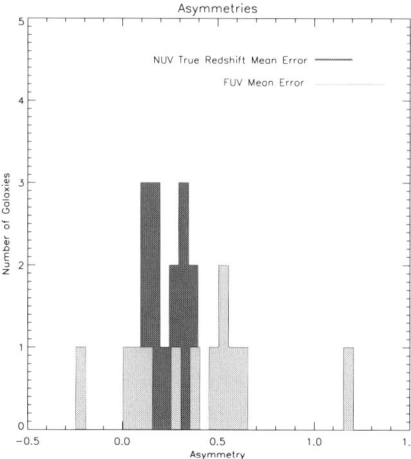

Figure 2. Distribution of asymmetries measured at Petrosian radius for both the H II galaxy (FUV) and LCBG (NUV) samples. The mean is roughly consistent with that of LBGs measured in the HDF by Conselice et al. (2003).

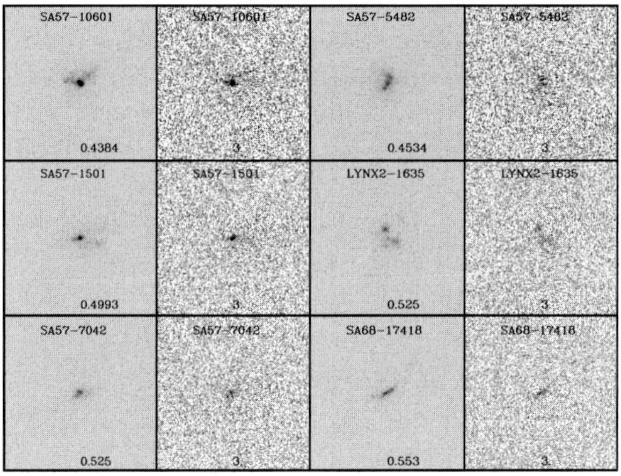

Figure 3. LCBGs shown at their true redshift (left panel of each pair) and simulated view as they would appear at redshift $z = 3$ observed in the HDF. All sources would be easily detected at $z = 3$, although faint, low surface brightness emission is lost to noise.

(Fig. 3). Despite the loss of low surface brightness features, the similarities to the appearances of real LBGs are striking.

We conclude that H II galaxy and LCBG morphologies, like their sizes, colors, and star-formation rates and intensities, are qualitatively and quantitatively similar to those of LBGs. They are therefore reasonable local testbeds for further comparative study of LBGs, including constraints on mass.

Our future plans include applying other morphological measures such as the Gini coefficient (Lotz et al. 2004) and combining these STIS UV images with WFPC2 and NICMOS images in hand to constrain stellar populations, dust content, and merger scenarios.

References

Conselice C.J., Bershady M.A., Dickinson M., Papovich C., 2003, ApJ, 136, 1183
Koo D.C., Guzmán R., Faber S.M., Illingworth G.D., Bershady M.A., Kron R.G., Takamiya M., 1995 ApJ, 440, L49
Guzmán R., Jangren A., Koo D.C., Bershady M.A., Simard L., 1998, ApJ, 495, L13
Lotz J.M., Primack J., Madau P., 2004, AJ, 128, 163
Meurer G., Heckman T., Lehnert M., Leitherer C., Lowenthal J.D., 1997, AJ, 114, 54.
Munn J.A., Koo D.C., Kron R.G., Majewski S.R., Bershady M.A., Smetanka J.J., 1997, ApJS, 109, 45
Pérez-González P.G., Gallego J., Zamorano J., Gil de Paz A., 2001, A&A, 365, 370
Pisano D.J., Kobulnicky H.A., Guzmán R., Gallego J., Bershady M.A., 2001, AJ, 122 1194

THE DISK WOLF-RAYET POPULATION OF THE NUCLEAR STARBURST GALAXY M83

Paul A. Crowther[1], Lucy J. Hadfield[1], Hans Schild[2], and Werner Schmutz[3]
[1]*Department of Physics & Astronomy, University of Sheffield, Hounsfield Road, Sheffield S3 7RH, UK*
[2]*Institut für Astronomie, ETH-Zentrum, CH 8092 Zürich, Switzerland*
[3]*Physikalisch-Meteorologisches Observatorium, CH-7260 Davos, Switzerland*
Paul.Crowther@sheffield.ac.uk

Abstract We present VLT imaging and spectroscopy of M83, revealing a large disk Wolf-Rayet population. The observed WC to WN ratio is ~ 1.3, extending previous Local Group results to higher metallicity, and is significantly higher than current evolutionary predictions at high metallicity. Late subtypes dominate the WC population, arguing in favour of metallicity-dependent WR winds, of application to the hardness of Lyman continuum fluxes from young starbursts. Finally, source #74 is identified as a super star cluster, has a large WR content of ~ 200 with $N(\text{WR})/N(\text{O}) \sim 1$, comparable to the WR cluster NGC3125-1.

1. Introduction

Wolf-Rayet (WR) galaxies are a subset of extragalactic nebular emission-line galaxies in which the characteristic signatures of Wolf-Rayet stars – the evolved descendants of the most massive stars ($\geq 20 M_\odot$ at solar metallicity) – are detected, such that they indicate massive star formation within the past $5-10$ Myr. Since the first detection 25 years ago (Allen et al. 1976), the number of known WR galaxies has grown rapidly to well over a hundred (Schaerer et al. 1999). The WR stellar content of such galaxies ranges from a few dozen in NGC 1569 (González Delgado et al. 1997) to 2×10^4 in Mrk 309 (Schaerer et al. 2000). WR signatures are most widely detected within unresolved knots or clusters of irregular, blue compact dwarf or spiral galaxies, and are seen in the combined rest-UV frame spectrum of Lyman Break galaxies (Shapley et al. 2003). In the case of nearby spirals, spectroscopy of bright HII regions obtained for nebular abundance studies often leads to the serendipitous discovery of WR clusters (Pindao et al. 2002, Bresolin & Kennicutt 2002).

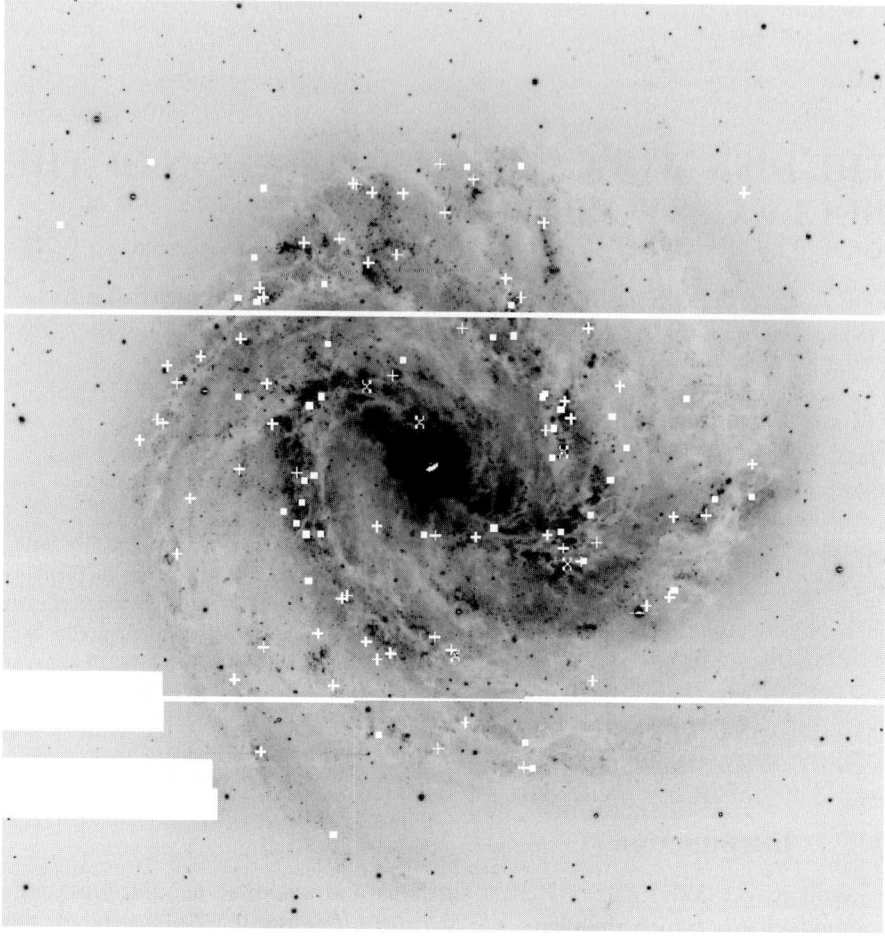

Figure 1. 12 × 12 arcsec^2 composite VLT FORS2 image (λ4684Å filter) of M83, indicating the location of the WR clusters (squares: WN, plus symbols: WC, crosses: WN+WC). North is up and east to the left. Regions to the south east are masked to avoid saturation by bright foreground stars.

2. Observations and determination of WR content

We have undertaken a systematic search for the WR content in the nearby metal-rich spiral galaxy M83 (NGC 5236). M83 is massive grand-design spiral at a distance of 4.5 Mpc (Thim et al. 2003) with on-going star formation in its spiral arms, plus an active nuclear starburst (Elmegreen et al. 1998). The favourable inclination and high metallicity (log O/H + 12 = 9.2; Bresolin & Kennicutt 2002) makes M83 an ideal candidate for studies of massive stellar populations at high metallicity.

We have used the ESO Very Large Telescope UT4 and Focal Reduced/Low Dispersion Spectrograph #2 (FORS2) to image M83 in April–June 2002 using narrow-band $\lambda4684$ (He II 4686 and C III 4650 emission lines), $\lambda4781$ (continuum) filters, plus Bessell B and Hα filters. The nucleus is extremely bright, such that we are unable to discuss the nuclear WR population in this study. A composite $\lambda4684$ image of M83 is presented in Fig. 1. Candidate WR sources were identified by constructing a difference image of the emission-line ($\lambda4684$) and continuum ($\lambda4781$) images, together with blinking the individual frames. In total, 283 candidates were identified, 198 of which were observed spectroscopically with FORS2 between April–June 2003 using the multi-object spectroscopy (MOS) mode and 300V grism (resolution 7Å).

We visually inspected the FORS2/MOS spectroscopy of individual sources to look for characteristic WR emission features, primarily He II $\lambda4686$ in WN stars, C III $\lambda4650$, $\lambda5696$ and/or C IV $\lambda5801$ in WC stars. We confirmed the presence of WR stars in 132 sources, corrected for slit losses via photometry, reddenings via Hα/Hβ nebular line fluxes, and calculated WR populations using average early/late WN and WC line fluxes from Schaerer & Vacca (1998). These are indicated in Fig. 1. In total we identify ~ 1000 WR stars in the disk of M83, neglecting remaining candidates without spectroscopic follow-up. These are fairly evenly distributed between WN (450 stars) and WC (560 stars) subtypes, representing surface abundances consistent with interior H (WN) or He (WC) burning. Several sources are in common with recent H II region (Bresolin & Kennicutt 2002) or cluster (Larsen 2004) studies, although the great majority of the sources are newly identified here.

3. Wolf-Rayet distribution and clusters

With regard to Local Group galaxies, the relative quiescent WC to WN population is known to increase monotonically with metal content (Massey & Johnson 1998), presumably due to the increase in wind strength during pre-WR phases at high metallicity. M83 continues this trend to higher metallicity with $N(\text{WC})/N(\text{WN}) \sim 1.3$. Recent bursts of star formation can cause strong deviations from this general trend via a short-lived enhancement of the WC population (Pindao et al. 2002), so galaxies where there has been a recent starburst episode may strongly deviate from this correlation, such as IC10 (Crowther et al. 2003). Since we are not studying the nuclear starburst of M83, these statistics should be reasonable – even omitting the massive cluster #74 (see later), $N(\text{WC})/N(\text{WN}) \sim 1.1$. This argues against a reduced upper-mass cutoff at high metallicity, in accord with other recent observational evidence (e.g., Pindao et al. 2002). Meynet & Maeder (2005) have recently constructed a set of evolutionary models for rotating massive stars. At low metallicity, observed statistics are well reproduced, but at twice solar metallicity far too few WC stars are predicted, i.e., $N(WC)/N(WN) \sim 0.36$ versus $\sim 1.1 - 1.3$ observed.

The most remarkable discovery of our spectroscopic survey is the dominant late-type WC population of M83, with C$_{III}$ λ5696 stronger than C$_{IV}$ λ5801. Over half the WR stars fall into the WC8–9 subtype, with the number of late to early WC stars ~ 9. In contrast, no such stars are observed in the SMC, LMC or M33. The total number of late WC stars in the Milky Way and M31 is less than ~ 50, with the number of late to early WC stars 0.9 and 0.2, respectively. Beyond the Local Group, C$_{III}$ λ5696 is observed in a small number of metal-rich WR galaxies (Pindao et al. 2002), although such populations are limited to exclusively integrated populations, versus the identification of individual late WC stars in M83.

Milky Way WC9 stars are universally observed towards the Galactic Centre, such that a metallicity role favouring their formation has long been recognised. Smith & Maeder (1991) argued that this trend was due to heavy mass-loss revealing WC subtypes at an earlier evolutionary phase, such that (C+O)/He decreases from early to late WC stars. Subsequent spectral analysis failed to confirm any systematic trend in C/He versus subtype. Instead, Crowther et al. (2002) claimed that WC subtypes resulted from metallicity-dependent wind strengths, with C$_{III}$ λ5696 scaling sensitively with wind density. Since WR winds appear to be radiatively driven (Gräfener & Hamann 2005), their wind strengths should increase with (heavy element) metallicity, analogous to OB stars, with later WC subtypes, as observed in M83. Consequently, the present observations favour metallicity-dependent winds amongst WR subtypes, of relevance to the Lyman continuum ionizing fluxes from young starbursts (Smith et al. 2002).

Super Star Clusters (SSCs) are well known in starburst and interacting galaxies, such as the Antennae (e.g., Mengel et al. 2002), but it has only recently been recognised that "quiescent" spiral galaxies also host massive, dense clusters (Larsen 2004). Within M83, WR source #74, located 30 arcsec north of the nucleus (Fig. 1) has an exceptional mixed WN+WC population of ≥ 200. From comparison with evolutionary synthesis models for a 5 Myr burst, we obtain a mass of $\sim 2 \times 10^5 M_\odot$, whilst archival *HST*/ACS images indicate FWHM~ 0.2 pc or ~ 4 pc at 4.5 Mpc, such that it is a SSC. We can estimate the O star population of this cluster from our continuum-subtracted Hα imaging, such that $N(WR)/N(O) \sim 1$. Amongst nearby galaxies, only the WR cluster NGC 3125-1 (Chandar et al. 2004) is comparable with #74. WR signatures *are* seen in rest-frame UV spectroscopy of Lyman Break galaxies (Shapley et al. 2003), but only a few (e.g., BX 418) exhibit strong WR signatures (Pettini, priv. comm.).

4. Summary

We have identified $\sim 1,000$ WR stars in the disk of the nuclear starburst galaxy M83, equally distributed amongst WN and WC subtypes, in conflict with recent evolutionary predictions for quiescent star formation at high metallicity. Late WC stars constitute half the entire WR population, in contrast with all Local Group galaxies. With regard to the nuclear starburst, Pellerin (2004) has carried out spectral synthesis of far-UV FUSE spectroscopy of the central 30×30 arcsec^2, suggesting a mass of $1.5 \times 10^6 M_\odot$ and age of 3.5 Myr, which indirectly infer a WR population of $\sim 1,700$. If such a large nuclear WR population is confirmed, the total WR population of M83 may exceed 3,000. Finally, we identify a SSC in M83, which has a similar mass to the Galactic SSC Westerlund 1 (Clark et al. 2005) but a WR content an order of magnitude higher.

More details on this study are presented in Crowther et al. (2004) and Hadfield et al. (in prep.). We are currently pursuing VLT FORS imaging / spectroscopy programs on other nearby WR galaxies, including NGC 3125 and NGC 1313.

References

Allen D.A., Wright A.E., Goss W.M., 1976, MNRAS, 177, 91
Bresolin F., Kennicutt R.C., 2002, ApJ, 572, 838
Chandar R., Leitherer C., Tremonti C., 2004, ApJ, 604, 153
Clark J.S., Negueruela I., Crowther P.A., Goodwin S., 2005, A&A, submitted
Crowther P.A., Dessart L., Hillier D., Abbott J., Fullerton A., 2002, A&A, 392, 653
Crowther P.A., Drissen L., Abbott J., Royer P., Smartt S., 2003, A&A, 404, 483
Crowther P.A., Hadfield L., Schild H., Schmutz, W., 2004, A&A, 419, L17
Elmegreen D., Chromey F., Warren A., 1998, AJ, 116, 2834
González Delgado R.M., Leitherer C., Heckman T.M., Cerviño M., 1997, ApJ, 483, 705
Gräfener G., Hamann W.-R., 2005, A&A, in press (astro-ph/0410697)
Larsen S., 2004, A&A, 416, 537
Massey P., Johnson O., 1998, ApJ, 505, 793
Mengel S., Lehnert M.D., Thatte N., Genzel R., 2002, A&A 383, 137
Meynet G., Maeder A., 2005, A&A, 429, 581
Pellerin A., 2004, Ph.D. thesis, Université Laval, Quebec, Canada
Pindao M. Schaerer D., González Delgado R.M., 2002, A&A, 394, 443
Schaerer D, Contini T., Pindao M., 1999, A&AS, 136, 35
Schaerer D. Vacca W., 1998, ApJ, 497, 618
Schaerer D., Guseva N. Izotov Y., Thuan T., 2000, A&A, 362, 53
Shapley A., Steidel C., Pettini M., Adelberger K., 2003, ApJ, 588, 65
Smith L.F., Maeder A., 1991, A&A, 241, 77
Smith L.J., Norris R., Crowther P.A., 2002, MNRAS, 337, 1309
Thim F., Tammann G.A., Saha A., Dolphi A., Sandage A., Tolstoy E., Labhardt L., 2003, ApJ, 590, 256

LASER ILLUMINATES COMPACT GALAXIES

J. Melbourne
UCO/Lick Observatory, Department of Astronomy and Astrophysics, University of California at Santa Cruz, 1156 High St., Santa Cruz, CA 95064, USA
jmel@ucolick.org

Abstract Lick laser adaptive optics (AO) images of 6 low-redshift luminous compact blue galaxies were obtained in the H and K bands. The near-IR morphologies range from disky to irregular and include one possible galaxy merger. For one of the irregular galaxies, we combine the AO data with *HST* far-ultraviolet (FUV) STIS imaging and do stellar population synthesis modeling of the sub-arcsec structure. We find a string of compact knots in the STIS image with the brightest knot roughly 10 times brighter than 30 Doradus, indicating a recent major star-forming event. This structure is seen in the near-IR AO images which also contain an even larger red concentration that is undetected in the FUV. The stellar population synthesis models indicate that the red concentration is significantly older than the blue knot. One possible explanation for the observations is a recurrent burst history for this object.

1. Introduction

Adaptive optics (AO) is a technique for correcting ground-based near-infrared (NIR) images for distortions from the turbulence in the Earth's atmosphere. Images taken with AO systems effectively overcome atmospheric seeing and recover the diffraction limit of the telescope. Until recently AO systems could only operate within 40 arcsec of $\sim 12^{\text{th}}$-magnitude guide stars, meaning most of the sky was not available for study. Now however, Lick and Keck observatories have laser AO systems that create their own artificial guide stars, thus opening up much larger portions of the sky (note: a tip-tilt guide star is still needed to correct the low-order aberrations, but this star can be much fainter, i.e., 15th mag at Lick or 18th mag at Keck, and you can typically work much farther from the guide star, up to 1 arcmin).

We take advantage of the high spatial resolution of the Lick AO system (~ 0.3 arcsec FWHM) to study the NIR structure of local counterparts to luminous compact blue galaxies (LCBGs). With typical radii smaller than 2.5 kpc, luminosities $\sim L^*$, and colors bluer than $(B - V) = 0.6$, LCBGs make up roughly 20% of galaxies at $z = 1$ (Phillips et al. 1997, Guzmán et al.

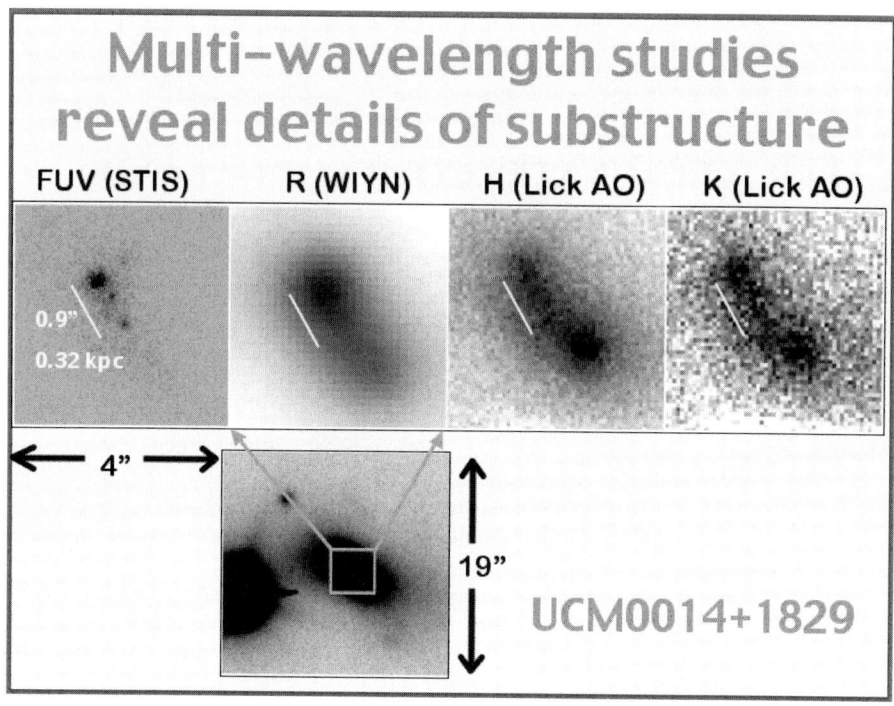

Figure 1. Images of UCM0014+1829 spanning the FUV (STIS) to the NIR (Lick). The optical image from WIYN is shown twice, once zoomed in on the core of the galaxy and once zoomed out to reveal the larger extent of the object.

1997) and may account for 45% of the SFR density at intermediate redshifts (Guzmán 1997). Studies of LCBGs with *HST* indicate that the rest-frame UV to optical morphologies include objects made up of single, double or multiple blue knots (Koo et al. 1994, Phillips et al. 1997, Guzmán et al. 1998). Our AO imaging of 6 low-redshift counterparts to the LCBGs also reveal complex sub-structures in the NIR. In this paper, we combine the AO imaging of one of these galaxies with *HST* STIS imaging in the far ultraviolet (FUV). The large wavelength coverage of high-resolution data allows us to do stellar population synthesis modeling of the subcomponents and deduce aspects of the star-formation history. Results for the 5 other galaxies will be presented in a future paper (Melbourne et al., in prep.).

2. The Images

Figure 1 shows the AO and STIS images of galaxy UCM0014+1829 ($z = 0.018$). This galaxy is similar in size, color and surface brightness to the LCBGs at higher redshift but it is about a magnitude fainter. The FUV image shows a large knot (hereafter, the blue knot) and a string of smaller knots along the semi-major axis of the galaxy. The blue knot has a FUV luminosity of $M_{\rm FUV} = -15.2$ (where $M_{\rm FUV} = -12.6$ for 30 Doradus; Smith et al. 1996), consistent with a region of intense star formation. The seeing-limited optical WIYN image is bright at the location of the blue knot and displays a tail in the direction of the smaller star-forming regions. The NIR AO images show two concentrations, one at the location of the blue knot (to the north east) and a brighter "red" concentration to the south west beyond the terminus of the string of star-forming regions. The red concentration does not appear in the FUV image.

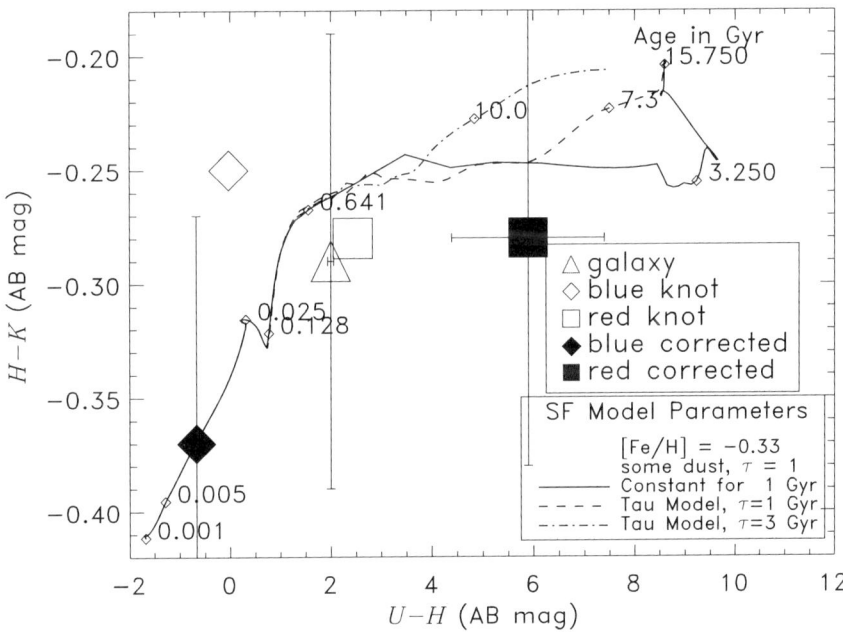

Figure 2. $(U-H)$, $(H-K)$ color-color diagram of the substructure in UCM0014+1829. The color of the entire galaxy is shown as a triangle. Colors of the blue knot are shown as diamonds, with the filled diamond showing the color of the blue knot after correcting for background light. Colors of the red knot are shown as squares, with the filled square being the color after correcting for background light. Bruzual & Charlot (2003) burst models are plotted as lines, with the model parameters given in the legend. Population ages are given in Gyr.

3. Stellar Populations

We perform aperture photometry on the high-resolution *HST* and AO images, measuring the galaxy as a whole, and the blue and red components separately. The results are plotted on the color-color diagram, Fig. 2. Also plotted on the figure are Bruzual & Charlot (2003) star-formation models including a continuous star formation model and two τ models. The numbers on the plot indicate the number of Gyr since the initial burst. We plot both a raw and background-corrected color for each component. We perform a background correction by selecting an annulus just outside our aperture to measure the background contamination. The aperture for the knots has a radius of 0.4 arcsec and the annulus a radius of 0.5 arcsec. We see that the blue knot is truly young, while the red concentration may be significantly older. One possible explanation is that the galaxy has undergone more than one starburst phase.

Acknowledgments

I would like to thank the Lick Observatory staff, especially Elinor Gates, for making these observations possible. I would also like to thank my advisors, David Koo and Claire Max, for their help with this work.

References

Bruzual G., Charlot S., 2003, MNRAS, 344, 1000

Guzmán R., Gallego J., Koo D.C., Phillips A.C., Lowenthal J.D., Faber S.M., Illingworth G.D., Vogt N.P., 1997, ApJ, 489, 559

Guzmán R., Jangren A., Koo D.C., Bershady M.A., Simard L., 1998, ApJ, 495, L13

Koo D.C., Bershady M.A., Wirth G.D., Stanford S.A., Majewski S.R., 1994, ApJ, 427, L9

Phillips A.C., Guzmán R., Gallego J., Koo D.C., Lowenthal J.D., Vogt N.P., Faber S.M., Illingworth G.D., 1997, ApJ, 489, 543

Smith D.A., et al., 1996, ApJ, 473, L21

FIRST SPECTROSCOPIC RESULTS FROM THE SPITZER INFRARED NEARBY GALAXIES SURVEY

John-David T. Smith, Robert C. Kennicutt, and the SINGS Team
Steward Observatory, University of Arizona, USA
jdsmith@as.arizona.edu, robk@as.arizona.edu

Abstract   We present first results of Spitzer spectroscopy from SINGS, the Spitzer Infrared Nearby Galaxies Survey. Spitzer's Infrared Spectrograph (IRS) is being used to spectrally map various-sized regions centered on the nucleus, and covering star-forming rings and extra-nuclear HII regions. We highlight the powerful diagnostic capabilities made possible by the sensitivity of the IRS for local samples of low, moderate, and active star-forming galaxies, including [OIV] as a tracer of very massive stars, a new PAH feature at 17.1 μm, and molecular hydrogen lines tracing hot photo-dissociation regions. We also demonstrate the versatility of the IRS spectral mapping mode for producing spatially resolved maps tracking variations of the physical parameters of the gas, dust and stars on kpc scales.

1.   Introduction

The Spitzer Infrared Nearby Galaxies Survey (SINGS) is a comprehensive survey of 75 nearby galaxies chosen to span the full range of morphological type, luminosity, and infrared activity present in the local Universe. Figure 1 demonstrates the large ranges spanned in this three-dimensional physical parameter space. Though not predominantly a starburst sample, SINGS does have a starburst component (e.g., M82), and, since it covers the full range of star-forming environments, is a natural complement to more focused investigations of extreme star formation.

SINGS is using all three Spitzer instruments to produce 3.6–160 μm imaging out to the D_{25} optical radius and 5–38 μm resolved spectroscopy of the nucleus and selected extra-nuclear targets. In addition, a large ancillary and complementary data set targeting the SINGS sample is being provided, including Chandra X-Ray, GALEX and UIT UV, optical $BVRI$, Hα, and drift-scan spectroscopy, NIR JHK and *HST* Paα imaging, SCUBA submm, BIMA SONG CO, HI, and radio continuum, and VLA HI maps. For more informa-

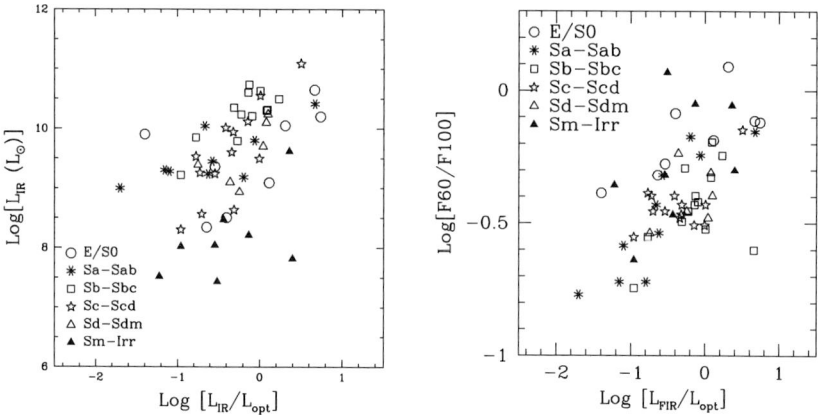

Figure 1. The SINGS sample, illustrating the large range of infrared to optical colors, total infrared luminosity, morphological type, and far-infrared color temperature spanned.

tion on the SINGS sample, data sets, and scientific objectives, see Kennicutt et al. (2003).

2. Spectral Mapping

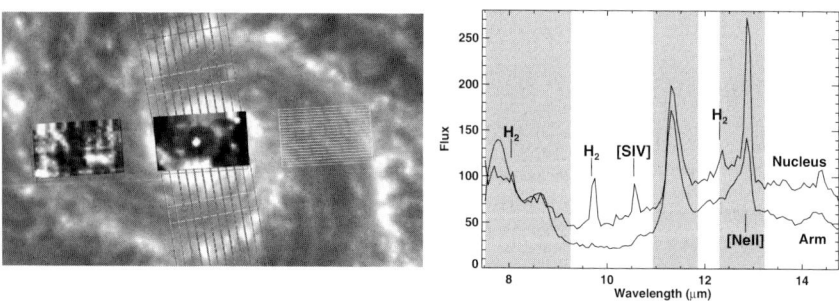

Figure 2. At left, the SINGS 8 μm image of the inner arm and nucleus of M51, showing the overlay areas of the short-low (5–15 μm, three regions left to right) and LL (15–38 μm, top to bottom) spectral maps. The spectral map of SL order 1 is overlaid, and each of the two segments is separately scaled to demonstrate the sensitivity to low-level features in the disk. At right, two representative spectra extracted from the nucleus and inner arm illustrate the band and feature variations occurring over small scales. The spectral map is shown in the three PAH bands 7.7, 11.3, and 12.7 μm, denoted on the spectra.

The SINGS spectroscopic program make exclusive use of spectral mapping mode, in which the instrument slits are rastered around the nuclei and selected extra-nuclear targets within the sample, to create 5–38 μm spectral data cubes,

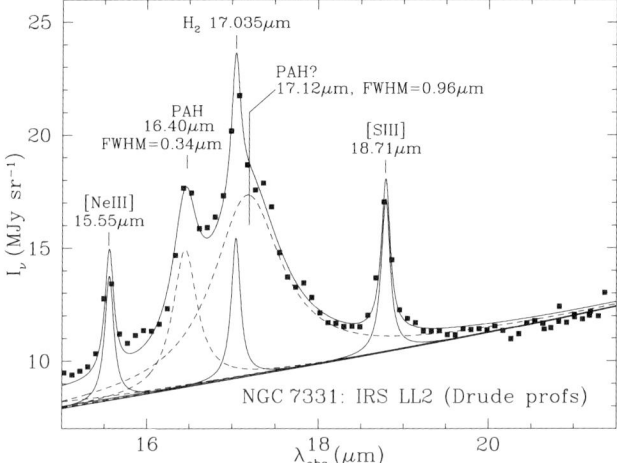

Figure 3. A fit to the 17.1 μm feature discovered in SINGS galaxy NGC 7331. The fit includes unresolved [NeIII], [SIII], and $H_2S(1)$, as well as two Drude-profile PAH components: the previously discovered 16.4 μm feature, and a new, much stronger 17.1 μm band.

including larger radial strips at 15–38 μm. An example illustrating the SINGS spectral mapping technique is shown in Fig. 2. The 8 μm inner ring and nucleus image of M51 is overlaid with the observed low-resolution spectral mapping regions, and with a PAH map created from one of the spectral cubes produced. Since the low-resolution slits are divided into two sub-slits, one for each order, an "outrigger" field is obtained as each sub-slit is separately mapped. The strong detection of features in the outrigger field demonstrates the sensitivity of the SINGS spectral maps to weak emission knots in the disk.

3. A new PAH feature

In the early spectra of SINGS galaxy NGC 7331, Smith et al. (2004) found a new, broad emission band, presumed to be due to PAHs, centered at 17.1μm. This band is associated with the 16.4μm band discovered in ISO spectra of several Galactic objects by Moutou et al. (2000), but much stronger. It was simultaneously seen in spectra of the Galactic reflection nebula NGC 7023 (Werner et al. 2004). Shown in Fig. 3, the feature is blended with [SIII], [NeIII], and $H_2S(1)$ at 17.04 μm. After removing the line contributions, the total band luminosity was found to be $\sim$ 50% of the 11.3μm PAH band. Tracking variations in inter-band strength ratios and equivalent widths to help constrain the interpretation of deep Spitzer survey counts is a major goal of SINGS.

4. MIR diagnostics

The spectral mapping capabilities of Spitzer make it possible to investigate variations in the PAH emission spectrum *within* individual galactic environments, and couple these variations using fine structure and molecular lines to the physical parameters of the ISM, HII and photo-dissociation regions (PDRs). Among the HII-region lines are two lines of [SIII], which together provide an excellent, temperature insensitive density diagnostic. Pure vibrational H_2 lines ranging from S(0) at $28.2\mu m$ to S(6) at $6.1\mu m$ are found in many SINGS spectra. They likely arise in PDRs surrounding or adjacent to the HII regions, and can be used to provide joint constraints on the pressure or density and starlight intensity there.

A nebular line of particular interest is [OIV] $25.9\mu m$, seen by ISO in many bright starbursts, and now found to be present at more than $10\times$ lower fluxes in almost all SINGS galaxies forming stars. The O^{+++} ionization potential is just above the He^+ Lyman limit of 54.4eV, and as a result [OIV] is rarely formed in HII regions. Any observed [OIV], therefore, requires shocks, the wind-enhanced far-ultraviolet emission of Wolf-Rayet stars (as seen for the first time in Spitzer spectra of Galactic WR star WR6; Morris et al. 2004), or an AGN, to provide sufficient high-energy photons. We are investigating the excitation mechanisms of this crucial diagnostic in a range of environments.

Acknowledgments

We are pleased to acknowledge the support of the Spitzer Science Center, and in particular the IRS IST. Funding for the SINGS project is provided by NASA through JPL contract 1224769.

References

Kennicutt R.C. Jr., et al., 2003, PASP, 115, 928
Morris P.W., Crowther P.A., Houck J.R., 2004, ApJS, 154, 413
Moutou C., Verstraete L., Léger A., Sellgren K., Schmidt W., 2000, A&A, 354, L17
Smith J.D.T., et al., 2004, ApJS, 154, 199
Werner M.W., Uchida K.I., Sellgren K., Marengo M., Gordon K.D., Morris P.W., Houck J.R., Stansberry J.A., 2004, ApJS, 154, 309

NEAR-IR SUPER STAR CLUSTERS IN STARBURST AND LUMINOUS INFRARED GALAXIES

Almudena Alonso-Herrero
Departamento de Astrofísica Molecular e Infrarroja, Instituto de Estructura de la Materia,
Consejo Superior de Investigaciones Científicas, Serrano 113b, 28006 Madrid, Spain

Abstract   NICMOS on the *HST* is playing a relevant role in understanding the properties of near-infrared super star clusters (SSCs) and the associated younger population of giant HII regions. Giant HII regions and SSCs represent the dominant mode of recent star formation – younger (a few million years old) and "older" ($\simeq 3-500$ Myr), respectively. In particular, Paα ($\lambda_{\rm rest} = 1.87\,\mu$m) observations together with radio observations offer us an unprecedented view of the truly youngest (high-mass) star-formation activity, as it is plausible that a large fraction of the youngest SSCs in galaxies are hidden by dust in their natal HII regions. We will summarize recent results on the properties of near-infrared SSCs and associated giant HII regions in nearby starburst galaxies and LIRGs.

1.   Introduction

Massive (young) star clusters, the so-called super star clusters (SSCs), were first discovered in galaxies with very active star formation (SF) in the optical (see, among others, Whitmore et al. 1993, Schweizer et al. 1996, Zepf et al. 1999), and more recently in the near-IR (Alonso-Herrero et al. 2000, 2001a, 2002 [AAH00, AAH01a, AAH02], Maoz et al. 2001, Scoville et al. 2000).

The general properties of near-IR SSCs are relatively well known from high-resolution *HST* imaging, and high-resolution near-infrared (IR) and optical spectroscopy. Their sizes are $1-5$ pc or larger, kinematic masses on the order of $10^5 - 10^6\,{\rm M}_\odot$ (e.g., McCrady et al. 2003 for SSCs in M82; Mengel et al. 2002 for SSCs in the Antennae), comparable to those of globular clusters (GCs), and a broad range of ages ($\simeq 3-200$ Myr), significantly younger than GCs. SSCs can contribute to up to 20% of the stars formed in the current episode of SF of galaxies (e.g., Zepf et al. 1999).

In this paper I will summarize recent results, as well as on-going research by our group, on the properties of near-IR SSCs and associated giant HII regions in nearby starburst galaxies and LIRGs.

2. The dwarf galaxy NGC 5253

Although initially discovered mainly in interacting galaxies and luminous and ultraluminous IR galaxies (LIRGs, ULIRGs), near-IR SSCs are also detected in ringed galaxies (e.g., Alonso-Herrero et al. 2001b [AAH01b], Maoz et al. 2001, and references therein), and dwarf galaxies (e.g., Aloisi et al. 2001, Alonso-Herrero et al. 2004 [AAH04]).

NGC 5253 is a nearby dwarf galaxy ($D = 4.1$ Mpc) whose central region is undergoing one of the youngest processes of SF in the local Universe. The *HST*/NICMOS images have revealed the presence of a double cluster (C1+C2) in the nucleus of the galaxy separated by $0.3 - 0.4$ arcsec or $6 - 8$ pc. This double cluster is also a bright double source of Paα ($\lambda_{\rm rest} = 1.87\,\mu$m) emission. The western cluster (C2) is almost entirely obscured ($A_V \simeq 11$ mag) at UV and optical wavelengths, but becomes the brightest source in the galaxy at $\lambda > 2\,\mu$m. The near-IR double cluster is coincident with the double radio nebula (Turner et al. 2000). Both clusters are extremely young, with ages of approximately $3 - 4$ Myr. C2 is more massive than C1 by a factor of 6 to 20 ($M_{\rm C2} = 7.7 \times 10^5 - 2.6 \times 10^6\,M_\odot$, for a Salpeter IMF in the mass range $0.1 - 100\,M_\odot$) putting them in the SSC category (see AAH04). Martín-Hernández et al. (2004) have obtained high-resolution mid-IR spectroscopy of the C2 SSC, and have shown that a very young ($3 - 4$ Myr) age for the SSC is only plausible if the C2 cluster is deficient in massive stars – implying a non-standard IMF. An older age ($5 - 6$ Myr) would require a longer "hidden" phase for SSCs than generally thought (e.g., Tan & McKee 2000).

In addition to the nuclear double cluster, we have identified 269 near-IR star clusters over the central 270 pc. Since the presence of hydrogen recombination line emission indicates a young stellar population, we have used the equivalent width (EW) of Paα and the absolute H-band magnitudes to study the ages and stellar masses of the youngest of these clusters (see Fig. 1). We find that $20-30\%$ of the detected clusters in the H-band have ages younger than 7 Myr, and (photometric) stellar masses of between $3 \times 10^3\,M_\odot$ and $3 \times 10^4\,M_\odot$. For older clusters – those without Paα emission – Harris et al. (2004) have estimated ages of up to 200 Myr using optical data.

3. Luminous Infrared Galaxies
Properties of SSCs and H_{II} regions

While a lot of effort is being devoted to understanding the properties of the SSCs, *HST*/NICMOS has only recently revealed a population of bright H_{II} regions in LIRGs (AAH00, AAH01a, AAH02). A large fraction show Hα luminosities in excess of that of 30 Doradus (Fig. 2), the prototypical giant H_{II} region. These exceptionally bright H_{II} regions are more common in LIRGs

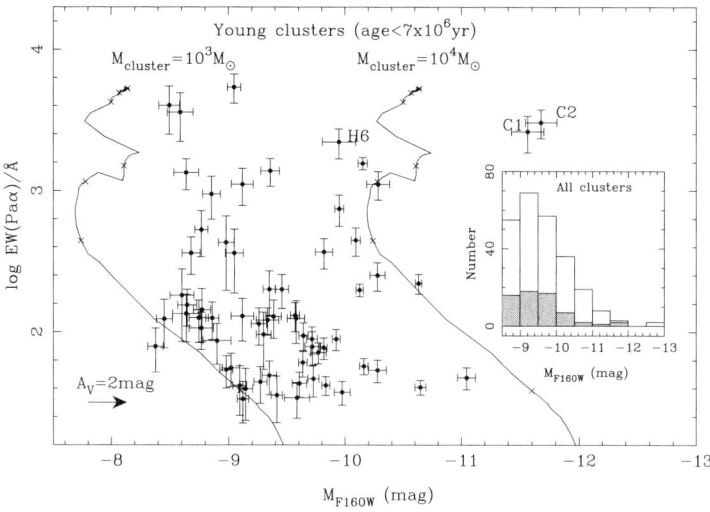

Figure 1. Absolute $M_{\rm F160W}$ magnitude vs. EW of the Paα emission line for the youngest ($\log {\rm EW(Pa}\alpha)/ \geq 1.5$ or age < 7 Myr) H-band selected clusters (filled dots) in the central region of NGC 5253. The H-band (F160W) magnitudes have not been corrected for extinction. The arrow shows what would be the effect of correcting the observed absolute $M_{\rm F160W}$ for $A_V = 2$ mag. The lines represent the time evolution (crosses on the curves are drawn at 1 Myr intervals, youngest ages at the top) of star clusters with masses $M = 10^3 {\rm M}_\odot$ and $M = 10^4 {\rm M}_\odot$ for instantaneous SF, a Salpeter IMF ($M_{\rm low} = 1 {\rm M}_\odot$ and $M_{\rm up} = 100 {\rm M}_\odot$), and solar metallicity, using STARBURST99 (Leitherer et al. 1999). Insert: The open histogram shows the distribution of the absolute $M_{\rm F160W}$ magnitudes of all star clusters selected in the H band, whereas the filled histogram shows the clusters with ages younger than 7 Myr, that is, those shown in the main panel of this figure.

than in normal galaxies, as illustrated in Fig. 2 where we show histograms of the Hα luminosities of HII regions identified in LIRGs, and normal galaxies (see Alonso-Herrero & Knapen 2001 and AAH02). The giant HII regions identified in LIRGs – with sizes of 80 to 200 pc – are often located close to, but in most cases not spatially coincident with, the near-IR SSCs. The giant HII regions detected in LIRGs are not confined to the nuclei, but are also found off-nucleus: at the interface of interacting galaxies, and in the spiral arms of interacting/merging systems, emphasizing that the effects of extreme SF can propagate throughout the galaxies and not only the nuclei.

The extraordinary luminosities of giant HII regions and SSCs in LIRGs may imply up to 10^6 M$_\odot$ in newly-formed stars (see AAH00, AAH01a, AAH02) per individual HII region/SSC. Such massive clusters and HII regions are rarely seen in normal galaxies (Fig. 2). There are a number of possible explanations. We may just be seeing the extended tail of a luminosity function, or the bright HII regions may represent aggregations of normal HII regions, or perhaps there is a truly unique population of SF regions in LIRGs.

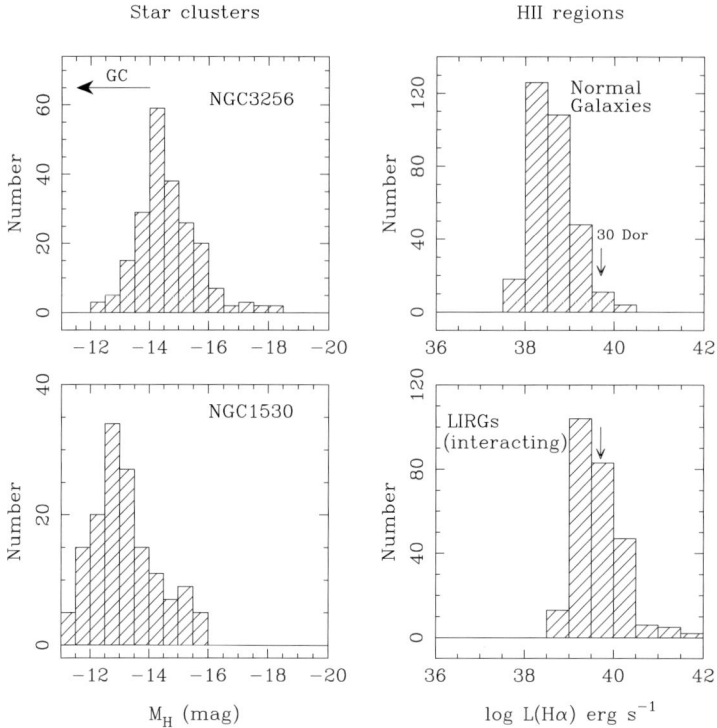

Figure 2. Left panels: Comparison of the absolute H-band magnitudes (from *HST*/NICMOS) observations) of star clusters detected in the LIRG NGC 3256 and the starburst galaxy NGC 1530. We also show the typical magnitudes of globular clusters (see, e.g., Kissler-Patig et al. 2002). Right panels: Hα luminosities (from NICMOS Paα observations, not corrected for reddening) of HII regions detected in LIRGs (NGC 3256 and Arp 299), and in normal galaxies. The arrow indicates the luminosity of 30 Dor, the prototypical giant HII region.

Theoretical and observational arguments suggest that these HII regions and SSCs represent the dominant mode of recent SF in LIRGs – younger (a few millions years old) and "older" ($\simeq 5 - 200$ Myr), respectively – in LIRGs (AAH00, AAH01a, AAH02, Mengel et al. 2001). Tan & McKee (2000) have proposed a model in which young star clusters within clouds with masses $5 \times 10^4 - 5 \times 10^5$ M$_\odot$ can survive for up to 3 Myr, and allow for the formation of gravitationally bound clusters (which, in turn, may evolve into GCs) in the presence of vigorous feedback (Wolf-Rayet winds and supernovae), before the gas will be dispersed. In AAH02 we showed that the apparent offsets between the location of the SSCs and the HII regions in LIRGs is an age effect, as the youngest SSCs are still embedded in the HII regions, and demonstrates that giant HII regions offer us an unprecedented view of the truly youngest (high-mass) SF activity, while the SSCs show a broader range of ages.

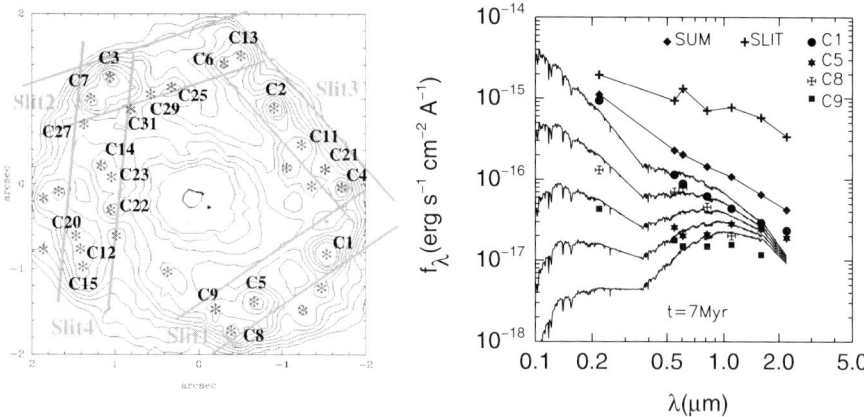

Figure 3. Contours of the *HST*/NICMOS 1.1 μm emission (left panel) of NGC 7469 showing the PA of the UKIRT slits and the near-IR (at 1.1 μm) selected SSCs (star symbols). The right panel shows the UV to near-IR SEDs (symbols) of the four SSCs included in Slit 1, as well as the SED of the sum of the four SSCs and that of the region covered by the slit. The lines are outputs of STARBURST99 (Leitherer et al. 1999) for an instantaneous burst with an age of 7 Myr, and different values of the extinction: $A_V = 0$ (top line), 0.5, 1, 1.5 and 2 mag (bottom line).

NGC 7469

NGC 7469 is a nearby ($D = 65$ Mpc) LIRG that contains a Seyfert 1 nucleus and a 1 kpc-diameter ring of SF (Wilson et al. 1991). The ring is detected in the UV, optical, near and mid-IR (Soifer et al. 2003), and radio (Colina et al. 2001). High spatial resolution *HST* data covering the spectral range from 2000Å to 2.2 μm reveal a large number of SSCs within the ring of SF (Fig. 3, left panel). We have also obtained UKIRT K-band spectroscopy of the SSCs, placing slits at four different position angles (PA; see Fig. 3), covering almost the entire SF ring of this galaxy. The measured EWs of Brγ ($\lambda = 2.166$ μm) and the CO index (at 2.3 μm) indicate a remarkably uniform range of ages for the four slit positions, 6 – 7 Myr after the peak of SF.

Genzel et al. (1995) modeled the integrated properties of the ring in NGC 7469 and concluded that either the SF has been progressing at a constant rate for the past 10 Myr, with few high mass stars, or has been decaying exponentially since the onset of a burst 15 Myr ago, with a fairly normal IMF. We have constructed UV to near-IR spectral energy distributions (SEDs) of near-IR selected clusters covered by one of the slits (Fig. 3, right panel, and Díaz-Santos et al. 2005). The SEDs of the individual clusters are compared to the SEDs of the sum of all of them, and that of the area covered by the slit. The SEDs of the individual SSCs show different degrees of extinction and a range of ages, whereas the SEDs of the sum and the slit resemble that of the brightest cluster, C1. Moreover, the SED of C1 is well reproduced with an age of 7 Myr and little extinction as inferred from the near-IR spectroscopy, whereas the less dominant SSCs do not necessarily show the same properties. We will characterize the properties (extinction, age, stellar mass) of individual SSCs, as well as determine the effects of spatial resolution when modeling the integrated properties of the ring of SF (Genzel et al. 1995) vs. those of individual SSCs.

Acknowledgments

I would like to thank all my collaborators (M.J. Rieke, G.H. Rieke, J. Knapen, S. Ryder, N. Scoville, L. Colina, T. Díaz-Santos, A. Baker, A. Quillen, C. Engelbracht, M. Imanishi, T. Toshinobu) for their help in making the research presented in this paper possible.

References

Aloisi A., et al., 2001, AJ, 121, 1425
Alonso-Herrero A., Rieke G.H., Rieke M.J., Scoville N.Z., 2000, ApJ, 532, 845 (AAH00)
Alonso-Herrero A., Engelbracht C.W., Rieke M.J., Rieke G.H., Quillen A.C., 2001a, ApJ, 546, 952 (AAH01a)
Alonso-Herrero A., Ryder S.D., Knapen J.H., 2001b, MNRAS, 322, 757 (AAH01b)
Alonso-Herrero A., Knapen J.H., 2001, AJ, 122, 1350
Alonso-Herrero A., Rieke G.H., Rieke M.J., Scoville N.Z., 2002, AJ, 124, 166 (AAH02)
Alonso-Herrero A., Takagi T., Baker A.J., Rieke G.H., Rieke M.J., Imanishi M., Scoville N.Z., 2004, ApJ, 612, 222 (AAH04)
Colina L., Alberdi A., Torrelles J.M., Panagia N., Wilson A.S., 2001, ApJ, 553, L19
Díaz-Santos T., Alonso-Herrero A., Colina L., et al., 2005, in prep.
Genzel R., Weitzel L., Tacconi-Garman L.E., Blietz M., Cameron M., Krabbe A., Lutz D., Sternberg A., 1995, ApJ, 444, 129
Harris J., Calzetti D., Gallagher III J.S., Smith D.A., Conselice C.J., 2004, ApJ, 603, 503
Kissler-Patig M., Brodie J.P., Minniti D., 2002, A&A, 391, 441
Leitherer C., et al., 1999, ApJS, 123, 3
McCrady N., Gilbert A.M., Graham J.R., 2003, ApJ, 596, 240
Maoz D., Barth A.J., Ho L.C., Sternberg A., Filippenko A.V., 2001, AJ, 121, 3048
Martín-Hernández N.L., Schaerer D., Sauvage M., 2004, A&A, in press (astro-ph/0408582)
Mengel S., Lehnert M.D., Thatte N., Tacconi-Garman L.E., Genzel R., 2001, ApJ, 550, 280
Mengel S., Lehnert M.D., Thatte N., Genzel R., 2002, A&A, 383, 137
Schweizer F., Miller B.W., Whitmore B.C., Fall S.M., 1996, AJ, 112, 1839
Scoville N.Z., et al., 2000, AJ, 119, 991
Soifer T., Bock J.J., Marsh K., Neugebauer G., Matthews K., Egami E., Armus L., 2003, AJ, 126, 143
Tan J.C., McKee C.F., 2001, in: Starburst Galaxies: Near and Far, Tacconi L., Lutz D., eds., Springer: New York, p. 188
Turner J.L., Beck S., Ho P.T.P., 2000, ApJ, 532, L109
Whitmore B.C., Schweizer F., Leitherer C., Borne K., Robert C., 1993, AJ, 106, 1354
Wilson A.S., Helfer T.T., Haniff C.A., Ward M.J., 1991, ApJ, 381, 79
Zepf S.E., Ashman K.M., English J., Freeman K.C., Sharples R.M., 1999, AJ, 118, 752

HIGH-RESOLUTION IMAGING OF THE SSCS IN NGC 1569 AND NGC 1705

M. Sirianni[1], G. Meurer[2], N. Homeier[2], M. Clampin[3], R. Kimble[3], and the ACS Science Team
[1]*European Space Agency – Research and Scientific Support Department, Space Telescope Science Institute, 3700 San Martin Drive, Baltimore, MD, 21218, USA*
[2]*Johns Hopkins University, USA*
[3]*NASA Goddard Space Flight Center, USA*
sirianni@stsci.edu

Abstract We observed the local starburst galaxies, NGC 1569 and NGC 1705 with the *HST/ACS* High Resolution Channel in the U, V, and I filters. The superb resolution of the images (0.018 arcsec pix^{-1}) shows, for the first time, the real morphology of the Super Star Clusters. We derive the structural parameters and total luminosity and recalculate, with the new parameters, the virial mass of the clusters.

1. Introduction

The *Hubble Space Telescope (HST)* has shown that young, compact, luminous star clusters ("super star clusters" or SSCs) may exist in different environments. Their average luminosity, size and spectroscopically inferred masses make them young analogues of classical globular star clusters. In order to evolve into globular clusters, the SSCs need to be more massive than $\sim 3 \times 10^4$ M$_\odot$ and of the right size (half-light radius $\sim 1-10$ pc). While the size of the cluster is, in principle, relatively easy to measure, an estimate of the mass is more complicated. If the total light of an object is known, then the mass can be estimated by assuming an initial mass function (IMF) or M/L ratio. If instead, the mass can be measured independently (for example, assuming complete virialization of the cluster), the observed M/L ratio can be compared with model conditions to constrain the IMF. We present new data on NGC 1569 and NGC 1705, two nearby dwarf galaxies. Each contains well-studied SSCs. The high resolution of the new *HST* images allows us to determine improved structural parameters for these clusters. We will present the analysis of the color profiles and of the stellar content of these SSCs in a separate paper (Sirianni et al., in prep.).

2. The Data

The images were obtained with the High Resolution Channel (HRC) of the Advanced Camera for Surveys (ACS) as part of the ACS/GTO program. HRC provides a 27×31 arcsec^2 field of view with a pixel scale of 0.025 arcsec pix^{-1}. The data consist of a 3-point dithered observation in the filters F330W, F555W and F814W. Short and long exposures were taken to ensure that information on saturated pixels in the long-exposure images could be recovered from the short exposures. The images were processed with the ACS-IDT pipeline (APSIS; Blakeslee et al. 2003) using a square kernel and re-sampling the point-spread function to a final scale of 0.018 arcsec pix^{-1}.

SSCs in NGC 1569

González Delgado et al. (1997) detected signatures of both Wolf-Rayet (W-R) stars and red supergiants (RSG) in ground-based data of the SSC NGC 1569A. Instantaneous burst models do not predict the simultaneous presence of the two types of stars. In *HST*/WFPC2 images, De Marchi et al. (1997) were able to identify two main components in cluster A, separated by ~ 0.2 arcsec, and suggested that the two different stellar populations could be associated with the two different clumps. However, the large overlap of the two components in WFPC2 images made it difficult to disentangle the two sub-clusters. Origlia et al. (2001) were able to isolate and age-date the hot and cool stellar component of NGC 1569-A with *HST* UV and IR imaging. The authors inferred an age of ~ 5 Myr for A2 and ~ 10 Myr for A1. Their conclusion has been confirmed by Maoz et al. (2001), who obtained STIS spectra of the region of NGC 1569-A. Their data show that the W-R emission is confined to cluster A2. The new HRC images clearly show the multiple nature of the SSC-A (see Figs. 1 and 2). We list the integrated magnitude and half-light core radius for SSCs A-1, A-2 and B in Table 1. Despite the strong evidence of a dual nature of SSC-A, most authors kept considering them as a single entity for the calculation of their M/L ratio and therefore to infer an IMF.

SSC in NGC 1705

The structure of the SSC NGC 1705-1 has recently been analyzed by Smith & Gallagher (2001), and its stellar content by Vázquez et al. (2004). Our results, listed in Table 1, confirm the findings of Smith & Gallagher (2001), with a slightly larger half-light radius.

3. Results

With the new determination of the luminosity and half-light core radius, we can recalculate the virial masses and the observed M/L ratio. If we assume

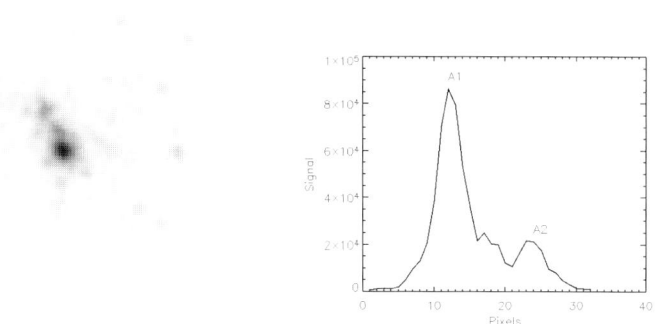

Figure 1. Negative image showing NGC 1569 A as observed through the F555W filter of the HRC.

Figure 2. Counts per pixel along the vector connecting the two components of NGC 1569A. The pixel scale is 0.018 arcsec pix^{-1}.

a distance of 5.1 Mpc for NGC 1705 and 2.2 Mpc for NGC 1569, and the velocity dispersion from Ho & Filippenko (1996) and Gilbert et al. (2003), we derive an L/M ratio of 48.5 for NGC 1705-1 and 59 and 43 for NGC 1569-A1 and NGC 1569-B, respectively. When we compare the observed L/M ratio vs. age with STARBURST99 models with different IMF (see for example Vázquez et al. 2004) we find that NGC 1569-A1, NGC 1569-B and NGC 1705-1 are very similar and consistent with a Kroupa-type or Salpeter IMF with a lower mass cutoff at ~ 1 M$_\odot$.

Table 1. Photometric and structural parameters

SSC	M_{F555W} (mag)	(F330W−F555W)$_0$ (mag)	(F555W−F814W)$_0$[1] (mag)	R_{05} (arcsec)	Age[2] (Myr)
NGC 1705-1	−13.8	−1.4	0.6	0.08	12
NGC 1705-2	−10.9	−1.5	0.1	0.11	...
NGC 1569-A1	−13.6	−1.30	−0.07	0.10	12
NGC 1569-A2	−11.9	−1.35	−0.06	0.17	5–7
NGC 1569-B	−12.6	−1.29	0.62	0.16	12

[2]The effects of the red halo in the F814W filter have not yet been removed.
[2]Ages for NGC 1705-1 from Vázquez et al. (2004), NGC 1569-A1 and NGC 1569-B from Anders et al. (2004), NGC 1569-A2 from Origlia et al. (2001).

Anders et al. (2004) determined the masses for the most SSCs in NGC 1569. Their masses are higher than those inferred by kinematic studies. Since the estimate of kinematical masses assume the complete virialization of the cluster, which is likely not to be the case for these massive and young clusters, the virial mass could be an underestimate. With a higher mass, the L/M ratio will be lower and both clusters and the major SSCs of NGC 1569 would be more consistent with a normal IMF with a 0.1 $M_\odot$ lower mass cut-off.

References

Anders P., de Grijs R., Fritze-v. Alvensleben U., Bissantz N., 2004, MNRAS, 347, 17
Blakeslee J.P., et al., 2003, in: Astronomical Data Analysis Software and Systems XII, ASP Conf. Ser., Payne H.E., Jedrzejewski R.I., Hook R.N., eds., 295, 257
De Marchi G., Clampin M., Greggio L., Leitherer C., Nota A., Tosi M. 1997, ApJ, 479, L27
González Delgado R.M., Leitherer C., Heckman T., Cerviño M., 1997, ApJ, 483, 705
Maoz D., Ho L.C., Sternberg A., 2001, ApJ, 544, L139
Origlia L., Leitherer C., Aloisi A., Greggio L., Tosi M., 2001, AJ, 122, 815
Vázquez G., Leitherer C., Heckman T.M., Lennon D.J., De Mello D.F., Meurer G.R., Martin C.L., ApJ, 600, 161

MASSIVE STAR CLUSTERS, FEEDBACK, AND SUPERWINDS

Andrea M. Gilbert[1] and James R. Graham[2]
[1] Max-Plank-Institut für extraterrestrische Physik, Garching, Germany
[2] Department of Astronomy, University of California, Berkeley, CA, USA
agilbert@mpe.mpg.de, jrg@astro.berkeley.edu

Abstract Young massive star clusters return matter and energy to the ISM via cluster winds that ultimately combine to power superwinds in high-density starbursts. We present high-resolution near-IR spectroscopy of such clusters that dominate the current star formation in the nearby merger, the Antennae. By fitting outflow models to their broad hydrogen lines, we measure the mass-loss rates and mechanical luminosities of these cluster outflows. They are strongly mass-loaded with high thermalization efficiencies, two properties that also characterize superwinds. While the Antennae do emit large-scale diffuse X-rays, their present star cluster density may be too low to generate an M82-like superwind, but it is perfect for providing a resolved look at feedback mechanisms in a young merger.

1.   Introduction

Star formation in starbursts creates massive ($10^5 - 10^6$ $M_\odot$) young super star clusters (SSCs) that are not often found in more quiescent environments like the Milky Way's disk. They are found in all starburst environments, from dwarfs to mergers, and one of the most spectacular examples of the latter is the Antennae Galaxies (NGC 4038/39, Fig. 1), which have a rich SSC population that is dominated by very young clusters (< 30 Myr; e.g., Whitmore et al. 1999). Whether young SSCs survive to become GCs or disperse into the field star population of a galaxy, they have a great influence on the energetics of its interstellar medium (ISM) because they harbor thousands of massive stars producing ionizing and FUV radiation. The Lyman continuum fluxes of the youngest SSCs range over $Q[H^+] = 10^{52-53}$ s^{-1}, dwarfing that of R136, which has a mass $\geq 10^{4.5}$ $M_\odot$ (Massey & Hunter 1998) and $Q[H^+] = 10^{51.4}$ s^{-1} (Walborn 1991). The radiation from OB stars excites H II regions (H IIRs) and photodissociation regions, and heats the ISM. Their winds and supernova (SN) ejecta stir and inject energy into the surrounding ISM. In the most extreme starbursts, the combined effects of a starburst drive large-scale galactic

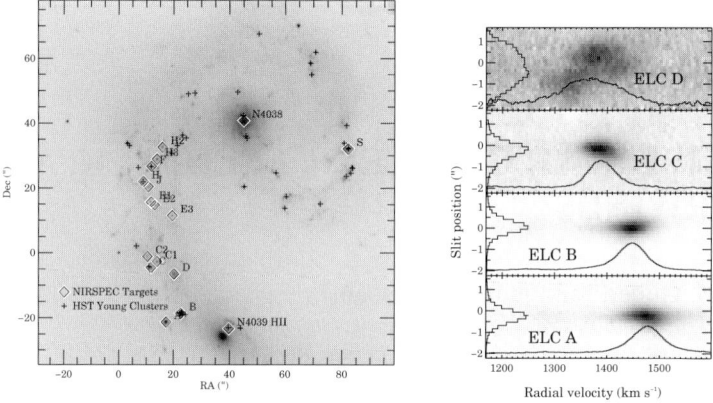

Figure 1. ISAAC narrow-band (2.25 μm) image (left) of the Antennae with our sample of ELCs and the brightest young optical clusters of Whitmore et al. (1999) marked; due to variations in extinction, the brightest optical and IR sources do not necessarily coincide. NIRSPEC spectra of the four brightest ELCs in the southern overlap region (right) feature broad Brγ emission that is spatially extended relative to the continuum; the spatially-resolved velocity gradients suggest non-spherical flows (Gilbert & Graham 2005a).

winds that can eject matter into the intergalactic medium. SSCs are concentrated power sources for feedback on both star-cluster and galactic scales.

2. ELCs in the Antennae: A New Class of HII Region

We observed signatures of the feedback from SSCs in a near-IR imaging and spectroscopic study of a sample of young Antennae SSCs. Most of the targets are emission-line clusters (ELCs) that have broad, spatially extended, non-Gaussian Brγ lines (Fig. 1) whose supersonic widths (60 – 110 km s^{-1}) exceed estimates of their virial line widths (Gilbert 2002). Thus the HIIRs of these clusters contain at least some high-velocity gas that is not bound.

Figure 2 (Gilbert & Graham 2005a) places the HIIRs that are excited by ELCs in context with other types of HIIRs. The range in $Q[\mathrm{H}^+]$ for ELCs exceeds but overlaps with that of top-ranked extragalactic giant HIIRs (GHIIRs). The nonthermal component of line width among all HIIR classes overlaps, though there is a well-known trend of increasing luminosity with line width. The HIIR classes are better separated in a diagram of mean electron density, n_e, versus size, which shows Galactic compact and ultracompact HIIRs ((U)CHIIs) at the highest n_e and smallest size, GHIIRs at the lowest n_e but largest size, and ELCs falling between these extremes in a region shared with the newly discovered class of deeply embedded ultra-dense HIIRs (UDHIIs; Kobulnicky & Johnson 1999). UDHIIs are radio and mid-IR-detected HIIRs with very high inferred n_e and radio sizes as small as a few parsec, while their

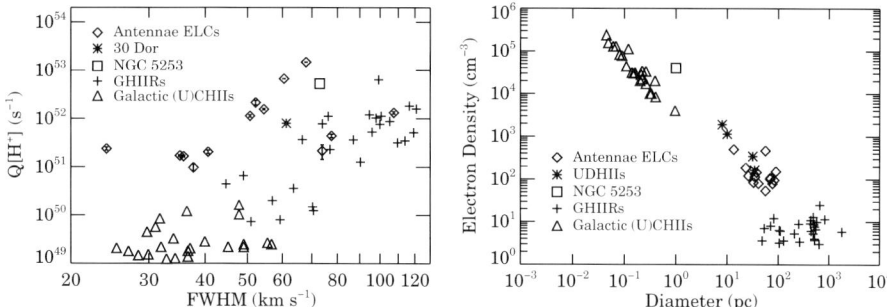

Figure 2. Comparison of properties of ELCs and other HIIRs. ELCs have extreme values in the continuum of $Q[H^+]$ vs. FWHM of the nonthermal component of H recombination lines *(left)*. ELCs and UDHIIs occupy a common region in the space *(right)* of mean n_e vs. diameter (of infrared emission for ELCs and UDHIIs, and of radio emission for other objects). Assuming the pc-scale radio-inferred sizes for UDHIIs or typical optical SSC sizes for ELCs increases their mean densities, moving them closer to the (U)CHIIs. References are: ELCs, Gilbert & Graham (2005a,b); 30 Dor, Walborn (1991), Chu & Kennicutt (1994); NGC 5253, Turner et al. (2000, 2003); UDHIIs in He 2-10, Kobulnicky & Johnson (1999), Vacca et al. (2002); top-ranked GHIIRs, Arsenault & Roy (1988); Galactic (U)CHIIs, Garay & Lizano (1999).

IR sizes can be ten times larger (Vacca, Johnson & Conti 2002), in the range occupied by the K-band sizes of ELCs (Gilbert & Graham 2005a). The very young (< 1 Myr) UDHIIs may evolve into ELCs ($3 - 7$ Myr) as they ionize and expel their ambient material.

3. Feedback from ELCs: Mass-Loaded Cluster Winds

Because some ELCs appear to remove gas from their HIIRs by driving cluster outflows, we employed a simple kinematic model of such an outflow in order to measure the mass-loss rate $\dot{M}_{\rm HII}$ and amount of energy and momentum in the flows (Gilbert 2002). The Brγ emission-line profile for a spherical flow described by a β-wind velocity law (à la Kudritzki & Puls 2000), together with the assumption of a constant $\dot{M}_{\rm HII}$, provides excellent fits to the observed line profiles. From the fits we derive cluster wind values of $\dot{M}_{\rm HII} = 0.01 - 1$ M$_\odot$ yr^{-1} and terminal velocities < 200 km s^{-1}, which is an order of magnitude below the velocities observed for individual O or WR stellar winds. This suggests that the combined stellar winds are slowed by interactions with each other and the surrounding medium. That picture is supported by the great extent to which the observed values of $\dot{M}_{\rm HII}$ exceed the amount of stellar ejecta predicted by STARBURST99 (Leitherer et al. 1996) for the populations of massive stars that are providing the observed $Q[H^+]$ (Fig. 3). The amount of energy in observed gas phases cannot account for all of the available input energy from ELCs (Fig. 3; see Gilbert & Graham 2005b); this is also the case for wind-blown superbubbles and galactic superwinds, and the unseen energy must be radiated away

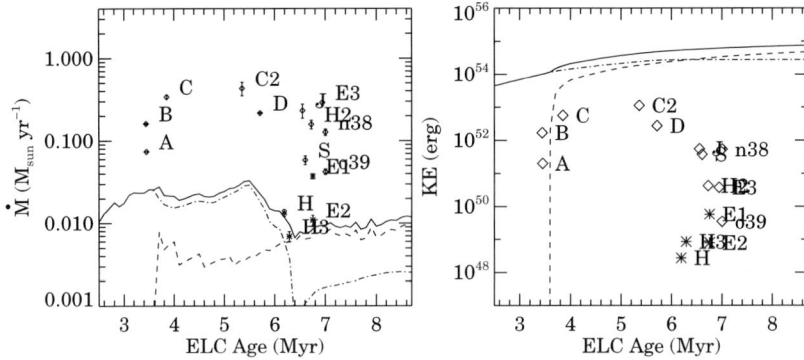

Figure 3. Mass-loss rates *(left)* and kinetic energy *(right)* measured for ELC outflows (points, normalized to a 10^6 M$_\odot$ cluster and extinction corrected where possible). Curves are predictions from STARBURST99 models for a cluster's stellar winds (dash-dotted), SNe (dashed), and their sum (solid). ELC outflows display much greater $\dot{M}_{\rm HII}$, and lower KEs than models predict are available from stellar ejecta alone, which suggests efficient mass loading and thermalization of KE.

or hidden in another phase. ELCs have pressures that are high enough to blow out of a Milky Way disk, and in the overlap region (~ 4 kpc) they satisfy the star-formation surface density threshold for creating superwinds (e.g., Heckman 2003). ELCs and SSCs disrupt the Antennae ISM on kpc scales, and may ultimately form a large-scale wind.

References

Arsenault R., Roy J.-R., 1988, A&A, 201, 199
Chu Y.-H., Kennicutt R.C., 1994, ApJ, 425, 720
Garay G., Lizano S., 1999, PASP, 111, 1049
Gilbert A.M., 2002, Ph.D. thesis, UC Berkeley, USA
Gilbert A.M., Graham J.R., 2005a,b, ApJ, subm.
Heckman T., 2003, in: Galaxy Evolution, Theory and Observations, Avila-Reese V., Firmani C., Frenk C.S., Allen C., eds., RMxAC, 17, 47
Kobulnicky H.A., Johnson K.E., 1999, ApJ, 527, 154
Kudritzki R.A., Puls J., 2000, ARA&A, 38, 613
Leitherer C., et al., 1999, ApJS, 123, 3
Massey P., Hunter D.A., 1998, ApJ, 493, 180
Turner J.L., Beck S.C., Ho P.T.P., 2000, ApJ, 532, L109
Turner J.L., Beck S.C., Crosthwaite L.P., Larkin J.E., McLean I.S., Meier D.S., 2003, Nature, 423, 621
Walborn N.R., 1991, in: The Magellanic Clouds, IAU Symp. 148, Haynes R., Milne D., eds., (Kluwer: Dordrecht), p. 145
Whitmore B.C., Zhang Q., Fall S.M., Schweizer F., Miller B.W., 1999, AJ, 118, 1551
Vacca W.D., Johnson K.E., Conti P.S., 2002, AJ, 123, 772

30 DORADUS – A TEMPLATE FOR "REAL STARBURSTS"?

Bernhard R. Brandl
Leiden Observatory, P.O. Box 9513, 2300 RA Leiden, The Netherlands
brandl@strw.leidenuniv.nl

Abstract 30 Doradus is the closest massive star-forming region and the best studied template of a starburst. In this conference paper, we first summarize the properties of 30 Doradus and its stellar core R136. We discuss the effects of insufficient spatial resolution and cluster density profiles on dynamical mass estimates of super star clusters, and show that their masses can be easily overestimated by a factor of ten or more. From a very simple model, with R136-like clusters as representative building blocks, we estimate typical luminosities on the order of 10^{11} L$_\odot$ for starburst galaxies.

1. Overview

At a distance of only about 53 kpc 30 Doradus is the closest massive star-forming region. To date, it has been studied and described in over 3000 papers. In this conference paper, we discuss the relevance of 30 Doradus as a local template for more luminous starburst systems. The characteristic dimensions for structures related to 30 Doradus are given in Table 1.

Table 1. Characteristic dimensions, from Walborn (1991)

Name	Class	angular Ø	linear Ø
LMC	galaxy	5°	5000 pc
30 Dor region	complex	1°	1000 pc
30 Dor nebula	H II region	15′	200 pc
NGC 2070	stellar cluster	3′	40 pc
R136	stellar core	10″	2.5 pc

The most relevant properties of its stellar content and interstellar medium (ISM) are summarized in Table 2. Figure 1a shows the complex nature of the ISM in 30 Doradus with its interplay between large-scale filaments and wind-blown cavities. According to Walborn & Blade (1997), NGC 2070 consists of several distinct, young stellar generations. The IMF is fully populated up to at least 100 $M_\odot$ (Massey & Hunter 1998), and there is no evidence for a truncated low-mass IMF, even within R136 (Andersen et al., in prep.). The stellar distribution near the center R136 is shown in Fig. 1b.

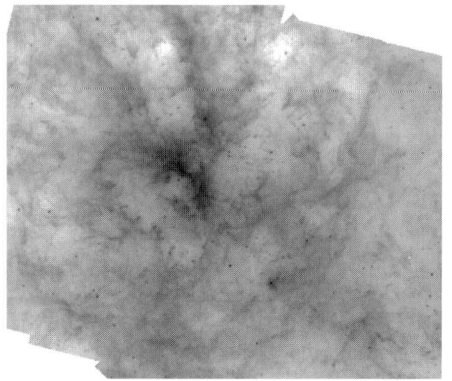
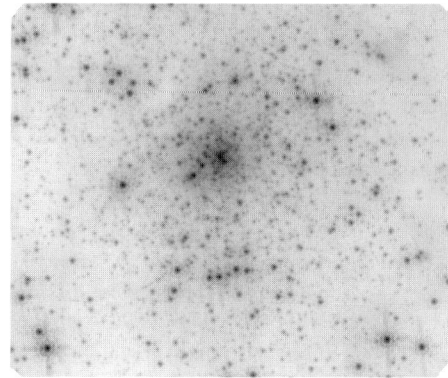

Figure 1a. Spitzer-IRAC image at 8μm, showing the distribution of warm dust and UV-excited PAH molecules in 30 Doradus. The size is 27×20 arcmin^2 (360×270 pc^2; Brandl et al., in prep.).

Figure 1b. HST/NICMOS H-band image of the 90×75 arcsec^2 (23×19 pc^2) around R136 (Andersen et al., in prep.); the faintest stars visible are about 2 $M_\odot$.

Table 2. Properties of the stars and the ISM in 30 Doradus

Quantity	Value	Reference
Hα luminosity	1.5×10^{40} erg s^{-1}	Kennicutt (1984)
Ly-cont flux (30 Dor)	1.1×10^{52} phot s^{-1}	Kennicutt (1984)
Ly-cont flux (NGC 2070)	4.5×10^{51} phot s^{-1}	Walborn (1991)
# OB stars (NGC 2070)	2400	Parker (1993)
H$_2$ mass	7×10^7 M$_\odot$	
H$_{II}$ mass	8×10^5 M$_\odot$	Kennicutt (1984)
E_{kin}^{gas}	$\geq 10^{52}$ erg	Chu & Kennicutt (1994)
L_{FIR}	4×10^7 L$_\odot$	Werner et al. (1978)

2. Is R136 a Super Star Cluster (SSC)?

With about 40 stars of spectral type O3, R136 is the densest concentration of very massive stars known (Massey & Hunter 1998). R136 is centrally condensed and displays a remarkably smooth exponential cluster profile (see, e.g., Malumuth & Heap 1994; their fig. 13). The stellar properties of R136 are summarized in Table 3.

Table 3. Stellar properties of R136

Quantity	Value	Reference
age	$\approx 2 \pm 1$ Myr	[a]
central density ρ_c	5.5×10^4 M$_\odot$ pc^{-3}	Hunter et al. (1995)
total mass m_{tot}	6.3×10^4 M$_\odot^b$	Hunter et al. (1995)
core radius r_c	0.12 pc (0.5 arcsec)	Brandl et al. (1996)
half-mass radius r_{hm}	1.2 pc (5 arcsec)	Brandl et al. (1996)
tidal radius r_t	5 pc (21 arcsec)	Meylan (1993)
IMF slope ξ	2.2^c	Andersen et al. (in prep.)

[a] the age of the cluster is still subject of controversy; [b] for $m \geq 0.1$ M$_\odot$; [c] the Salpeter (1955) slope is $\xi = 2.35$ in this notation.

However, there are numerous examples of massive young clusters that do *not* show a density profile like R136. For instance, NGC 604, the most luminous H II region in M33, looks very similar to R136 in Hα (Hunter et al. 1996) both in structure and luminosity, but the stellar distribution is completely different, given by numerous smaller clusters – a structure sometimes referred to as a scaled OB association (SOBA; Maíz-Apellániz et al. 2004).

In recent years, so-called super star clusters, such as in the Antennae galaxies, have received a lot of attention (e.g., Mengel et al. 2002). The dynamical mass can be estimated via $m_{\rm dyn} = \eta \frac{\sigma^2 r_{\rm hl}}{G}$, where $\eta \approx 10$. However, this is based on the assumption that the cluster is well resolved, i.e., that $r_{\rm hl}$ can be accurately determined. Fig. 2a shows R136 in the upper left as observed with NICMOS (same as Fig. 1b), and then progressively at 2× lower resolution. The FWHM derived from the same cluster but at different resolution is plotted as a function of distance in Fig. 2b. Because of spatial undersampling and the light from bright stars near the cluster core, the half-light radius $r_{\rm hl}$ – and thus $m_{\rm dyn}$ – can easily be overestimated by a factor of ten or more at distances beyond a few Mpc. To complicate matters, the spectroscopically measured velocity dispersion may be significantly affected by the orbital velocities of massive binary stars (Bosch et al. 2001). Furthermore, at distances like the Antennae, a centrally condensed cluster like R136 is indistinguishable from a non-virialized, NGC 604-like SOBA. Given the large uncertainties, it remains to be seen how much more "super" than R136 super star clusters really are.

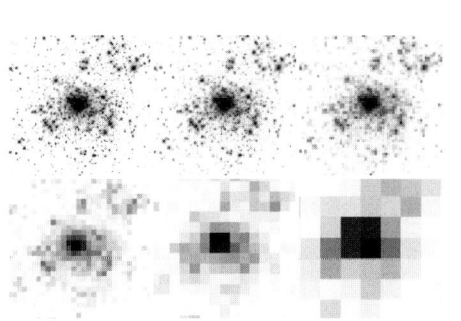

Figure 2a. R136 at the observed NICMOS resolution (upper left) and at larger distances that are increasing by factors of 2, 4, 8, 16, and 32.

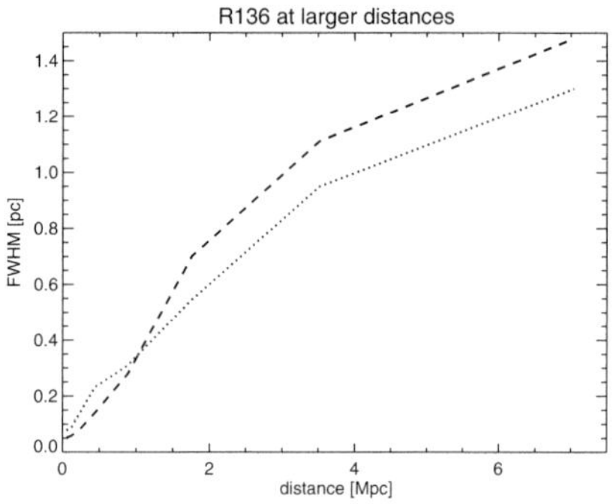

Figure 2b. The FWHM derived from 2D Gaussian fits to the light profiles of R136 in Fig. 2a.

3. Are luminous starbursts made of 30 Doradus complexes?

30 Doradus does not experience a strong gravitational field from its host galaxy; the entire complex can almost freely expand, and its center has no continuous supply from a larger gas reservoir. Under different boundary conditions, which may be prevalent in a circumnuclear starburst, it is conceivable that the densities of gas, dust, and embedded star clusters are much higher – without necessarily requiring different unit cells than R136. Using a very simplified model of identical clusters, spherical geometry, and constant gas density, we can estimate what the total luminosity of such a starburst might be.

Assuming 200 pc for the linear extent of a "typical" extragalactic starburst region, and taking R136's tidal radius r_t as the (half-)size of a unit cell, there could be the equivalent of as many as 8000 R136-like clusters within the given volume. If only a quarter (10^7 $L_\odot$) of the total 30 Doradus far-IR luminosity is being produced within the volume defined by r_t, and if $L_{\rm FIR}$ is the sum of the reprocessed UV radiation and (about the same amount) of shock-heated gas from supernovae, the total far-infrared luminosity is

$$L_{\rm FIR}^{\rm tot} = 8000 \times 2 \times 10^7 = 1.6 \times 10^{11} L_\odot \qquad (1)$$

This number is well within the ballpark of luminous starburst galaxies, although it falls short of the luminosity of ultraluminous infrared galaxies. At any rate, if starbursts have such porous, inhomogeneous structures, they represent a big challenge for accurate starburst modelling.

References

Bosch G., Selman F., Melnick J., Terlevich R., 2001, A&A, 380, 137
Brandl B.R., et al., 1996, ApJ, 466, 254
Chu Y.-H., Kennicutt R.C., 1994, ApJ, 425, 720
Hunter D.A., Shaya E.J., Holtzman J.A., Light R.M., O'Neill E.J. Jr., Lynds R., 1995, ApJ, 448, 179
Hunter D.A., Baum W.A., O'Neill E.J. Jr., Lynds R., 1996, ApJ, 456, 174
Kennicutt R.C., 1984, ApJ, 287, 116
Maíz-Apellániz J., Pérez E., Mas-Hesse J.M., 2004, AJ, 128, 1196
Malumuth E.M., Heap S.R., 1994, AJ, 107, 1054
Massey P., Hunter D.A., 1998, ApJ, 493, 180
Mengel S., Lehnert M.D., Thatte N., Genzel R., 2002, A&A, 383, 137
Meylan G., 1993, in: The globular clusters – galaxy connection, Smith G.H., Brodie J.P., eds., ASP Conf. Ser., (ASP: San Francisco), vol. 48, p. 588
Parker J.W., 1993, AJ, 106, 560
Salpeter E.E., 1955, ApJ, 123, 666
Walborn N.R., 1991, in: The Magellanic Clouds, Haynes R., Milne D., eds., IAU Symp., 148, p. 145 (Kluwer: Dordrecht)
Walborn N.R., Blades J.C., 1997, ApJS, 112, 457
Werner M.W., Becklin E.E., Gatley I., Ellis M.J., Hyland A.R., Robinson G., Thomas J.A., 1978, MNRAS, 184, 365

Session II

The Initial Mass Function in Starburst Regions: Environmental Dependences?

THE INITIAL MASS FUNCTION IN STARBURSTS

Bruce G. Elmegreen
IBM Research Division, T.J. Watson Research Center, P.O. Box 218, Yorktown Heights, NY 10598, USA
bge@watson.ibm.com

Abstract   The history of the initial mass function (IMF) in starburst regions is reviewed. The IMFs are no longer believed to be top-heavy, although some super star clusters, whether in starburst regions or not, could be. General observations of the IMF are discussed to put the starburst results in perspective. Observed IMF variations seem to suggest that the IMF varies a little with environment in the sense that denser and more massive clusters produce more massive stars, and perhaps more brown dwarfs too, compared to intermediate-mass stars.

1.   Introduction: History of Starburst IMFs

Early starburst observations suggested that the luminous mass from young massive stars is comparable to the dynamical mass from the rotation curve (see reviews in Telesco 1988, Scalo 1990, Zinnecker 1996, Leitherer 1999). This implied that there was a deficit in low-mass stars. The nearby starburst galaxy, M82, was one of the best cases. Rieke et al. (1980, 1993) modeled M82's gas mass, luminous-star mass, rotation-curve mass, 2.2μm flux, Lyman continuum flux, CO index for supergiants, and the Brα/Brγ ratio, giving an extinction $A_V = 25$ mag. They concluded that only the usual IMF models with lower mass limits $M_L > 3 - 6$ M$_\odot$ worked. Kronberg et al. (1985) used the Lyman-continuum flux from radio emission in M82 to determine the star-formation rate, and used the total gas mass combined with an efficiency estimate to conclude that the IMF is top-heavy compared to the Miller & Scalo (1979) IMF. Bernlohr (1992) fit the same M82 properties as Rieke et al., plus the heavy element abundance and FIR line ratios, and concluded that either the IMF slope is shallower than the Scalo (1986) IMF by 1, or there is a lower mass cut-off greater than $1.5 - 2$ M$_\odot$. Independent models of M82 by Doane & Mathews (1993), emphasizing the supernova (SN) rate and the total dynamical mass, led to the same lower cut-off even for the Salpeter IMF, which is shallower than both the Miller-Scalo and Scalo functions.

Early observations obtained truncated IMFs in other starbursts, too. Wright et al. (1988) found $M_L > 3-6$ M$_\odot$ from low M/L ratios in 12 starburst galaxies assuming a Miller-Scalo IMF. For the merger remnant NGC 3256, Doyon, Joseph & Wright (1994) modeled the Brγ equivalent width, HeI $\lambda 2.06\mu$m/Brγ, CO index, and N(Lyc)/LIR ratio with different IMFs and star-formation histories. The HeI $\lambda 2.06\mu$m/Brγ index suggested an upper mass limit, $M_U = 30$ M$_\odot$, the rotation curve and CO molecular cloud observations gave $M_{\rm gas} \sim M_{\rm total}$, and all constraints gave either an IMF slope shallower than the Salpeter slope by ~ 0.5 or $M_L > 3$ M$_\odot$. For the merger remnant UGC 8387, Smith et al. (1995) suggested $M_L > 8$ M$_\odot$ for a Miller-Scalo IMF using the low ratio of $1 - 500\mu$m flux to Lyα, as obtained from the 5 GHz flux. Smith, Herter & Haynes (1998) got the same result from IR excesses in 20 starburst galaxies, suggesting that the IMF slope for $M > 10$ M$_\odot$ was -2.7 ± 0.2 ($\Gamma = -1.7$, see below), shallower than the Miller-Scalo slope of -3.3 ($\Gamma = -2.3$) in this mass range, but steeper than Salpeter (-2.35; $\Gamma = -1.35$).

At about the same time as these observations were suggesting the IMF was top-heavy in starburst galaxies, several other observations suggested that it was normal. Devereux (1989) observed 20 nearby starbursts like M82, and using 2.2μm and FIR fluxes, and estimates for the central dynamical masses, found acceptable fits to the Miller-Scalo IMF from $0.09 - 30$ M$_\odot$ (the Miller-Scalo function is the least top-heavy of the main IMF models). He also suggested that extinction corrections in M82 made by others were too high, and this made it appear like M82 had a truncated IMF when really it did not. Satyapal (1995) indeed found low extinction in M82 from Paβ/Brγ, which gave $A_V = 2 - 12$ mag, compared to 25 mag in Rieke et al. (1980). Satyapal then got a K-band luminosity 3 times lower than Rieke et al., and saw no need for IMF truncation. Satyapal (1997) also found an age gradient in the center of M82 and fit an IMF with the Salpeter slope from $0.1 - 100$ M$_\odot$, accounting for only 36% of the dynamical mass.

In other observations of starburst IMFs, Schaerer (1996) applied evolutionary models to the WR/O star ratios and found a Salpeter IMF slope, although M_L could not be determined. Stasińska & Leitherer (1996) modeled the emission-line spectra of giant HII regions and starburst galaxies, having a factor-of-ten range in metallicities, and also found a Salpeter IMF up to 100 M$_\odot$, with no information about the lower mass limit. Calzetti (1997) modeled multiwavelength spectroscopy and broad-band infrared photometry of 19 starburst galaxies to derive reddening values and found a general consistency with the Salpeter IMF between 0.1 and 100 M$_\odot$.

Finally, returning to M82, Förster Schreiber et al. (2003) modeled it with 25 pc resolution using near-IR integral-field spectroscopy and mid-IR spectroscopy. An upper stellar mass limit greater than 50 M$_\odot$ was derived from the $L_{\rm bol}/L_{\rm Lyc}$ and [NeIII]/[NeII] ratios; short decay times were observed for star formation locally (1 − 5 Myr), and the models were insensitive to the shape or slope of IMF at intermediate to high mass. An IMF turnover somewhere below 1 $M_\odot$ was concluded.

There were also suggestions that top-heavy IMFs would cause problems with stellar populations or metallicities. Charlot et al. (1993) suggested that an inner-truncated IMF would produce a very red population of red giants, without the corresponding main sequence stars, after the turn-off age reaches the stellar lifetime at the truncation mass. Wang & Silk (1993) suggested that the truncated model gives an oxygen abundance that is too high when the star-formation process is over. These red populations or elevated oxygen abundances have not been observed.

At the present time, the observations suggest that the Salpeter IMF with a lower-mass flattening somewhere between 0.5 $M_\odot$ and 1 $M_\odot$ is a reasonable approximation to the IMF in large integrated regions of starburst galaxies.

2. Local IMFs

The Field

Starburst IMFs are useful for understanding star formation only in comparison to local IMFs or IMFs in non-starburst regions, where many of the star-formation processes can be observed in more detail. There are many such observations of non-starburst IMFs, as reviewed in the conference proceedings "The Stellar Initial Mass Function" edited by Gilmore, Parry & Ryan (1998) or in Chabrier (2003). We give a brief summary here.

In 1955, Salpeter showed that if 10% of a star's mass is converted into Helium on the main sequence, if the star-formation rate in the local Milky Way disk is constant over time, and if the present-day mass function is that given by the available catalogs (some of which dated back several decades – even into the 1920's), then the mass function of stars at birth has a slope of $\Gamma = -1.35$ on a log-log plot.

More recent derivations of this field-star IMF generally give steeper slopes. Miller & Scalo (1979) fitted the observations to a log-normal IMF, which has about the Salpeter slope near one solar mass, but an increasingly steep slope toward higher masses, reaching $\Gamma \sim -2.3$. Scalo (1986) found a field-star IMF with a slope between -1.5 and -1.7 in the range from 1 to 10 $M_\odot$, and a slightly shallower slope, between -1.35 and -1.5, at higher mass. Rana (1987) derived a field-star IMF with somewhat different data than Scalo (1986) and found $\Gamma = -1.8$ for $M > 1.6\,M_\odot$.

The difference between these field-star IMFs and Salpeter's IMF is significant: for a slope difference of 0.5, the number of high-mass stars between 10 $M_\odot$ and 100 $M_\odot$ compared to the number of intermediate-mass stars between 1 $M_\odot$ and 10 $M_\odot$ is three times larger in the Salpeter IMF than in the others. This excess factor of 3 for nearby high-mass stars can easily be ruled out. However, Salpeter did not have observations that extended to the high-mass range. In the region of overlap, which is near 1 $M_\odot$, the modern field-star IMF slope is comparable to Salpeter's value. The big question for starburst regions is how the IMF near 1 $M_\odot$ extrapolates to OB stars.

IMF measurements outside starburst regions give a wide range of slopes. Parker et al. (1998) did photometry on 37,300 stars in the LMC and SMC, and found slopes near $\Gamma = -1.0, -1.6$, and -2.0 for the Davies, Elliot & Meaburn (1976) HII regions, and $\Gamma = -1.80 \pm 0.09$ for all the field stars, considering only stars with $M > 2\ M_\odot$. The IMFs near the HII regions are probably too shallow as a result of inadequate corrections for background and foreground stars (Parker et al. 2001), but the field-star IMF appears to be free of this systematic effect. Note that the statistical accuracy is very high for this measurement.

Massey et al. (1995) and Massey (2002) surveyed the remote field in the LMC and SMC, defining these to be regions more than 30 pc from a Lucke & Hodge (1970) or Hodge (1986) association. The survey was complete down to 25 $M_\odot$ and included 450 stars, which should give a statistical uncertainty of $\Delta\Gamma \sim \pm 0.15$ (Elmegreen 1999a). By assuming a constant star-formation rate over the last 10 Myr, Massey et al. found Γ significantly steeper in the remote field than in clusters, having a value between -3.6 and -4.

One could imagine several systematic effects that make this slope artificially steep. First, note that runaway O stars could not do this, because the field has too few O stars compared to intermediate-mass stars. Other likely processes could do it, however: (i) Selective evaporation of cluster envelopes into the field, considering that some cluster envelopes have $\Gamma \sim 4$ already (de Grijs et al. 2002). (ii) Greater migration into the field of the longer-lived, low-mass stars compared to high-mass stars. (iii) Greater self-destruction of low-pressure clouds in the field by OB star formation, compared to high-pressure clouds in associations (Elmegreen 1999a). Hoopes, Walterbos & Bothun (2001) also explained the steep mass function required for diffuse interstellar ionization in nearby galaxies with the differential drift of low-mass stars into the field. Tremonti et al. (2002) found a Salpeter IMF for clusters and a steeper IMF for the field in the dwarf starburst galaxy NGC 5253, and explained this difference as a result of cluster dispersal after 10 Myr, when the most massive stars have disappeared.

Another observation of a systematically steep IMF was by Lee et al. (2004). They fit the high M/L in low surface brightness galaxies with population synthesis models requiring low metallicity, recent (1 – 3 Gyr) star formation, and a steep IMF: $\Gamma = -2.85$ from $0.1 - 60\ M_\odot$. These galaxies have a low pressure like the extreme field regions in Massey et al.

The Taurus region in the nearby field may have a steep IMF at high mass too (Luhman 2000). The pre-stellar condensations there are peculiar anyway, showing more extended structures, like isothermal spheres, than those in Perseus or Ophiuchus which appear truncated (Motte & André 2001).

Clusters

The IMF in the Sco-Cen OB association is best fit with a slope of $\Gamma \sim -1.8$ (Preibisch et al. 2002). Another Galactic region, W51, has 4 subgroups, all with $\Gamma \sim -1.8$ at intermediate to high mass, but two of the subgroups have a statistically significant excess, by a factor of ~ 3, of stars in the highest mass bin (~ 60 M$_\odot$; Okumura et al. 2000). There are other anomalies like this, too. Scalo (1998) suggested that the IMF varied significantly from cluster to cluster, but Elmegreen (1999a) and Kroupa (2001) showed that most of these variations could be statistical in origin, given the small number of stars usually observed.

Most clusters have Salpeter IMFs. A good example is the R136 cluster in the 30 Dor region of the LMC. The slope is $\Gamma \sim 1.3 - 1.4$ out to stellar masses greater than 100 M$_\odot$ (Massey & Hunter 1998). In addition, h and χ Persei (Slesnick, Hillenbrand & Massey 2002), NGC 604 in M33 (González Delgado & Pérez 2000), NGC 1960 and NGC 2194 (Sanner et al. 2000), NGC 6611 (Belikov et al. 2000) and many other clusters have Salpeter IMFs (e.g., Sakhibov & Smirnov 2000, Sagar, Munari & de Boer 2001). Massey & Hunter (1998) concluded that the Salpeter IMF occurs in star-forming regions spanning a factor of 200 in density.

Whole galaxies are often observed to have the Salpeter slope too. There were many studies in the 1990's using Hα equivalent widths, spectrophotometry, metallicity, and galaxy evolution models (see the review in Elmegreen 1999b). More recently, Baldry & Glazebrook (2003) derived the Salpeter IMF from the cosmic star-formation rate, Rejkuba, Greggio & Zoccali (2004) obtained it for the halo of NGC 5128 (Cen A), and Pipino & Matteucci (2004) fit the photochemical evolution of elliptical galaxies with a Salpeter IMF.

If whole galaxies have the same average IMF as clusters, and if most stars form in clusters, which is believed to be the case (Lada & Lada 2003), then the mass of any star cannot depend on the cluster mass. That is, any type of star can form in any type of cluster (as long as the cluster mass is larger than the stellar mass). If this were not the case, then the summed IMFs would differ from the cluster IMFs. For example, if low-mass clusters were able to form only low-mass stars, and high-mass clusters formed all types of stars, as observed, then the sum of the low and high-mass clusters would produce far more low-mass stars than each cluster's IMF (Elmegreen 1999b).

3. Top-heavy IMFs in Super Star Clusters

Some super star clusters (SSCs) apparently have "top heavy" or "bottom light" IMFs. Sternberg (1998) found a high L/M ratio in NGC 1705-1 and concluded that either $|\Gamma| < 1$ or there is an inner-mass cut-off. Smith & Gallagher (2001) got a high L/M in M82F and inferred an inner cut-off at $2-3$ M$_\odot$ for $\Gamma = -1.3$; they also confirmed the inner truncation for NGC 1705-1 found by Sternberg. Alonso-Herrero et al. (2001) observed a high L/M in the starburst galaxy NGC 1614, suggesting a top-heavy IMF. McCrady et al. (2003) found that MGG-11 in M82 is deficit in low-mass stars. Mengel et al. (2002) found the same for the Antennae, NGC 4038/9, and noted that the clusters in the high-pressure regions had more normal IMFs, as if the Jeans mass were lower there.

Other SSCs have normal IMFs, however. This is the case for NGC 1569-A (Ho & Filippenko 1996, Sternberg 1998), NGC 6946 (Larsen et al. 2001), and M82/MGG-9 (McCrady et al. 2003).

Measuring the IMF in SSCs is subject to many uncertainties. It requires observations of the velocity dispersion and radius to get the mass, and observations of the luminosity. One problem is that Δv can vary inside a cluster (i.e., it may not be isothermal – e.g., NGC 6946) and it is often measured with large uncertainties. The value of R is uncertain too if the core is unresolved or the outer part of the cluster is blended with field stars. The luminosity is uncertain because of possible field star blending. Mass segregation makes the IMF vary with radius (de Grijs et al. 2002), so the cluster colors vary with radius, giving another uncertainty about the core radius. The average IMF depends on where the outer cut-off is placed. The cluster could also be evaporating or be out of radial equilibrium, in which case the usual expressions for cluster mass in terms of Δv and R do not apply. Several SSCs are observed to have sub-clusters inside their halos, giving irregular and asymmetric light profiles.

Implications

These observations suggest a correlation between Γ and star-formation density. In the extreme field, $\Gamma \sim -4$; in low surface brightness galaxies, $\Gamma \sim -2.85$; in the Milky Way and LMC fields, $\Gamma \sim -1.8$; in many clusters, $\Gamma \sim -1.35$; in some SSCs, $|\Gamma| < 1.35$ or there is an inner-mass truncation, and in starburst regions as a whole, $\Gamma \sim -1.35$ – with or without inner-truncation (this is uncertain). We should probably remove the local field from this list because it is a mixture of dispersed clusters and cluster envelopes integrated over time; both the mixing process and the local star-formation history are uncertain. Aside from this, the trend suggests significantly denser regions have slightly shallower IMF slopes at intermediate to high mass. More observations are needed to confirm this. If true, it could imply that enhanced gas accretion and protostellar coalescence are important for high-mass stars in the densest environments (see more extensive reviews of this point in Stahler, Palla, & Ho 2000, Elmegreen 2004, Shadmehri 2004).

References

Alonso-Herrero A., Engelbracht C.W., Rieke M.J., Rieke G.H., Quillen A.C., 2001, ApJ, 546, 952
Baldry I.K., Glazebrook K., 2003, ApJ, 593, 258
Belikov A.N., Kharchenko N.V., Piskunov A.E., Schilbach E., 2000, A&A, 358, 886
Bernlohr K., 1992, A&A, 263, 54
Calzetti D., 1997, AJ, 113, 162
Chabrier G., 2003, PASP, 115, 763
Charlot S., Ferrari F., Matthews G.J., Silk J., 1993, ApJ, 419, L57
Davies R.D., Elliot K.H., Meaburn J., 1976, Mem. RAS, 81, 89
de Grijs R., Gilmore G.F., Johnson R.A., Mackey A.D., 2002, MNRAS, 331, 245
Devereux N.A., 1989, ApJ, 346, 126
Doane J.S., Mathews W.G., 1993, ApJ, 419, 573
Doyon R., Joseph J.D., Wright G.S., 1994, ApJ, 421, 101
Elmegreen B.G., 1999a, ApJ, 515, 323
Elmegreen B.G., 1999b, in: The Evolution of Galaxies on Cosmological Timescales, Beckman J.E., Mahoney T.J., eds., ASP Conf. Ser., (ASP: San Francisco), vol. 187, p. 145
Elmegreen B.G., 2004, MNRAS, 354, 367
Förster Schreiber N.M.F., Genzel R., Lutz D., Sternberg A., 2003, ApJ, 599, 193
Gilmore G., Parry I., Ryan S., 1998, eds., The Stellar Initial Mass Function, (CUP: Cambridge)
González Delgado R.M., Pérez E., 2000, MNRAS, 317, 64
Ho L.C., Filippenko A.V., 1996, ApJ, 466, L83
Hodge P., 1986, PASP, 98, 1113
Hoopes C.G., Walterbos R.A.M., Bothun G.D., 2001, ApJ, 559, 878
Kronberg P.P., Bierman P., Schwab F.R., 1985, ApJ, 291, 693
Kroupa P., 2001, MNRAS, 322, 231
Lada C.J., Lada E.A., 2003, ARA&A, 41, 57
Larsen S.S., Brodie J.P., Elmegreen B.G., Efremov Y.N., Hodge P.W., Richtler T., 2001, ApJ, 556, 8011
Lee H.-C., Gibson B.K., Flynn C., Kawata D., Beasley M.A., 2004, MNRAS, 353, 113
Leitherer C., 1999, in: Galaxy Interactions at Low and High Redshift, IAU Symp. 186, Barnes J.E., Sanders D.B., eds., (Kluwer: Dordrecht), p. 243
Lucke P.B., Hodge P.W., 1970, AJ, 75, 171
Luhman K.L., 2000, ApJ, 544, 1044
Massey P., 2002, ApJS, 141, 81
Massey P., Lang C.C., DeGioia-Eastwood K., Garmany C.D., 1995, ApJ, 438, 188
Massey P., Hunter D.A., 1998, ApJ, 493, 180
McCrady N., Gilbert A., Graham J.R., 2003, ApJ, 596, 240
Mengel S., Lehnert M.D., Thatte N., Genzel R., 2002, A&A, 383, 137
Miller G.E., Scalo J.M., 1979, ApJS, 41, 513
Motte F., André P., 2001, A&A, 365, 440

Okumura S., Mori A., Nishihara E., Watanabe E., Yamashita T., 2000, ApJ, 543, 799
Parker J.W., et al., 1998, AJ, 116, 180
Parker J.W., Zaritsky D., Stecher T.P., Harris J., Massey P., 2001, AJ, 121, 891
Pipino A., Matteucci F., 2004, MNRAS, 347, 968
Preibisch T., Brown A.G.A., Bridges T., Guenther E., Zinnecker H., 2002, AJ, 124, 404
Rana N.C., 1987, A&A, 184, 104
Rejkuba M., Greggio L., Zoccali M., 2004, A&A, 415, 915
Rieke G.H., Lebofsky M.J., Thompson R.I., Low F.J., Tokunaga A.T., 1980, ApJ, 238, 24
Rieke G.H., Loken K., Rieke M.J., Tamblyn P., 1993, ApJ, 412, 99
Sagar R., Munari U., de Boer K.S., 2001, MNRAS, 327, 23
Sakhibov F., Smirnov M., 2000, A&A, 354, 802
Salpeter E.E., 1955, ApJ, 121, 161
Sanner J., Altmann M., Brunzendorf J., Geffert M., 2000, A&A, 357, 471
Satyapal S., et al., 1995, ApJ, 448, 611
Satyapal S., Watson D.M., Pipher J.L., Forrest W.J., Greenhouse M.A., Smith H.A., Fischer J., Woodward C.E., 1997, ApJ, 483, 148
Scalo J.S., 1986, Fund. Cos. Phys, 11, 1
Scalo J.M., 1990, in: Windows on Galaxies, Renzini A., Fabbiano G., Gallagher J.S., eds., (Kluwer: Dordrecht), p. 125
Scalo J.S., 1998, in: The Stellar Initial Mass Function, Gilmore G., Parry I., Ryan S., eds., (CUP: Cambridge), p. 201
Schaerer D., 1996, ApJ, 467, L17
Shadmehri M., 2004, MNRAS, 354, 375
Slesnick C.L., Hillenbrand L.A., Massey P., 2002, ApJ, 576, 880
Smith D.A., Herter T., Haynes M.P., Beichman C.A., Gautier T.N. III, 1995, ApJ, 439, 623
Smith D.A., Herter T., Haynes M.P., 1998, ApJ, 494, 150
Smith L.J., Gallagher J.S., 2001, MNRAS, 326, 1027
Stahler S.W., Palla F., Ho P.T.P., 2000, in: Protostars and Planets IV, Mannings V., Boss A.P., Russell S.S., eds., (Univ. Arizona Press: Tucson), p. 327
Stasińska G., Leitherer C., 1996, ApJS, 107, 427
Sternberg A., 1998, ApJ, 506, 721
Telesco C.M., 1988, ARA&A, 26, 343
Tremonti C.A., Calzetti D., Leitherer C., Heckman T.M., 2002, ApJ, 555, 322
Wang B., Silk J., 1993, ApJ, 406, 580
Wright G.S., Joseph R.D., Robertson N.A., James P.A., Meikle W.P.S., 1988, MNRAS, 233, 1
Zinnecker H., 1996, in: The Interplay between Massive Star Formation, the ISM, and Galaxy Evolution, Kunth D., Guiderdoni M., Heydari-Malayeri M., Thuan T.X., eds., (Editions Frontieres: Gif-sur-Yvette), p. 151

DYNAMICAL MODELS OF STAR FORMATION AND THE INITIAL MASS FUNCTION

Ian A. Bonnell
School of Physics and Astronomy, University of St Andrews, Scotland, KY16 9SS, UK
iab1@st-andrews.ac.uk

Abstract Numerical simulations highlight that star formation is a dynamical process in which stars interact during their formation. Of particular interest is that accretion in a clustered environment is non-uniform in that stars located in the centre of the potential accrete more and become more massive. This competitive accretion process can explain the initial mass function of stars, including a shallow slope for low-mass stars and a steeper Salpeter-like slope for higher-mass stars. Numerical simulations of the fragmentation and formation of a stellar cluster show that the final stellar masses are due to competitive accretion and that this results in a realistic IMF. Competitive accretion also naturally results in a direct correlation between the richness of a cluster and the mass of the most massive star therein. Shallower IMFs are possible if the Jeans mass is significantly higher than one $M_\odot$.

1.   Introduction

Our understanding of the star-formation process has recently undergone a paradigm shift from the older quasi-static models where stars form singly and in isolation (Shu, Adams & Lizano 1987) to one where we recognise that stars form predominantly in groups (Lada & Lada 2003), and that dynamics play an important role in determining the stellar products (Clarke, Bonnell & Hillenbrand 2000, Larson 2003, Maclow & Klessen 2004). The advent of large-scale numerical simulations of star formation now allows us to study not just the dynamics of turbulent molecular clouds (Vázquez-Semadeni et al. 2000) but crucially, the development of self-gravitating cores and subsequent star formation therein (Bate, Bonnell & Bromm 2003, Bonnell, Bate & Vine 2003). Here, I review how dynamical models for star formation lead to an understanding of the origin of the initial mass function (IMF) for stars.

2. Turbulence and the IMF

One possible origin for the IMF is from the direct fragmentation of molecular clouds due to their internal "turbulent" motions (Padoan & Nordlund 2002). These motions are highly supersonic, and thus induce shocks in the clouds. The resulting structure can form the basis for the IMF if we associate a Jeans mass with the gas at a given density (for an isothermal gas): $M_J \propto \rho^{-1/2}$. The inferred mass function is then of a log-normal variety, mimicking the density distribution. This mechanism is particularly attractive as observations of starless cores in Ophiuchus appear to follow a stellar-like IMF (Motte et al. 1998) although this does not appear to be true in some regions of Orion (Coppin et al. 2000).

One problem with a fragmentation model for the IMF is that the more massive cores must be less dense if they contain just one Jeans mass, and are thus not going to fragment into several smaller pieces. This implies that more massive stars should be well separated at distances greater than the Jeans radius, the minimum radius for an object to be gravitationally bound: $R_J \propto \rho^{-1/2}$. This would result in an inverse mass segregation where the more massive stars are in low-density regions, in direct opposition to observations (Clarke et al. 2000). Another potential problem with a turbulently driven origin for the IMF is that simulations of such clouds shows that only a small subset of the generated cores are actually bound, and that the masses of these cores are approximately the mean Jeans mass of the cloud (Clark & Bonnell 2005). Furthermore, these bound cores commonly fragment to form multiple systems. There is therefore no one-to-one mapping of the pre-stellar core mass distribution to the stellar IMF.

3. Competitive accretion and the IMF

An alternative explanation is that the IMF originates due to competitive accretion in stellar clusters. Competitive accretion arises as many stars, or other sources of gravity, compete to accrete from the same reservoir of gas. This occurs in any system where both the stars and gas are free to move under their combined gravitational influence. In systems where the communal reservoir contains significant mass, competitive accretion can determine the final distribution of stellar masses.

Competitive accretion takes two forms depending on whether the gas or the stars dominate the local stellar potential (Bonnell et al. 2001a). In gas-dominated potentials, the stars and gas have similar velocities and the accretion rate onto an individual star is limited by its tidal radius relative to the other stars and the cluster as a whole. In such cases, the stars need not be moving, as it is their ability to attract the gas which provides the competition. In contrast, when stars dominate the cluster potential, as inevitably occurs due to the accretion, the stars virialise and have high velocities relative to the infalling gas. The accretion then follows the classic Bondi-Hoyle formalism.

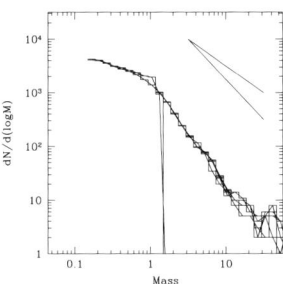

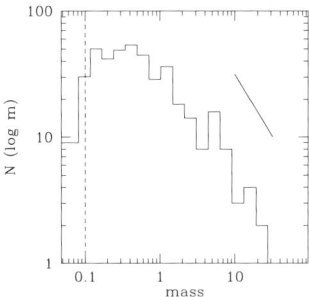

Figure 1. The two power-law IMF is shown that results from competitive accretion in a model cluster (Bonnell et al. 2001b).

Figure 2. The IMF from a simulation of the formation of a 419-star cluster is compared to a slope of $\gamma = -2$ (Bonnell et al. 2003).

We can use the above formulation of the accretion rates, with a simple model for the stellar cluster in the two physical regimes, in order to derive the resultant mass functions (Bonnell et al. 2001b). The primary difference is the power of the stellar mass in the accretion rate, $M_*^{2/3}$ for tidal-accretion and M_*^2 for Bondi-Hoyle accretion. Starting from a gas-rich cluster with equal stellar masses, tidal accretion results in higher accretion rates in the centre of the cluster, where the gas density is highest. This results in a spread of stellar mass and a mass-segregated cluster. The lower dependency of the accretion rate on the stellar mass results in a fairly shallow IMF of the form (where Salpeter [1955] has $\gamma = -2.35$)

$$\mathrm{d}N/\mathrm{d}M_* \propto M_*^{-3/2}. \tag{1}$$

Once the cluster core enters the stellar-dominated regime where Bondi-Hoyle accretion occurs, the higher dependency of the accretion rate on the stellar mass results in a steeper mass spectrum. Zinnecker (1982) first showed how a Bondi-Hoyle type accretion results in a $\gamma = -2$ IMF. In a more developed model of accretion into the core of a cluster with a pre-existing mass segregation, the resultant IMF is of the form

$$\mathrm{d}N/\mathrm{d}M_* \propto M_*^{-5/2}. \tag{2}$$

This steeper IMF applies only to those stars that accrete the bulk of their mass in the stellar-dominated regime, i.e., the high-mass stars in the core of the cluster (see Fig. 1).

An additional result of competitive accretion is that it produces an "initial" mass segregation in that stars near the centre of the potential have the highest accretion rates and therefore become the more massive stars (Bonnell et al. 1997, 2001a). This is an important result, as young stellar clusters are observed to be mass segregated even when they are too young to have undergone dynamical mass segregation (Bonnell & Davies 1998).

4. The formation of stellar clusters

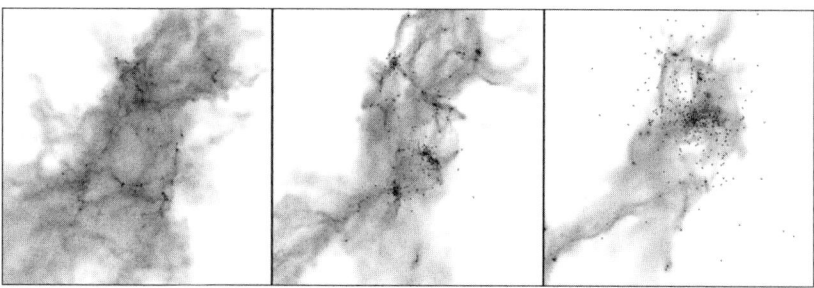

Figure 3. The fragmentation of a 1000 M$_\odot$ molecular cloud and the formation of a stellar cluster containing 419 stars (Bonnell et al. 2003).

Numerical simulations of cluster formation reproduce the above competitive accretion process. Figure 3 shows the evolution of a 1000 M$_\odot$ cloud, with a 0.5 pc radius and initially supported by supersonic turbulent motions, which fragments into > 400 stars due to the filamentary structure generated by the shocks (Bonnell, Bate & Vine 2003). The cluster forms in a hierarchical manner with many subclusters forming before eventually merging into one larger cluster. In all, 419 stars form in 5×10^5 years (Fig. 3). The simulation produces a field-star IMF with a shallow slope for low-mass stars, steepening to a Salpeter-like slope for high-mass stars (Fig. 2). The stars all form with low masses and accrete up to their final masses. The stars that are in the centres of the subclusters accrete more gas and thus become higher-mass stars. A careful dissection of the origin of the more massive stars reveals the importance of competitive accretion in setting the IMF. Using the Lagrangian nature of the SPH simulations, we can trace the mass from which a star forms. This shows that the mass of the more massive stars does not arise from either the initial fragment mass or from the mass of a contiguous envelope around this fragment extending to the next forming star. Instead, the vast majority of the mass falls in onto a pre-existing cluster of stars where it is therein accreted by the developing massive star.

5. The formation of massive stars

One of the predictions that we can extract from the simulations is that there should be a strong correlation between the mass of the most massive star and the number of, and mass in, stars in the cluster, (Fig. 4; Bonnell, Vine & Bate 2004 [BVB 2004]). This occurs due to the simultaneous accretion of stars and gas into a forming stellar cluster. Given an effective initial efficiency of fragmentation, for every star that falls into the cluster a certain amount of gas also enters the cluster. This gas joins the common reservoir from which the most massive star takes the largest share in this competitive environment. Thus, the mass of the most massive star increases as the cluster grows in numbers of stars.

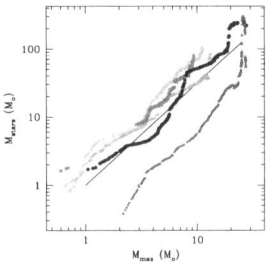
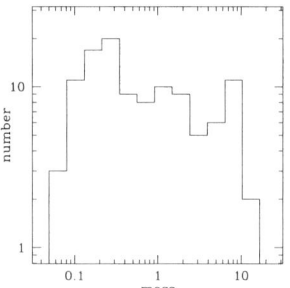

Figure 4. The total stellar mass in a subcluster is plotted against the mass of the most massive star therein (BVB 2004).

Figure 5. The IMF that results from a fragmentation of a 1000 $M_\odot$ cloud where the Jeans mass is 5 $M_\odot$.

6. Variations in the IMF

The IMF appears to be remarkably universal in the vast majority of stellar systems (Elmegreen, these proceedings). This tells us that either the conditions for star formation are remarkably universal or that the physics that determines the IMF is very robust. There are a number of possible ways in which the IMF could depend on local properties. This is evident from Fig. 5, which shows the resultant IMF for a cluster formation simulation where the Jeans mass is increased from 1 to 5 $M_\odot$. The IMF is fairly shallow up to masses of $\approx 5 - 10$ $M_\odot$, corresponding roughly to the Jeans mass. In Fig. 2 we see that the slope of the IMF becomes steeper at ≈ 1 $M_\odot$, the Jeans mass in this simulation. We can thus deduce that the Jeans mass helps set the *knee* in the IMF. Thus, one way of producing a shallow IMF is to have a relatively high Jeans mass.

Another potential variation in the upper-mass IMF arises due to the effects of radiation pressure from high-mass stars on dust grains (Wolfire & Casinnelli 1986). This effect, neglected in the above simulations, can halt accretion and thus set an upper mass limit. If disc accretion is unable to overcome this difficulty (Yorke & Sonnhalter 2002), then an alternative is that massive stars could form through stellar mergers in the dense cores of clusters (Bonnell, Bate & Zinnecker 1998, Bonnell & Bate 2002). In this case, an additional feature in the IMF somewhere near the critical mass (10–20 $M_\odot$) is naively expected as the physical process changes.

7. Conclusions

Competitive accretion in a clustered environment can naturally explain the stellar initial mass function. Numerical simulations of star cluster formation show that the high-mass stars gain their mass due to mass accretion into the cluster and that many stars compete from the same mass reservoir. This process naturally results in a two power-law IMF with a shallow $\gamma = -3/2$ slope for low-mass stars that accrete most of their mass in a gas-dominated potential and a steeper $\gamma = -5/2$ slope for high-mass stars that accrete most of their mass in a stellar-dominated core of a cluster. In addition to producing a mass-segregated cluster, competitive accretion also predicts a correlation between the mass of the most massive star and the cluster properties (numbers and total mass in stars). Variations in a local IMF can result if the Jeans mass is increased, producing a shallower IMF to higher masses. Stellar mergers may also produce a knee in the high-mass end of the IMF.

References

Bate M.R., Bonnell I.A., Bromm V., 2003, MNRAS, 339, 577
Bonnell I.A., Bate M.R., Clarke C.J., Pringle J.E., 1997, MNRAS, 285, 201
Bonnell I.A., Bate M.R., Clarke C.J., Pringle J.E., 2001a, MNRAS, 323, 785
Bonnell I.A., Bate M.R., Vine S.G., 2003, MNRAS, 343, 413
Bonnell I.A., Bate M.R., Zinnecker H., 1998, MNRAS, 298, 93
Bonnell I.A., Clarke C.J., Bate M.R., Pringle J.E., 2001b, MNRAS, 324, 573
Bonnell I.A., Davies M.B., 1998, MNRAS, 295, 691
Bonnell I.A., Vine S.G., Bate, M.R., 2004, MNRAS, 349, 735 (BVB 2004)
Clark, P.C., Bonnell I.A., 2005, MNRAS, subm.
Clarke C.J., Bonnell I.A., Hillenbrand L.A., 2000, in: Protostars and Planets IV, Mannings V., Boss A.P., Russell S.S., eds., (UofA Press: Tucson), p. 151
Coppin K.E.K., Greaves J.S., Jenness T., Holland W.S., 2000, A&A, 356, 1031
Lada C.J., Lada E., 2003, ARA&A, 41, 57
Larson R.B., 2003, Rep. Prog. Phys., 66, 1651
Mac Low M.M., Klessen R.S., 2004, RvMP, 74, 125
Motte F., André P., Neri R., 1998, A&A, 336, 150
Padoan P., Nordlund A., 2002, ApJ, 576, 870
Salpeter E.E., 1955, ApJ, 123, 666
Vázquez-Semadeni E., Ostriker E.C., Passot T., Gammie C.F., Stone J.M., 2000, in: Protostars and Planets IV, Mannings V., Boss A.P., Russell S.S., eds., (UofA Press: Tucson), p. 3
Zinnecker H., 1982, New York Academy of Sciences Annals, 395, 226

RED SUPERGIANTS, MASS SEGREGATION AND M/L RATIOS IN YOUNG STAR CLUSTERS

Ariane Lançon and Christian Boily
Observatoire de Strasbourg (UMR 7550), 11 rue de l'Université, F-67 000 Strasbourg, France

Abstract Dynamical masses of star clusters are important constraints on the slope or lower-mass cut-off of the stellar IMF. The measurements of dynamical masses rest on model-based relationships between mass, line-of-sight velocity dispersion and projected half-light radius. We have used dynamical models that account for stellar evolution to compute the evolution of this relationship, depending on the wavelengths at which radii and velocity dispersions are observed. The conversion factor varies significantly over a few $\times 10^7$ yr. In reddened starburst clusters that are observed at near-IR wavelengths, red supergiants are the dominant sources of light. We report on progress in the synthesis of stellar populations with strong red supergiant contributions. These stars are initially among the most massive stars, but they lose mass rapidly and the resulting dynamical evolution is complex.

1. Introduction

The now standard diagnostic plot used to investigate the stellar initial mass function (IMF) in star clusters displays the mass-to-light ratio versus age. It shows values derived from cluster observations superimposed on the IMF-dependent curves predicted by population synthesis models. In order to position a given cluster on this diagram, it is necessary to estimate its age (based on its spectrum), its intrinsic luminosity (i.e., the distance, the integrated flux and extinction), and its mass. The clusters found in starburst galaxies are often highly reddened and therefore best observed at near-IR wavelengths. There, red supergiant stars are the dominant sources of light. In the following, we briefly summarize difficulties in the synthesis of red supergiant populations, and ongoing work to improve the reliability of these models. We then describe new dynamical simulations of star clusters that account for stellar evolution and mass segregation. The significant effects of these evolutionary processes on mass estimates are highlighted. They may occur even on the short timescales relevant to the clusters found in starburst galaxies.

2. Red supergiants, cluster ages and cluster luminosities

Current limitations in the population synthesis models that predict near-IR spectra for young clusters (8 Myr to 100 Myr) include:

- the shortcomings of spectral libraries for red supergiants (insufficient sampling of the energy distribution; poor understanding of the surface abundances);

- the uncertain locations of the library stars in the theoretical HR diagram (luminosity sub-classes not taken into account in effective temperature scales; atmosphere models only in approximate agreement with the low-resolution energy distributions and molecular band spectra);

- the known inadequacy of standard evolutionary tracks to reproduce the temperatures and relative lifetimes of the red supergiant phases at different metallicities (mixing processes due to rotation or other mechanisms must be envisaged).

Within an extended collaboration[1], we have recently obtained new red supergiant spectra ($\lambda/\Delta\lambda \sim 1000$) with Spex on IRTF, Hawaii, and Caspir on the 2.3m ANU Telescope at Siding Spring. They cover optical and near-IR wavelengths *continuously* out to 2.4 μm, which is essential for the assessment of the energy distribution and the broad molecular bands. We found systematic differences between classes Ia, Ib and II that would be worth exploiting in the future. Using standard effective temperature scales, we found that an extinction correction of order $A_V \sim 1.5$ is necessary for most M1+ supergiants *if* one attempts to reproduce the semi-empirical colours of the spectra collected in the Basel library (Lejeune et al. 1998). Atmosphere models (provided on the one hand by P. Hauschildt, Hamburg, Germany, and on the other hand by B. Plez, Montpellier, France), reproduce the general aspect of the warmer M supergiants, but discrepancies in the molecular bands become more important for objects with later spectral types. In collaboration with P. Hauschildt, we will investigate models with non-solar abundances of C, N and O, in order to account for the internal mixing that evolutionary models predict. We hope to improve reddening and effective temperature estimates, and to constrain the initial mass and evolutionary status of the library stars.

Having incorporated these spectra in the population synthesis code PÉGASE (initially described in Fioc & Rocca-Volmerange 1997), and using the stellar evolution tracks of Bressan et al. (1993), we find that the spectra of various clusters of M82 between 0.8 and 2.4 μm can be reproduced well, but at ages that tend to be younger than those of Smith & Gallagher (2001) and with smaller amounts of extinction.

3. Cluster masses: results from dynamical models

The mass, M, can be derived from the integrated velocity dispersion along the line of sight, σ_v, and the projected half-light radius, $r_{\rm ph}$, each observed over a specific range of wavelengths (λ), via:

$$M = \eta \, \sigma_v^2(\lambda) \, r_{\rm ph}(\lambda)/G \qquad (1)$$

This relationship reflects virial equilibrium. In spherical Plummer or King models, η takes values between about 6 and 10, and such values have been used in the literature to discuss the IMF, in clusters with ages ranging from a few $\times 10^6$ to several $\times 10^7$ years (Smith & Gallagher 2001, Mengel et al. 2002, McCrady et al. 2003). However, as mentioned as a caveat by some of these authors, King models do not account for mass segregation, which does *not* occur on the relaxation timescale ($t_{\rm rel} \sim 10^8$ yr or more for a cluster of $\sim 10^5$ M$_\odot$), but on a generally shorter time, that scales (among others) with the ratio of the mean stellar mass to the maximum stellar mass present in the cluster (Farouki & Salpeter 1982, Spitzer 1987).

We have conducted numerical simulations to determine the evolution of η in spherically symetric clusters, using the code GASTEL (Louis & Spurzem 1991, Spurzem & Takashi 1995). The dynamical evolution is described by moments of the Boltzmann equation (as in fluid dynamics), and closure of this hierarchical set of equations is achieved with a prescription for diffusion of kinetic energy as a function of the *local* density and velocity dispersion. The stellar mass spectrum is sampled with 7 to 14 logarithmic mass bins. We checked that the mass-segregation timescales for 2 and 3 mass bins agree quantitatively with published results from N-body or Fokker-Planck integrators. Stellar evolution follows the Cambridge tracks (Hurley et al. 2000) at solar metallicity, and for a given initial mass and stellar age the flux in a specified waveband is derived from the Basel stellar library (Lejeune et al. 1998). Note that the dynamical mass bins are subdivided finely for these flux calculations. This allows us to determine $\sigma_v(\lambda)$ and $r_{\rm ph}(\lambda)$ for direct comparison with observations.

We ran simulations with a Salpeter IMF from 0.15 to about 17 M$_\odot$, for clusters with a total mass of 2×10^5 M$_\odot$, and with an initial value of 6 for the King parameter Ψ_0/σ^2 (Boily et al. 2004). Stellar evolution works against the effects of mass segregation: by the time the most massive stars would sink to the center in models with no stellar evolution, they may in fact have lost most of their mass and have started drifting outwards. In low-density environments, the mass segregation timescale is longer than the lifetime of massive stars, and over 10^8 yr η remains approximately constant. However, in dense clusters such as R136 in 30 Doradus or the luminous clusters of M82, mass segregation wins. η is found to increase by a factor of about 2 within 20 Myr in our densest model (representative of R136), mainly as a result of a decrease of $r_{\rm ph}$. The value of η is found to depend only little on λ in the optical/near-IR range, because similar mass bins provide the light in all cases. While the profile of the gravitational potential remains relatively close to a King 6 model, the apparent concentration (ratio of tidal to core radius) obtained from fits of King models to the light profile increases by almost a factor of 3. This bias in observations will have a strong impact on predictions of survival probabilities of clusters.

A campaign of simulations has been started to sample possible initial conditions (J.-J. Fleck et al., in prep.). Non-spherical cases will be investigated with N-body codes. Uncertainties due to the stochastic nature of the population of massive stars will be studied as well, as a function of total cluster mass.

4. Conclusions

Studies of the IMF in the young, dense and reddened clusters found in starburst galaxies remain challenging. Their near-IR light is produced by red supergiants for which we hope to provide better population synthesis models soon, based on new spectra with a broad spectral coverage and a careful comparison between these observations and atmosphere models. Dynamical modelling of star clusters including the evolution of stars will help us improve empirical cluster masses, and constrain the shape of their gravitational potential. In the case of the luminous cluster F in M82, our preliminary results (near-IR based age and extinction, dynamical evolution) weaken the case for a non-standard IMF made by Smith & Gallagher (2001), but without solving the discrepancy completely. The ellipticity of this object (confirmed by McCrady et al., this meeting) suggests that more numerical work will be needed before its status and its survival probability can be assessed in a final way.

Notes

1. P. Wood, ANU, Australia; L.J. Smith, UCL, UK; J.S. Gallagher, Univ. of Wisconsin at Madison, USA; M. Mouhcine, Univ. of Nottingham, UK; W.D. Vacca, NASA ARC, USA; R. de Grijs, Univ. of Sheffield, UK; R.W. O'Connell, Univ. of Virginia, USA

References

Boily C.M., Lançon A., Deiters S., Heggie D.C., 2004, ApJ, 620, L27
Bressan A., Fagotto F., Bertelli G., Chiosi C., 1993, A&AS, 100, 647
Farouki R.T., Salpeter E.E., 1982, ApJ, 253, 512
Fioc M., Rocca-Volmerange B., 1997, A&A, 326, 950
Hurley J.R., Pols O.R., Tout C.A., 2000, MNRAS, 315, 543
Lejeune T., Cuisinier F., Buser R., 1998, A&AS, 130, 65
Louis P.D., Spurzem R., 1991, MNRAS, 251, 408
McCrady N., Gilbert A.M., Graham J.R., 2003, ApJ, 596, 240
Mengel S., Lehnert M.D., Thatte N., Genzel R., 2002, A&A, 383, 137
Smith L.J., Gallagher J.S., 2001, MNRAS, 326, 1027
Spitzer L. Jr., 1987, The Dynamics of Star Clusters (Princeton Univ. Press: Princeton)
Spurzem R., Takahashi K., 1995, MNRAS, 272, 772

IMF VARIATION IN M82 SUPER STAR CLUSTERS

A Population Study

Nate McCrady[1], James R. Graham[1], and William D. Vacca[2]
[1]*University of California, Berkeley, USA*
[2]*NASA-Ames Research Center, USA*

Abstract The nuclear starburst in M82 is host to over 20 infrared-bright, dense, young super star clusters (SSCs). We use high-resolution near-infrared Keck/NIRSPEC echelle spectroscopy to measure the stellar velocity dispersions. The SSCs are resolved in *Hubble Space Telescope* images, from which we measure half-light radii and integrated luminosities. We calculate virial masses for the SSCs, and compare the observed light-to-mass ratios to population synthesis models to constrain the initial mass function. There are apparent variations of the IMF within this single starburst galaxy. We present evidence for mass segregation despite the young ages, and discuss implications for the interpretation of the IMF.

Star formation in starburst galaxies can be resolved into young, dense, massive "super star clusters" (SSCs) that represent a substantial fraction of new stars formed in a burst event (Meurer et al. 1995, Zepf et al. 1999). The nuclear starburst in M82 is host to numerous SSCs, as detailed in McCrady, Gilbert & Graham (2003). The clusters suffer from varying degrees of extinction in the dusty, highly inclined disk of M82; most SSCs in the nuclear starburst are, in fact, optically invisible. Infrared observations penetrate the pervasive dust and facilitate detailed study of the clusters. In particular, we measure cluster virial masses based on stellar velocity dispersions and cluster sizes. These direct measurements of a cluster's mass enable us to characterize the present-day mass distribution of its stars and to apply population synthesis modelling to constrain the initial mass function (IMF).

We have obtained spectra of 20 SSCs in the nuclear starburst of M82, using the near-IR echelle spectrometer NIRSPEC on the 10m Keck II telescope (Fig. 1). The spectra of young (10^7 to 10^8 Myr old) clusters are dominated by the light of red supergiant stars (Lançon & Boily, these proceedings). Ro-vibrational CO band-head features are prominent, as well as numerous strong metal absorption features characteristic of cool, evolved stars. NIRSPEC attains high spectral resolution of ~ 13 km s^{-1} in the H band. For a 13th magnitude cluster, NIRSPEC observations achieve a signal-to-noise ratio of ~ 40 in a 10-minute exposure – infrared observing is very efficient.

To determine the stellar velocity dispersion of an SSC, we use echelle spectra of red supergiant stars (RSG) as templates. We have prepared an atlas of H-band spectra for spectral types G2 through M5. Direct comparison shows that the spectrum of an SSC resembles supergiant light with the features washed out by the velocity dispersion of the constituent stars (Fig. 1). By cross-correlating an SSC with a template RSG, we measure the stellar velocity dispersion within the cluster. The abundance of features in cool star spectra aid in the cross-correlation analysis, but the signal-to-noise ratio of the input spectra remains a key source of error.

Derivation of the virial mass requires measurement of a cluster's size. We fit elliptical King functions to cluster light profiles and measure the half-light radius, typically 2–3 pc in M82. Integration over the fitted profile provides photometry for the cluster. With the velocity dispersion and half-light radius as inputs, the virial theorem may be used to determine the mass of a cluster (Spitzer 1987). The masses of SSCs in M82 are a few $\times 10^5$ or 10^6 $M_\odot$, packed into volumes typical of Galactic globular clusters. SSCs are a massive, dense mode of star formation.

Evolved high-mass stars, with their enormous luminosities, dominate the light output of young SSCs. Measurement of the virial mass is the only available means to detect the low-mass stars in a cluster. Our technique determines the virial mass independent of any assumptions about the IMF, thus enabling us to characterize the IMF cluster-by-cluster and seek variations within the nuclear starburst SSC population. Specifically, we use population synthesis models to compare the measured light-to-mass (L/M) ratio to predictions for various IMF forms. Changes to either the IMF slope or the lower-mass cut-off affect the expected cluster L/M ratio in the H band.

Cluster age and line-of-sight extinction are the primary uncertainties that affect our mass measurements. The cluster spectra provide a few indications of age, including the presence or absence of Wolf-Rayet or AGB-star features and nebular emission. For a coeval population, a single spectral type will dominate the integrated cluster light at a given time. Our cross-correlation analysis determines the dominant spectral type, which can be used in concert with stellar evolution models to further constrain the age of the cluster. Knowledge of the dominant spectral type also tells us the expected color of the cluster. By measuring the observed color and applying an extinction law to the derived color excess, we can estimate the extinction.

To date, we have examined three of the nuclear SSCs in M82: highly-reddened clusters MGG-9 and MGG-11 (McCrady et al. 2003) and the optically-bright cluster M82-F (McCrady, Graham & Vacca 2005). These clusters have ages of 10–50 Myr, based on their dominant spectral types. We apply extinction corrections to the photometry to derive the luminosity of each cluster, and derive the virial mass to determine the L/M ratio. To investigate the IMF of a cluster, we use population synthesis models from STARBURST99 (Leitherer et al. 1999) to generate the time evolution of the L/M ratio. Figure 2 shows the H-band L/M ratios for the three M82 SSCs versus two fiducial IMFs: a standard Kroupa (2001) IMF over a full range of masses and a "top-heavy" Salpeter IMF with no stars of $M > 2$ $M_\odot$. MGG-9 is consistent with a Kroupa IMF, whereas the other clusters are too luminous for their mass relative to the Kroupa IMF. Although we cannot distinguish changes to the IMF slope from changes to the lower-mass cut-off, it is apparent that no single IMF fits all three clusters. The IMF appears to vary from cluster to cluster within the M82 starburst, perhaps systematically with location.

There is, however, a complication. In *HST*/ACS and NICMOS images, the cluster half-light radius is progressively larger at shorter wavelengths. This is characteristic of mass segregation, as discussed by Sirianni (these proceedings). The near-infrared light is dominated by red supergiants. When we measure the virial mass based on near-IR light, we are insensitive to mass outside the volume occupied by the RSG stars. In a coeval stellar population, the RSG stars are the most massive stars still present, as they leave the main sequence first. If the cluster is mass segregated, we understate the total cluster mass and thus overstate the L/M ratio. In a mass-segregated cluster, the IMF varies with radius (Brandl et al. 1996). The apparently top-heavy IMFs for some clusters may, in fact, reflect mass segregation.

Our work on the SSCs in the M82 nuclear starburst is continuing. We have NIRSPEC spectra of approximately 20 SSCs, enabling us to look for IMF variations within the cluster population and quantify the effect of mass segregation. This data set represents the largest virial mass study of SSCs in a single galactic environment, providing a good test of environmental dependence of the IMF.

References

Brandl B., et al., 1996, ApJ, 466, 254
Kroupa P., 2001, MNRAS, 322, 231
Leitherer C., et al., 1999, ApJS, 123, 3
McCrady N., Gilbert A.M., Graham J.R., 2003, ApJ, 596, 240
McCrady N., Graham J.R., Vacca W.D., 2005, ApJ, in press (astro-ph/0411256)
Meurer G.R., Heckman T.M., Leitherer C., Kinney A., Robert C., Garnett D.R., 1995, AJ, 110, 2665
Spitzer L. Jr., 1987, Dynamical Evolution of Globular Clusters (Princeton Univ. Press: Princeton)
Zepf S.E., Ashman K.M., English J., Freeman K.C., Sharples R.M., 1999, AJ, 118, 752

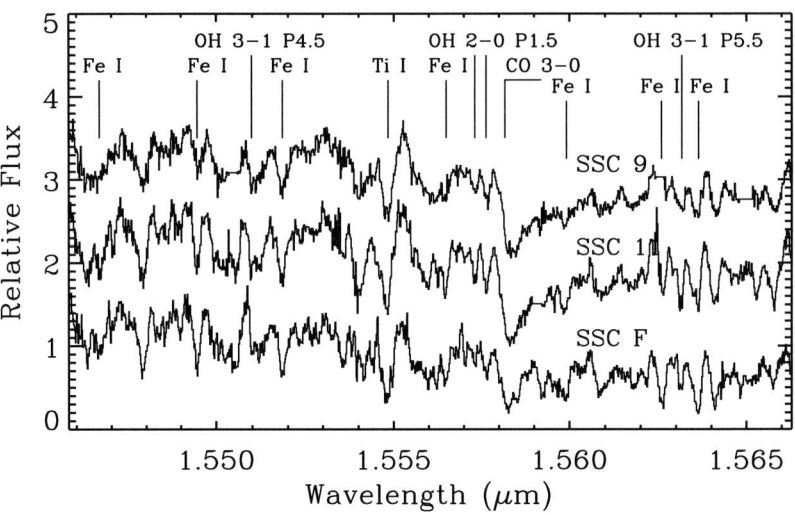

Figure 1. NIRSPEC echelle spectra of three SSCs in M82. The spectra resemble red supergiant spectra, with features "washed out" by the stellar velocity dispersion of the clusters.

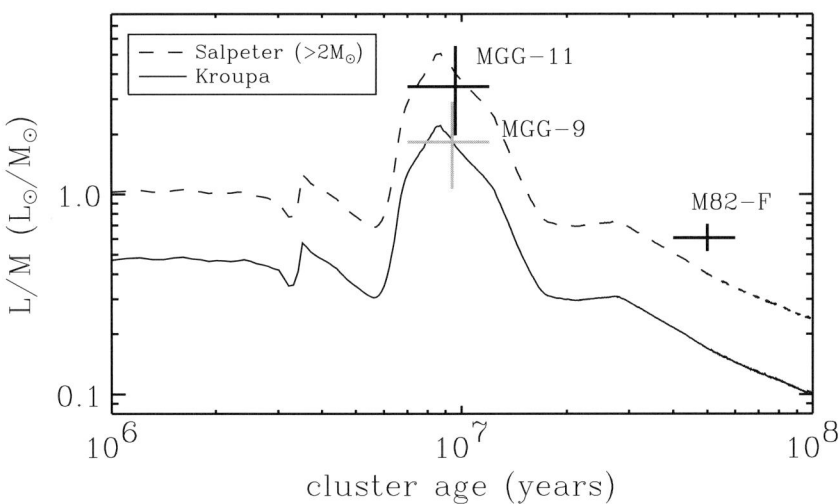

Figure 2. Comparison of observed H-band L/M ratios with population synthesis models. MGG-9 is consistent with a Kroupa IMF, while the other two clusters appear to have "top-heavy" IMFs deficient in low-mass stars.

Session III

Starbursts as a Function of Wavelength

COLOURFUL STARBURSTS

Jürgen Knödlseder
Centre d'Étude Spatiale des Rayonnements, 9 avenue du Colonel-Roche, B.P. 4346, 31028 Toulouse Cedex 4, France
knodlseder@cesr.fr

Abstract This paper provides a concise review of the multi-wavelength properties of starburst galaxies, from the radio domain up to high-energy gamma-rays. A few selected topics will be addressed in some depth, such as correlations between wavebands, the high-energy emissivity of starburst galaxies, and their particular supernova activity.

1. Introduction

Starburst galaxies are characterised by star-formation rate (SFR) densities (i.e., SFR per unit area) considerably exceeding the values found in normal galaxies (see Heckman, these proceedings). About ~ 10–20% of the global star formation in the local Universe occurs in starbursts, so starburst activity is, even today, an important and common mode of star formation (see Kennicutt, these proceedings). Starburst activity implies large dust column densities, large radiant energy densities, large supernova rates per unit volume, and large gas pressures. Consequently, the starburst activity of a galaxy manifests itself in many different ways across the electromagnetic spectrum, and only a multi-wavelength view of the phenomenon allows one to catch all aspects of this activity.

The spectral energy distribution of a typical starburst galaxy is shown in Fig. 1. The radio emission of a starburst galaxy is a combination of non-thermal synchrotron emission, arising from ultra-relativistic cosmic-ray electrons, accelerated in supernova shocks, and thermal free-free emission, produced by Bremsstrahlung of electrons in the interstellar medium (ISM) that has been ionised by the hot massive stars of the starburst. At frequencies of about 1 GHz, 90% of the radio emission is non-thermal; only in a small frequency range, around $30-200$ GHz, does thermal emission dominate, yet the intensity of this emission is generally too low to allow for detailed studies of distant starburst galaxies (Condon 1992). For nearby starbursts, however, observations at centimetre wavelengths provide a powerful tool to study the ionising power of

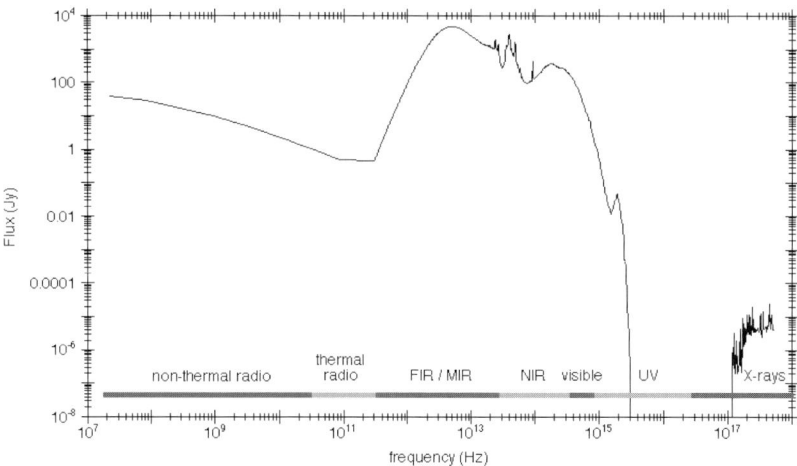

Figure 1. The observed radio to X-ray spectral energy distribution (SED) of the starburst galaxy M82. Starburst galaxies emit most of their radiant luminosity in the far-infrared.

hot stellar populations, that may be invisible in the optical and near-infrared domains due to dust obscuration (see Ulvestad, these proceedings).

Above frequencies of a few 100 GHz, in the far-infrared (FIR) domain, the radio emission gets swamped by thermal re-radiation of starlight from dust grains. This dust emission dominates the spectrum also in the mid-infrared (MIR) range, and the shape of the spectrum carries valuable information about the dust temperature and grain size distribution. Strong FIR and NIR emission is one of the characteristics of starburst galaxies, with FIR luminosities considerably exceeding the blue luminosities of the galaxies. In the infrared colour-colour diagram, starburst galaxies occupy a well-defined region, with colours similar to galactic H II regions, indicating that the starburst phenomenon is related to the young populations of hot and massive stars.

For wavelengths shorter than $\sim 10\mu$m, in the near-infrared (NIR) domain, starlight starts to dominate the emission from the starburst galaxy. Although most of the NIR emission arises from old stellar populations (in particular from red giants, supergiants, and asymptotic giant branch stars), some starburst diagnostics can be derived using NIR emission lines (H and He recombination lines, [Fe II], H_2 vibrational lines; see Förster Schreiber, these proceedings; van der Werf, these proceedings).

The optical emission of starburst galaxies is characterised by narrow nebular emission lines that arise from the ionisation of the ISM by the energetic photons of the hot stars. Young starbursts may eventually also show broad emission lines originating in the strong stellar winds of Wolf-Rayet stars. The nebular emission lines carry information about the SFR, metallicity, dust reddening, and the ionisation conditions of the ISM. However, dust obscuration heavily affects the emission in this wavelength band and makes the interpretation of the observations complex (see Bergvall, these proceedings; Calzetti, these proceedings).

The ultraviolet spectrum of a starburst galaxy is dominated by the hottest stars (O-type or Wolf-Rayet stars), showing characteristic photospheric absorption lines and stellar wind emission lines. Interstellar absorption is most severe in this wavelength domain, making studies of starburst characteristics extremely difficult. In particular, the majority of the stars is hidden from view in this waveband, and the small number of them that contribute to the emission needs not to be representative for the entire population of objects that participate in the starburst (see Leitherer, these proceedings).

In the X-ray domain, the emission originates from an extremely hot (up to $\sim 10^6$ K) thermal plasma that manifests itself through a thermal Bremsstrahlung continuum with atomic emission lines superimposed. The thermal plasma emits strongest in the soft X-ray domain ($\lesssim 1$ keV), and is generally interpreted as the hot component of the ISM heated by the passage of supernova shock waves. Interstellar absorption is still important in the soft X-ray band, and the emission morphology is generally determined by the local gas pattern. While the thermal plasma is intrinsically diffuse, point-like sources with a considerably harder spectrum also contribute to the X-ray emission. They are generally identified with accreting binary systems hosting either a neutron star or a black hole. Their emission extends above 2 keV, where the ISM becomes transparent to X-rays, and their study provides an unbiased view of the underlying source populations.

2. Correlations

The distinguishing feature of starburst galaxies is the presence of a large number of young, hot, and massive stars. These populations manifest themselves throughout large parts of the electromagnetic spectrum and it is therefore, first of all, not surprising that the emergent spectrum shows some correlations between the different wavelength domains.

Probably the tightest and most universal correlation amongst global properties of galaxies is that between FIR and non-thermal radio continuum emission that holds for all types of normal galaxies (i.e., galaxies without AGN activity), including starburst galaxies (Fig. 2; Condon 1992). This linear correlation spans ~ 5 orders of magnitude with less than 50% dispersion, and applies to redshifts of at least $z \sim 1.3$ (Yun et al. 2001, Garrett 2002).

The correlation is generally explained by the effects of the young massive stars on the surrounding medium. The FIR emission arises from the re-radiation of FUV photons that have been absorbed by dust grains, while the

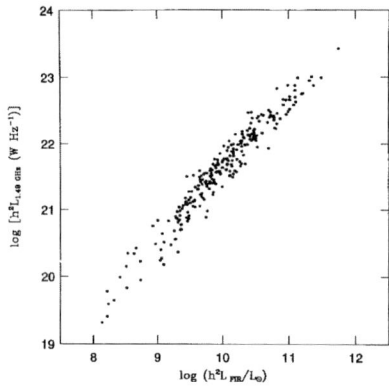 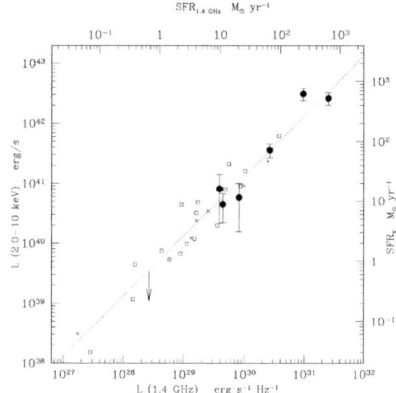

Figure 2. Correlation between FIR and non-thermal radio emission for a large variety of galaxies, including starburst galaxies (Condon 1992).

Figure 3. Correlation between hard X-ray (2–10 keV) and non-thermal radio emission (Ranalli et al. 2003).

radio synchrotron emission originates from ultrarelativistic cosmic-ray (CR) electrons that have been accelerated by supernova shocks. Since both processes rely on the presence of massive stars, and since massive stars are short-lived, the FIR and synchrotron luminosities are assumed to be indicators of the global SFR.

In view of the indirect link between FIR and radio emission, the tightness and the linearity of the relation comes as a surprise, and is, on theoretical grounds, not fully understood (Groves et al. 2003, and references therein). Most theories require some form of coupling between the magnetic field and gas density, and it is usually asserted that equipartition provides the coupling. Yet, the way in which equipartition can arise is left open. Recent ISO observations suggest that the linearity of the correlation only holds for the warm dust component, while the cold dust component obeys a slightly non-linear correlation (Pierini et al. 2003). Bell (2003) even suggests that neither the FIR nor the radio luminosity is a linear tracer of the SFR, and that the observed radio-FIR correlation is simply a conspiracy of Nature.

In any case, on small size and time-scales, the radio-FIR correlation should break down, since FUV photons are produced early on during the evolution of a stellar population by the most massive members ($M_{\rm ini} \gtrsim 20 \, M_\odot$), while CRs are accelerated only a few $\times 10^6$ yr later following the explosion of stars covering a much wider initial mass range ($M_{\rm ini} \gtrsim 8 \, M_\odot$). Indeed, spatially resolved investigations of nearby galaxies indicate that the correlation becomes more complex inside individual galaxies than between galaxies (Gordon et al. 2004).

And also on short time-scales, the correlation seems to break down: some starburst galaxies appear to be deficient in synchrotron emission, which could be explained by nascent starbursts (< 1 Myr old), where most of the UV photons are absorbed (as in compact H II regions) and no supernova has so far exploded to accelerate CRs (see Roussel, these proceedings).

Another correlation has recently been suggested, between hard X-ray ($2-10$ keV) and non-thermal radio continuum emission (Fig. 3; Ranalli et al. 2003). The $2-10$ keV band, which is essentially free from extinction, is dominated by the combined emission of X-ray binaries, which are subdivided into two classes (e.g., Gilfanov 2004): low-mass X-ray binaries (LMXBs), consisting of a low-mass star ($\lesssim 1$ M$_\odot$) from which matter accretes via Roche-lobe overflow onto a compact object (either a neutron star or a black hole), and high-mass X-ray binaries (HMXBs), consisting of a massive star ($\gtrsim 8$ M$_\odot$) from which stellar wind material is accreted onto a compact object. Only HMXBs are related to massive stars, and therefore, to the local SFR, while the number of LMXBs, with their long evolutionary time-scales of $\sim 10^9 - 10^{10}$ yr, is believed to be proportional to the total mass of the galaxy (Gilfanov 2004).

In quiescent galaxies, such as the Milky Way, the hard X-ray luminosity is believed to be dominated by LMXBs, while in starburst galaxies HMXBs become dominant (Grimm et al. 2002). Assuming that the integrated HMXB and radio luminosities are both proportional to the SFR, a linear correlation between hard X-ray and radio emission would indeed be expected. Yet, for lower SFRs, where LMXBs start to dominate, this correlation should break down. Figure 3 shows, however, that the linear correlation holds for a wide range of SFRs (and in particular for low SFRs), making the existence of the correlation puzzling. In addition, the shallow slopes of the X-ray binary luminosity functions imply that the integrated hard X-ray emission of a galaxy is dominated by the $\sim 5 - 10$ most luminous sources, and that variability of individual sources or an outburst of a bright transient source can increase the overall luminosity by as much as a factor of ~ 2 (Grimm et al. 2002). It is intriguing that such a small number of objects can create such a tight correlation between processes that are only linked very indirectly. It therefore remains to be seen if the observed correlation reflects an underlying physical relation or if it is just another conspiracy of Nature.

3. Supernova factories

Starburst galaxies are literally factories of core-collapse supernovae (SNe). With SFRs of $10-1000$ M$_\odot$ yr^{-1}, core-collapse SN rates in the range $0.05-5$ SNe yr^{-1} are expected (Mattila & Meikle 2001). Record holders in the (visually) observed number of SNe in a single host galaxy are the starburst galaxies NGC 6946 (7 SNe since 1917), M83 (6 SNe since 1923), and Arp 299 (4 SNe

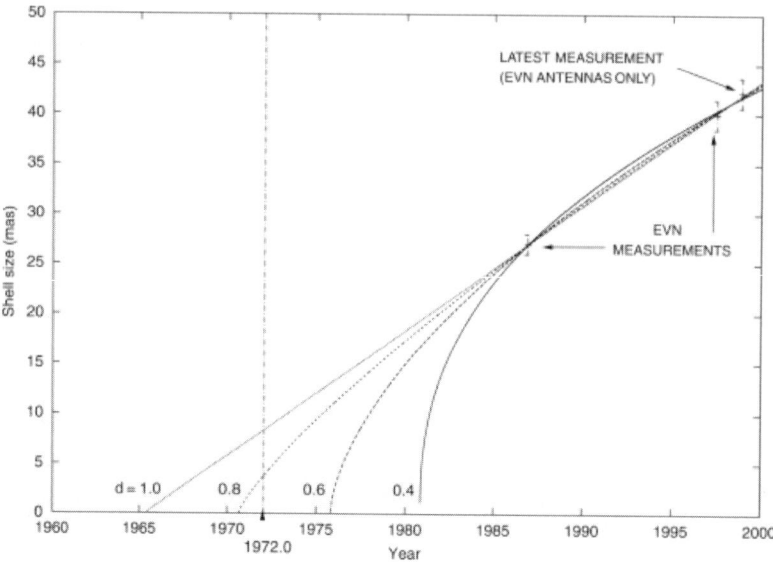

Figure 4. Expansion of the radio SNR 43.31+592 in M82 (McDonald et al. 2001). The curves indicate expansion models with different deceleration parameters, d, where $D \propto (t-t_0)^d$, with D being the shell size, t the epoch, and t_0 the date of explosion; $d = 1.0$ corresponds to free expansion. The vertical line indicates the upper limit for the explosion date t_0.

between 1992-1999), confirming the high SN rate estimates for starburst galaxies. The most recent and most nearby supernova in a starburst galaxy occurred this year in the active starbursting core of M82 (SN2004am).

Dust obscuration may prevent the detection of many of the SNe in the optical. Several groups have started major efforts to detect SNe in the less obscured NIR regime, with the first encouraging results having been reported recently (Mannucci et al. 2003, Mattila et al. 2004). The observed SN rates, however, are still somewhat low compared to the expectations, indicating that the extreme dust extinction that exists in nuclear regions of starburst galaxies may even prevent SN detections in the NIR domain.

As alternative to the optical and NIR surveys, the SN activity of starburst galaxies is also studied in the radio domain. In M82, for example, at least 61 compact radio sources have been identified, of which most show the spectral characteristics of young supernova remnants (SNRs; McDonald et al. 2002). Regular high-resolution VLBI observations of these SNRs allow us to directly determine the expansion velocity and diameter of the remnants, which, in turn, can be converted into a remnant age. Application to a large number of SNRs allows then the determination of the SN rate of the host galaxy.

Examples of recently studied objects in M82 are SNR 41.95+575 (Riley et al. 2004), the most compact radio source in this galaxy, for which an expansion velocity of $v_{\text{exp}} = 2500 \pm 1200$ km s^{-1} has been measured, and SNR 43.31+592 (McDonald et al. 2001), a typical shell-type SNR, with an expansion velocity of $v_{\text{exp}} = 7350 \pm 2100$ km s^{-1}. Assuming free expansion, explosion dates of ~ 1970 and ~ 1965 are inferred for these SNRs, respectively. That free expansion is a good assumption for such young SNRs is illustrated in Fig. 4: SNR 43.31+592 has been observed during 3 epochs, using VLBI, which are consistent with a wide variety of deceleration parameters, d. However, the remnant clearly existed already in 1972, as is shown in the 8.1 GHz map of Kronberg & Wilkinson (1975), indicating that no substantial deceleration occurred so far.

From a statistical analysis of the size distribution of 23 radio SNR in M82, and under the assumption of a mean (free) expansion velocity of 5000 km s^{-1}, Muxlow et al. (1994) estimated a supernova rate of 0.05 per year. This determination is in good agreement with estimates based on the present SFR in the core of M82 (Mattila & Meikle 2001), illustrating the potential of high-resolution radio observations in constraining the star-formation activity in nearby starburst galaxies.

4. Gamma-ray emission

With the advent of more sensitive instrumentation in the field of gamma-ray astronomy, starburst galaxies have now proven to be also sources of the most energetic photons that prevail in the Universe. Recently, Itoh et al. (2002) reported the first detection of TeV gamma-rays from the nearby starburst galaxy NGC 253, using the CANGAROO-II 10m imaging atmospheric Čerenkov telescope. The emission appears spatially extended and temporally steady, and has been interpreted as a halo of non-thermal emission due to inverse Compton scattering of ultrarelativistic CR electrons of cosmic microwave background and FIR photons (Itoh et al. 2003). If confirmed, this would be another manifestation of CR particle acceleration by the ubiquitous supernovae in starburst galaxies.

Starburst galaxies are also increasingly becoming of interest for the understanding of the most violent phenomenon known in the Universe, gamma-ray bursts (GRBs). Since 1997, the improvement of instrumentation in the hard X-ray and soft gamma-ray domain has allowed the identification of counterparts of these mysterious cosmic explosions in almost all wavelength bands. There is now growing evidence that at least the long-duration GRBs (duration $\lesssim 2$s) are associated with Type Ic supernova explosions, which are believed to arise from the collapse of Wolf-Rayet stars. That GRBs may be correlated to star-forming regions has been suggested since the first identifications of GRB host galaxies (Paczynski 1998) and there is now increasing evidence that (perhaps all) GRB host galaxies are undergoing starbursts (Sokolov et al. 2001).

If GRBs are intimately connected with the deaths of massive stars, they are potential probes of star formation in the early Universe. Since GRBs are extremely luminous they can be traced to extreme redshifts of $z \sim 20$, providing that GRBs existed already at this early epoch. Because gamma-rays are not attenuated by intervening columns of gas and dust, GRBs offer a unique perspective to probe star formation in the early Universe, independently of the biases associated with dust extinction. In addition, early-time spectroscopy of GRB optical afterglows allows one to measure the metallicity and dust content of the progenitor environment through absorption line studies. These afterglow studies of distant GRBs will undoubtedly advance our understanding of the first generations of stars, as well as of the chemical evolution in the early Universe.

References

Bell E.F., 2003, ApJ, 586, 794
Condon J.J., 1992, ARA&A, 30, 575
Garrett M.A., 2002, A&A, 384, 19
Gilfanov M., 2004, MNRAS, 349, 146
Gordon K.D., et al., 2004, ApJS, 154, 215
Grimm H.-J., Gilfanov M., Sunyaev R., 2002, A&A, 391, 923
Groves B.A., Cho J., Dopita M., Lazarian A., 2003, PASA, 20, 252
Itoh C., Enomoto R., Yanagita S., Yoshida T., Tsuru T.G., 2003, ApJ, 584, L65
Itoh C., et al., 2002, A&A, 396, L1
Kronberg P.P., Wilkinson P.N., 1975, ApJ, 300, 430
Mannucci F., et al., 2003, A&A, 401, 519
Mattila S., Meikle W.P.S., Greimel R., 2004, NewAR, 48, 595
Mattila S., Meikle W.P.S., 2001, MNRAS, 324, 325
McDonald A.R., Muxlow T.W.B., Wills K.A., Pedlar A., Beswick R.J., 2002, MNRAS, 334, 912
McDonald A.R., Muxlow T.W.B., Pedlar A., Garrett M.A., Wills K.A., Garrington S.T., Diamond P.J., Wilkinson P.N., 2001, MNRAS, 322, 100
Muxlow T.W.B., Pedlar A., Wilkinson P.N., Axon D.J., Sanders E.M., de Bruyn A.G., 1994, MNRAS, 266, 455
Paczynski B., 1998, ApJ, 494, 45
Pierini D., Popescu C.C., Tuffs R.J., Völk H.J., 2003, A&A, 409, 907
Ranalli P., Comastri A., Setti G., 2003, A&A, 399, 39
Riley J.D., Pedlar A., Muxlow T.W.B., McDonald A.R., Beswick R.J., Wills K.A., 2004, in: Supernovae, Marcaide J.M., Weiler K.W., eds., IAU Coll. 192, (Springer: New York), on accompanying CD-ROM (astro-ph/0405114)
Sokolov V.V., et al., 2001, A&A, 372, 438
Yun M.S., Reddy N.A., Condon J.J., 2001, ApJ, 554, 803

A FAR-ULTRAVIOLET VIEW OF STARBURST GALAXIES

Claus Leitherer
Space Telescope Science Institute, 3700 San Martin Drive, Baltimore, MD 21218, USA
leitherer@stsci.edu

Abstract Recent observational and theoretical results on starburst galaxies related to the wavelength regime below 1200Å are discussed. The review covers stars, dust, as well as hot and cold gas. This wavelength region follows trends similar to those seen at longer wavelengths, with several notable exceptions. Even the youngest stellar populations show a turn-over in their spectral energy distributions, and line-blanketing is much more pronounced. Furthermore, the O VI line allows one to probe gas at higher temperatures than possible with lines at longer wavelengths. Molecular hydrogen lines (if detected) provide a glimpse of the cold phase. I cover the crucial wavelength regime below 912Å, and the implications of recent attempts to detect the escaping ionizing radiation.

1. Background

The astrophysically important wavelength region below ~ 1200Å is still relatively unexplored, at least at low redshift where rest-frame observations must be obtained from space. Prior to the launch of FUSE (Moos et al. 2000), far-ultraviolet (far-UV) studies were limited to bright objects. The earliest spectral data for bright stars were obtained by Copernicus (Rogerson et al. 1973) and ORFEUS (Grewing et al. 1991), and with the UV spectrometers on the Voyager 1 and 2 spacecraft (Longo et al. 1989). Voyager 2 also succeeded in recording a far-UV spectrum of M33 (Keel 1998). HUT (Davidsen 1993) was the first instrument sensitive enough to collect spectra of faint galaxies below Lyα. HUT was flown on two missions and generated a rich database of far-UV spectra of actively star-forming and starburst galaxies. Subsequently, FUSE with its superior resolution and sensitivity fully opened the far-UV window to starburst galaxies. Most of this review deals with results obtained with FUSE and, to a smaller degree, with HUT.

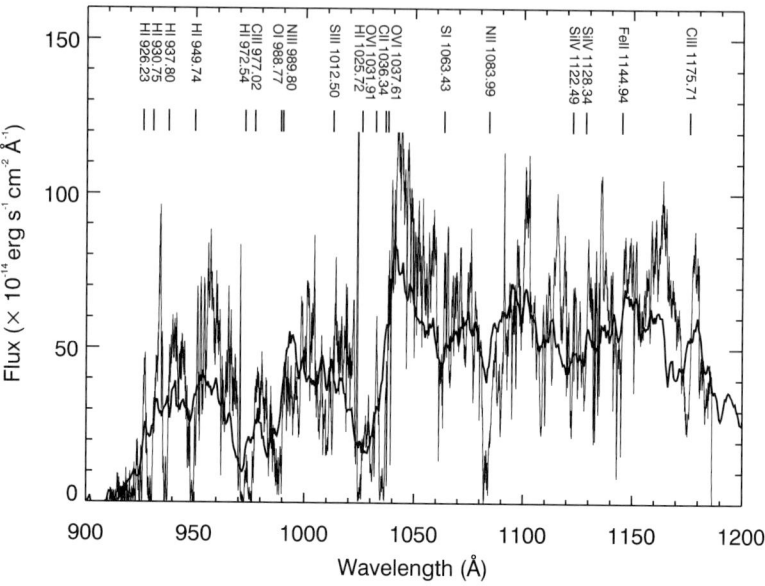

Figure 1. Spectral region between 900 and 1200Å for M83 (= NGC 5236). Major lines are labeled. Thick line: HUT; thin: FUSE (Leitherer et al. 2002).

2. Stellar Populations

The far-UV spectrum of the archetypal starburst galaxy M83 is reproduced in Fig. 1 (Leitherer et al. 2002). The wavelength region shown covers 900 – 1200Å, where blanketing is most severe. M83 has a supersolar oxygen abundance. Therefore, line-blanketing effects due to stellar wind, stellar photospheric, and interstellar lines are particularly strong in this galaxy. The stellar features generally originate from higher ionization stages than the features found above 1200Å. The most prominent transition is the O VI resonance doublet at 1032.38Å, which displays a spectacular P Cygni profile over a broad range of spectral types. At the resolution afforded by FUSE, the blueshifted absorption component of the P Cygni profile is resolved from nearby Lyβ and can be distinguished from the narrow interstellar C II at 1036Å. The (redshifted) emission component of its P Cygni profile is relatively unaffected by interstellar lines and provides additional diagnostic power. The C IIIλ1176 line is at the long-wavelength end of the spectral range covered and can also be observed with spectrographs optimized for wavelengths longward of Lyα. Surprisingly, the line has received relatively little attention in the earlier literature although it is a very good diagnostic of the properties of hot stars. C III is not a resonance transition, and consequently does not suffer from contamination by an interstellar component. CIII, like most other stellar lines, has a pronounced metallicity dependence, either directly via opacity variations, or indirectly via metallicity-dependent stellar wind properties.

Quantitative modeling of the stellar far-UV lines by means of evolutionary synthesis was done by Robert et al. (2003). In Fig. 2 I show an evolutionary sequence based on an empirical FUSE library of hot stars (Pellerin et al. 2002). The computed spectra are a good match to the M83 spectrum in Fig. 1. The CIIIλ1176 line is an excellent age diagnostic, mirroring the behavior of the well-studied SiIVλ1400 line: when luminous supergiants appear around 3 Myr, wind recombination raises the emission flux (Leitherer et al. 2001). The OVI line, in contrast, is largely decoupled from stellar parameters over a wide range of ages. This line is powered by shock heating and remains constant for stellar temperatures above $\sim$30,000 K. Combining lines with different optical depths, excitation, and ionization parameters allows age and metallicity estimates from far-UV spectra, analogous to methods calibrated at longer wavelengths (e.g., Keel et al. 2004).

Apart from their sensitivity to metallicity, the far-UV *lines* are affected the well-known age vs. initial mass function (IMF) degeneracy. In the absence of additional constraints, age and IMF can always be traded. This applies to the far-UV *continuum* as well, which in addition suffers from an age-reddening degeneracy. In contrast to wavelengths above 1200Å, the intrinsic stellar spectra below 1200Å are outside the Rayleigh-Jeans regime, and age effects are no longer negligible for the continuum slope generated by an instantaneous population. Alternatively, for a population of continuously forming stars, the region between 912 and 1200Å becomes even less age sensitive to population variations than the near-UV, because an equilibrium between star formation and stellar death is reached earlier in time (see Leitherer et al. 1999). Ages and star-formation rates in starburst galaxies derived from far-UV spectra are consistent with the results from longer wavelengths.

3. Dust Obscuration

If the age and IMF can be constrained independently, the observed far-UV spectral energy distribution is mostly a measure of the dust attenuation. The continua of star-forming galaxies are known to obey a well-defined average obscuration curve above 1200Å (Calzetti 2001). The curve accounts for the total absorption and encompasses the net effects of dust/star geometry, absorption, scattering, and grain-size distribution. Extension of this curve down to the Lyman limit using HUT and FUSE observations of starburst galaxies was done by Leitherer et al. (2002) and Buat et al. (2002), respectively. Their results are compared to stellar data and to theoretical predictions in Fig. 3. The reddening curve of Sasseen et al. (2002) applies to individual stars; it is signif-

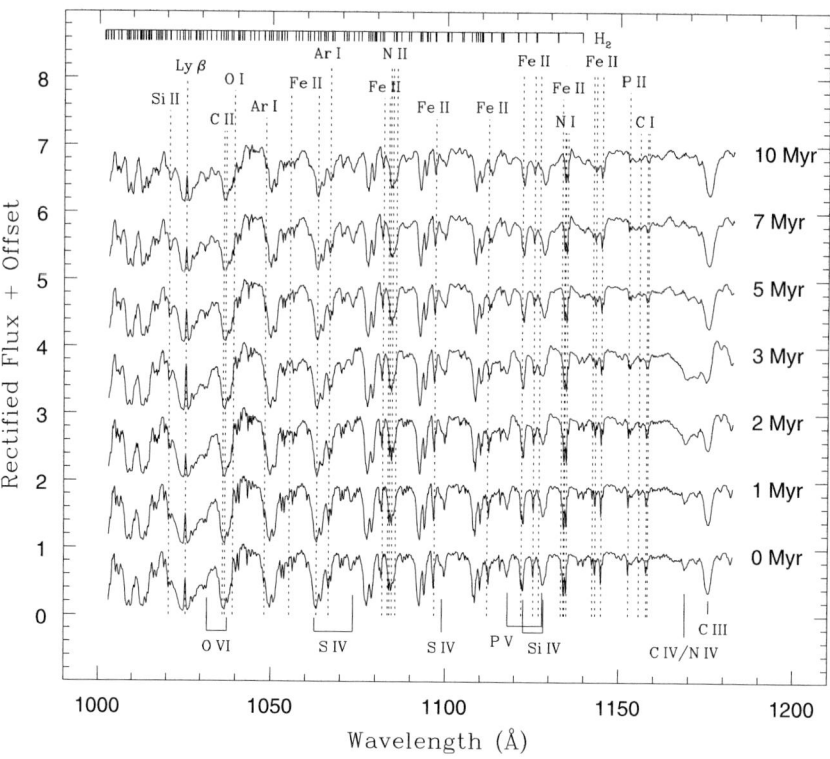

Figure 2. Evolution of a synthetic far-UV spectrum following a Salpeter IMF between 0 and 10 Myr. Stellar features (bottom), H$_2$ lines (top), and other interstellar absorptions (top; vertical dotted lines) are labeled. The emission line at 1026Å is geocoronal (Robert et al. 2003).

icantly steeper than the curves derived for galaxies. The physical interpretation of the "grayer" starburst reddening curve is non-uniform attenuation. Most of the stars are totally hidden from view, and the observed UV light is emitted by those few stars which happen to have low attenuation. This effect becomes progressively more important for shorter and shorter wavelengths. The implication is that far-UV observations sample only the tip of the iceberg and could be severely biased. For instance, if there were an age or IMF-dependence of the reddening, even the interpretation of spectral *lines* in the far-UV would be compromised with the assumption of a simple foreground dust screen. Circumstantial evidence for this effect to play a role has been presented by Chandar et al. (2004).

Witt & Gordon (2000) found that the empirical starburst attenuation law is most closely reproduced by a clumpy shell model with SMC-like dust and a

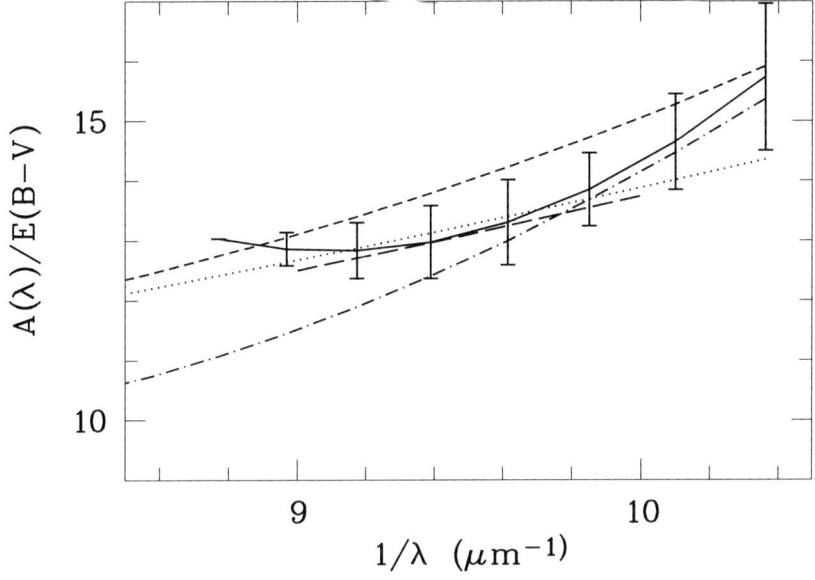

Figure 3. Comparison of attenuation curves. Solid: Buat et al. (2002); dotted: Leitherer et al. (2002); short-dashed: Calzetti et al. (2000); dot-dashed: Sasseen et al. (2002); long-dashed: model of Witt & Gordon (2000) for a shell distribution, a clumpy dust and an optical depth in the V band equal to $\tau_V = 1.5$ (Buat et al. 2002).

dust column density equivalent to $\tau_V = 1.5$. This corresponds to a far-UV attenuation correction factor of order 10 (Fig. 3).

4. Gas: Hot and Cold Phases

The far-UV spectral region includes line transitions that are sensitive to both the very hot ($\sim 10^5$K) and very cold ($\sim 10^2$K) interstellar material. Coronal gas with temperatures of several 100,000 K can be probed with the OVI line whose corresponding ionization potential is 114 eV. On the other hand, the rotational/vibrational transitions of H_2 trace cool molecular gas.

The combined effects of multiple stellar winds and supernovae are capable of heating the interstellar medium (ISM) and initiating large-scale outflows. Such outflows are a significant sink for the gas reservoir. They have been known for some time from optical and X-ray imagery and have recently been analyzed with *absorption*-line spectroscopy (Heckman 2005). Heckman et al. (2001) obtained FUSE far-UV spectra of the nearby dwarf starburst NGC 1705, probing the coronal ($10^5 - 10^6$K) and the warm (10^4K) phases of the outflow.

The kinematics of the warm gas are compatible with a simple model of the adiabatic expansion of a superbubble driven by the supernovae in the starburst. Radiative losses are negligible so that the outflow may remain pressurized over a characteristic flow time scale of 10^8 to 10^9yr, as estimated from the size and velocity. The same conclusion was reached for M82 by Hoopes et al. (2003), who used FUSE to search for O VI *emission* in the starburst superwind of M82. No O VI emission was detected at any of the pointings. These observations limit the energy lost through radiative cooling of coronal-phase gas to the same magnitude as that lost in the hot phase through X-ray emission, which has been shown to be small.

The total mass transported out of the starburst region via galactic superwinds is hard to constrain, given the uncertain ionization corrections and the strength of the observable spectral lines. Attempts were made by Johnson et al. (2000) and Pettini et al. (2000) for the nearby dwarf starburst galaxy He 2-10 and the luminous Lyman-break galaxy MS1512-cB58, respectively. In both cases, the mass-loss rate of the ISM is rather similar to the star-formation rate. Taken at face value, this suggests that the available gas reservoir will not only be depleted by the star-formation process but, more importantly, by removal of interstellar material. Starbursts may determine their own fate by their prodigious release of kinetic energy into the ISM.

The spectral range of FUSE includes numerous transitions of molecular hydrogen, making it possible to study H_2 in diffuse interstellar environments directly through absorption measurements, rather than relying on the indirect CO technique. Hoopes et al. (2004) searched for H_2 absorption in the five starburst galaxies NGC 1705, NGC 3310, NGC 4214, M83, and NGC 5253. Weak H_2 absorption was detected in M83 and NGC 5253, and upper limits on the H_2 column density were derived in the other three galaxies. The upper limits on the mass of molecular gas are orders of magnitude lower than the H_2 mass inferred from CO emission measurements. The interpretation is that almost all of the H_2 is confined to clouds with column densities in excess of 10^{20} cm^{-2} that are opaque to far-UV light and cannot be detected in the FUSE spectra. The far-UV continuum seen with FUSE originates from sightlines between the dense clouds, which have a covering factor < 1. This morphology is consistent with that of the interstellar dust, which is thought to be clumpy. The complex observational biases related to varying extinction across the extended UV emission in the FUSE apertures make it difficult to characterize the diffuse H_2 in these starburst galaxies.

5. Gas: Transparency of the Lyman Continuum

Star-forming galaxies are the dominant contributor to the non-ionizing UV radiation field in the Universe. Are they a significant component of the ioniz-

Table 1. Determinations of the Lyman escape fraction in galaxies

Reference	Instrument	Objects	$f_{\rm esc}$
Leitherer et al. (1995)	HUT	4 galaxies; $z \simeq 0$	$< 3\%$
Hurwitz et al. (1997)	HUT	4 galaxies; $z \simeq 0$	$< 19\%$
Deharveng et al. (1997)	Hα/UV	local luminosity function	$< 1\%$
Deharveng et al. (2001)	FUSE	Mrk 54; $z \simeq 0$	$< 5\%$
Giallongo et al. (2002)	FORS2	$z = 2.96, 3.32$	$< 16\%$
Fernández-Soto et al. (2003)	WFPC2	HDF; $1.9 < z < 3.5$	$< 4\%$
Malkan et al. (2003)	STIS	$1.1 < z < 1.4$	$< 1\%$
Steidel et al. (2001)	Keck/LRIS	29 galaxies; $z \simeq 3.4$	$\sim \mathbf{100\%}$

ing background as well? Simple arguments might suggest otherwise. An H I column density of $\sim 1 \times 10^{18}$ cm^{-2} is sufficient to absorb essentially all of the ionizing radiation. Since the measured extinctions imply column densities that are three or four orders of magnitude higher than this, it might appear that essentially no ionizing radiation can escape. However, the porosity of the ISM seen in the non-ionizing continuum ($\lambda > 912$Å) could very well extend below the Lyman edge and may dominate the shape of the emergent spectrum. The situation is sufficiently complex that the only way to determine the escape fraction $f_{\rm esc}$ of the ionizing radiation is via a direct measurement.

Attempts to measure $f_{\rm esc}$ fall into two categories: observations of local galaxies with a far-UV detector, and measurements using galaxies at cosmological redshifts, which are accessible from the ground with 8m-class telescopes. Either technique has its advantages and disadvantages. The "local" approach faces the obvious challenge of extreme UV observations, whereas the "distant" measurement is affected by the radiative transfer in the intergalactic medium. In addition, a somewhat less direct method is to determine the Lyman continuum opacity from a comparison of the Hα and UV luminosity functions in the local Universe. Table 1 gives a summary of recent results. Except for the last entry in this table, all studies quoted find more or less stringent upper limits on $f_{\rm esc}$, both in the low and high-redshift Universe. The ISM in the observed galaxies is highly opaque, and very little stellar ionizing radiation leaks out.

Steidel et al. (2001) detected significant Lyman-continuum flux in the composite spectrum of 29 Lyman-break galaxies with redshifts $z = 3.40 \pm 0.09$. If the inferred escaping Lyman-continuum radiation is typical of galaxies at $z \approx 3$, then these galaxies produce about 5 times more H-ionizing photons per unit co-moving volume than quasars at this redshift, with the obvious cosmological implications. Haehnelt et al. (2001) fitted the composite spectrum by a standard stellar population and no intrinsic H I opacity. Therefore, f_{esc} must be close to 100% for the observed sample. Confirmation or rejection of this striking result will be a major objective of observational cosmology.

References

Buat V., Burgarella D., Deharveng J.-M., Kunth D., 2002, A&A, 393, 33
Calzetti D., 2001, PASP, 113, 1449
Calzetti D., Armus L., Bohlin R.C., Kinney A.L., Koornneef J., Storchi-Bergmann T., 2000, ApJ, 533, 682
Chandar R., Leitherer C., Tremonti C.A., 2004, ApJ, 604, 153
Davidsen A.F., 1993, Science, 259, 327
Deharveng J.-M., Buat V., Le Brun B., Milliard B., Kunth D., Shull J.M., Gry C., 2001, A&A, 375, 805
Deharveng J.-M., Faisse S., Milliard B., Le Brun V., 1997, A&A, 325, 1259
Fernández-Soto A., Lanzetta K.M., Chen H.-W., 2003, MNRAS, 342, 1215
Giallongo E., Cristiani S., D'Odorico S., Fontana A., 2002, ApJ, 568, L9
Grewing M., et al., 1991, in: Extreme Ultraviolet Astronomy, Malina R.F., Bowyer S., eds., (Pergamon: New York), p. 437
Haehnelt M.G., Madau P., Kudritzki R., Haardt F., 2001, ApJ, 549, L151
Heckman T.M., 2005, in: Astrophysics in the Far Ultraviolet: Five Years of Discovery with FUSE, Sonneborn G., Moos H.W., Andersson B.G., eds., (ASP: San Francisco), in press (astro-ph/0410383)
Heckman T.M., Sembach K.R., Meurer G.R., Strickland D.K., Martin C.L., Calzetti D., Leitherer C., 2001, ApJ, 554, 1021
Hoopes C., Heckman T.M., Strickland D., Howk J.C., 2003, ApJ, 596, L175
Hoopes C., Sembach K.R., Heckman T.M., Meurer G.R., Aloisi A., Calzetti D., Leitherer C., Martin C.L., 2004, ApJ, 612, 825
Hurwitz M., Jelinsky P., Dixon W.V.D., 1997, ApJ, 481, L31
Johnson K.E., Leitherer C., Vacca W.D., Conti P.S., 2000, AJ, 120, 1273
Keel W.C., 1998, ApJ, 506, 712
Keel W.C., Holberg J.B., Treuthardt P.M., 2004, AJ, 128, 211
Leitherer C., Ferguson H., Heckman T.M., Lowenthal J., 1995, ApJ, 454, L19
Leitherer C., Leão J.R.S., Heckman T.M., Lennon D.J., Pettini M., Robert C., 2001, ApJ, 550, 724
Leitherer C., Li I.-H., Calzetti D., Heckman T.M., 2002, ApJS, 140, 303
Leitherer C., et al., 1999, ApJS, 123, 3
Longo R., Stalio R., Polidan R.S., Rossi L., 1989, ApJ, 339, 474
Malkan M., Webb W., Konopacky Q., 2003, ApJ, 598, 878
Moos H.W., et al., 2000, ApJ, 538, L1
Pellerin A., et al., 2002, ApJS, 143, 159
Pettini M., Steidel C.C., Adelberger K.L., Dickinson M., Giavalisco M., 2000, ApJ, 528, 96
Robert C., Pellerin A., Aloisi A., Leitherer C., Hoopes C., Heckman T.M., 2003, ApJS, 144, 21
Rogerson J.B., Spitzer L., Drake J.F., Dressler K., Jenkins E.B., Morton D.C., York D.G., 1973, ApJ, 181, L97
Sasseen T.P., Hurwitz M., Dixon W.V., Airieau S., 2002, ApJ, 566, 267
Steidel C.C., Pettini M., Adelberger K.L., 2001, ApJ, 546, 665
Witt A.N., Gordon K.D., 2000, ApJ, 528, 799

LOCAL STARBURSTS: PERSPECTIVES FROM THE OPTICAL

Daniela Calzetti
Space Telescope Science Institute, 3700 San Martin Drive, Baltimore, MD 21218, USA
calzetti@stsci.edu

Abstract The optical regime is historically the best-studied wavelength range. Gas ionized by massive stars produces optical emission lines that have been used to derive indicators of star-formation rate, metallicity, dust reddening, and the ionization conditions of the interstellar medium. Absorption lines have been used to measure velocity dispersions, and the 4000 Å break has been shown to be a useful indicator of the mean age of stellar populations. I briefly summarize some recent work done on, specifically, star-formation rate indicators, in view of their importance for understanding star-forming galaxies at high redshift.

1. Introduction

In recent years, the advent of improved infrared instrumentation on large, 8–10m class, telescopes has opened a window on rest-frame optical observations of high-redshift galaxies, and has revived interest for this historically well-studied wavelength regime.

The easy access from the ground to the optical emission from local celestial bodies (stars, HII regions, galaxies, etc.) has led to the development and definition of a series of tracers of the physical and chemical conditions of these objects. Of immediate relevance for complex systems like galaxies are the nebular emission lines, stellar and interstellar absorption lines, and broad-band features like the 4000 Å break.

The 4000 Å break is an estimator of a stellar population's mean age (e.g., Kauffmann et al. 2003), while absorption lines have been widely employed to measure velocity dispersions within the stellar systems. Among all optical tracers of physical and chemical conditions, however, the lion's share goes to the nebular emission lines. The gas ionized by young, massive stars produces optical emission lines from a number of chemical elements, and with a fairly large range of intensities; these have been "calibrated" to "measure":

- star-formation rates (SFR; from [OII], Hα, ...);

- gas chemical abundances (O, N, S, ...);

- dust reddening (e.g., from the Balmer series);

- diagnostics of star-formation feedback, and, in general, ionization conditions ([OI], [NII], [SII], ...).

The applicability of any such indicator for investigations at cosmological distances depends on the redshift range under consideration, and the number of lines that can be accessed. The [OII](λ3727 Å) emission line is sufficiently blue to be observable in the optical window up to redshifts less than about 1.5; however, this line alone (with no other information at shorter or longer wavelengths) can only provide a highly uncertain estimate of the distant galaxies' SFRs (see next section, and, e.g., Hammer et al. 1997, Hogg et al. 1998, Rosa-González et al. 2002, Hippelein et al. 2003, Teplitz et al. 2003, Kewley et al. 2004). New infrared instruments are providing more leverage, by allowing investigators to access a larger suite of rest-frame optical emission lines, from [OII](λ3727 Å) to [NII](λ6584 Å), up to, for some lines, redshift $z \sim 3$. Multiple emission lines from the same cosmological object afford better estimates of dust reddening, gas chemical abundances, etc. (e.g., Teplitz et al. 2000, Pettini et al. 2001, Lemoine-Busserolle et al. 2003). Last, but not least, once the James Webb Space Telescope is in orbit, from its vantage point above the atmosphere it will provide an unobstructed view of the earliest galaxies, detecting Hα up to $z \approx 6.5$, and, potentially, as high as $z \approx 40$ (depending on sensitivity and whether ionizing objects exist at such high redshifts). Redshift $z = 6.5$ corresponds to an epoch when the Universe was 6% of its current age (for a ΛCDM cosmology).

Given that the high-redshift "frontier" employs tools derived from the more accessible low-redshift Universe to understand the evolution of galaxies, it is worth revisiting the strengths and limitations of some of these tools, and whether more investigation is needed in some areas. For instance, recent studies have reiterated the limitations of the well-known and well-established "strong-line method" for chemical abundance measurements in metal-rich environments (Garnett et al. 2004).

Here, I concentrate on the SFR indicators accessible in the optical regime, highlighting recent progress in the area; I connect these optical indicators to those at other wavelengths, suggesting where additional investigation may be needed.

2. Star-Formation Rate Indicators in the Optical

The basic questions that come to mind when using SFR indicators at optical or other wavelengths are: are they consistent with one another? What level of "uncertainty" does each of them carry, and which factors produce such uncertainty?

At optical wavelengths, the two most widely employed SFR tracers are the Balmer emission lines (Hα, Hβ, ...), and the [OII](λ3727 Å) doublet emission lines. Both are measures of "instantaneous" star formation, as the gas is excited by the ionizing photons of the short-lived O and early-B stars. The intensity of both Balmer and [OII] lines is affected by dust extinction – the Hα less so than the bluer Hβ and [OII] – and by changes in the mass of the upper end of the stellar initial mass function (IMF, which affects the number of ionizing photons available to excite the gas). The intensity of the Balmer lines is additionally affected by the stellar absorption of the underlying galactic stellar population – again Hα at a lower level than Hβ. In contrast, [OII] is affected by the gas metallicity and, potentially, by its ionization conditions (but see Kewley et al. [2004], who find no such effect for galaxies with O/H > 8.5). A non-exhaustive list of studies addressing such effects and/or deriving calibrations for SFR estimates from line measurements includes: Gallagher et al. (1989), Kennicutt (1992, 1998), Charlot et al. (2002), Rosa-González et al. (2002), Kewley et al. (2002, 2004), Pérez-González et al. (2003), and Hopkins (2004).

Combined, the above effects impact SFR estimates from factors of a few to orders of magnitude, depending on the regime where the measurement is performed. This is easily seen in the case of dust extinction. Local starburst galaxies cover a wide range of dust attenuation values; even UV-selected starbursts can show extinctions as high as $A_V \sim 4.5$ mag, with a loose trend for more actively star-forming galaxies to have higher extinctions (Fig. 1, left; see also Wang & Heckman 1996, Sullivan et al. 2001, Pérez-González et al. 2003). If such a highly extincted galaxy is mistakenly corrected for a lower extinction value, e.g., $A_V \sim 1$ mag (the value derived for disks, Kennicutt 1983), the derived SFR(Hα) will underestimate the actual SFR by a factor of 14!

How common are such galaxies? From the Hα luminosity function of Pérez-González et al. (2003), an L^*(Hα) galaxy in the local Universe has $A_V^* \approx 2-3$ mag; thus, galaxies with heavily extincted ionized gas are not a rarity in our cosmological neighbourhood. How this translates to higher redshifts is still unclear, although there are suggestions that overall extinctions are decreasing at constant SFR in high-redshift galaxy populations (Adelberger & Steidel 2000).

The SFR–extinction correlation carries down to the "quiescent" star-forming galaxy regime, with a trend that resembles that of the starburst galaxies (Fig. 1, right). The fact that more actively star-forming galaxies and regions have, on average, higher dust extinction values is a consequence of the Schmidt-Kennicutt law (galaxies/regions with higher gas surface densities show larger specific SFRs, e.g., Kennicutt 1998) combined with the mass–metallicity relation (see also Heckman, these proceedings).

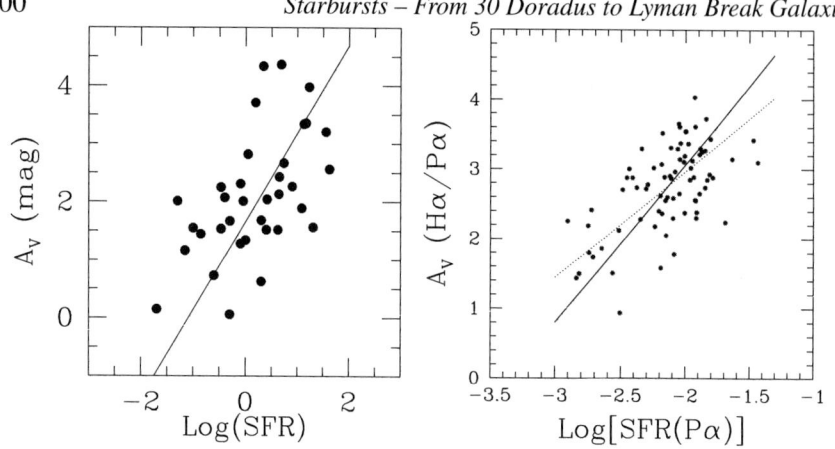

Figure 1. *Left:* The dust attenuation at V for the ionized gas, A_V, vs. SFR for the starburst galaxy sample of Calzetti et al. (1994). The SFR is calculated from the extinction-corrected Hα flux, while A_V is derived from various hydrogen emission lines, from Hβ in the optical to Brγ in the infrared. The continuous line is the best linear fit through the data points. *Right:* A_V vs. SFR for HII-emitting complexes in the central ~ 6 kpc of M51 (Calzetti et al., in prep.). A_V is derived from the Hα/Pα line ratio, and the SFR is calculated from the extinction-corrected P$\alpha(\lambda 1.8756~\mu$m). The continuous line is the best fit through the points, while the dotted line is the fir to the starburst galaxies (left panel), vertically rescaled by a factor of $10^{4.35}$.

Measurements of the stellar IMF are ridden with controversy (see the review by Brandl & Andersen 2004). Amid all of this, the high end of the stellar IMF appears to be fairly constant from galaxy to galaxy in the local Universe, and, possibly, has not changed from high redshift to the present day (Stasińska & Leitherer 1996, Wyse et al. 2002, Elmegreen, these proceedings). If variations do exist, they are likely a second-order effect, at least for what concerns SFR measurements. The comparison between SFR(IR) and SFR(Hα) for a sample of local galaxies by Kewley et al. (2002) supports such a statement. SFR(IR) is the star formation calculated from the galaxies' far-infrared emission, which is due to dust heated mainly by the stellar non-ionizing radiation (UV, optical, etc.). Thus, two different aspects of the stellar bolometric luminosity, the non-ionizing and the ionizing luminosities, determine SFR(IR) and SFR(Hα), respectively. The tight agreement between SFR(IR) and SFR(Hα), within $\sim 10\%$, over 4 orders of magnitude (Kewley et al. 2002) is supportive of a relatively constant upper end of the IMF.

The impact of the underlying stellar absorption can be quite significant for the Balmer emission lines, for two reasons: (i) because $EW_{abs}(H\alpha) \approx EW_{abs}(H\beta)$, the intrinsically weaker Hβ line emission will be proportionally more "depressed" by the underlying absorption than Hα, altering measurements of dust reddening; (ii) in galaxies where the star-formation intensity is proportionally a small fraction of the overall stellar emission, the $EW_{abs}(H\alpha)$ can be a significant fraction of the $EW_{em}(H\alpha)$, thus leading to underestimates of the emission fluxes. Measurements of local star-forming galaxy populations indicate values of $EW_{abs}(H\alpha) \sim 3$–6 Å (Kennicutt 1992, Calzetti et al. 1994, Charlot et al. 2002, Rosa-González et al. 2002), with the smaller value more commonly applicable to starburst galaxies.

These studies demonstrate that once the effects of dust extinction and the underlying stellar absorption are controlled, the Hα emission is a reliable indicator of instantaneous SFR, and existing calibrations (Kennicutt 1998) are sufficiently accurate for most purposes. For SFR([OII]), there are additional effects to consider: the dependence on metallicity and, potentially, on the ionization conditions, but these can be "calibrated" using samples of local galaxies (Kewley et al. 2004), or empirical approaches can be adopted (Kennicutt 1998, Rosa-González et al. 2002). Problems arise when only a limited amount of information, e.g., just one or two adjacent emission lines, is available, as is often the case for samples at cosmological distances. In these cases, the unknown dust extinction and underlying stellar absorption corrections can lead to two effects: (i) underestimates of the actual SFRs by a factor of a few, up to an order of magnitude, depending on the nature of the sample and the rest-frame wavelength region investigated; (ii) increase in the dispersion of the SFR distribution of the sample by at least a factor of 2 (Rosa-González et al. 2002).

3. Star Formation Rate Indicators at Other Wavelengths

The optical SFR indicators discussed above measure the presence and the amount of ionizing stars in galaxies, which are typically short-lived ($t_{life} < 10^7$ yr). Some of the other widely used SFR indicators at shorter or longer wavelengths probe star formation over longer timescales, typically 100 Myr or longer.

This is the case, for instance, of the rest-frame UV emission (where UV is intended here as the stellar emission between 1000 Å and 3000 Å). While the UV emission is well correlated with the ionized line emission in starburst galaxies (and thus both represent reliable tracers of current SFR), there is increasing evidence that the UV of quiescent star-forming galaxies traces recent, but not current, star formation, and cannot be directly used as an "instantaneous" SFR indicator (Kong et al. 2004, Calzetti et al., in prep.).

Another popular SFR indicator is the one derived from a galaxy's far-infrared emission, SFR(IR). As mentioned in the previous section, the far-infrared emission in a galaxy is from dust heated mainly by the non-ionizing stellar radiation. One of the standing questions is how much of the IR radiation from a galaxy is contributed by current or recent star formation, and how much by the general stellar radiation field. The answer to this question can affect the empirical calibration of SFR(IR).

Addressing some of those questions is one of the purposes of projects like SINGS (the Spitzer Infrared Galaxies Survey, one of the Spitzer Legacy Projects; P.I. R. Kennicutt; see J.D. Smith, these proceedings). The unprecedented angular resolution afforded by Spitzer together with observations at multiple wavelengths of local galaxies enables this project to shed light on the merits and problems of common (and uncommon) SFR indicators, for use on galaxies at cosmological distances.

References

Adelberger K., Steidel C.C., 2000, ApJ, 544, 218
Brandl B.R., Andersen M., 2004, in: IMF@50: The Initial Mass Function 50 years later, Corbelli E., Palla F., Zinnecker H., eds., (Kluwer: Dordrecht), in press (astro-ph/0410513)
Calzetti D., Kinney A.L., Storchi-Bergmann T. 1994, ApJ, 429, 582
Charlot S., Kauffmann G., Longhetti M., Tresse L., White S.D.M., Maddox S.J., Fall S.M., 2002, MNRAS, 330, 876
Gallagher J.S., Hunter D.A., Bushouse H., 1989, AJ, 97, 700
Garnett D.R., Kennicutt R.C., Bresolin F., 2004, ApJ, 607, L21
Hammer F., et al., 1997, ApJ, 481, 49
Hippelein H., et al., 2003, A&A, 402, 65
Hogg D.W., Cohen J.G., Blandford R., Pahre M.A., 1998, ApJ, 504, 622
Hopkins A.M., 2004, ApJ, 615, 209
Kauffmann G., et al., 2003, MNRAS 341, 33
Kennicutt R.C., 1983, ApJ, 272, 54
Kennicutt R.C., 1992, ApJ, 388, 310
Kennicutt R.C., 1998, ARA&A, 36, 189
Kewley L., Geller M.J., Jansen R.A., 2004, AJ, 127, 2002
Kewley L., Geller M.J., Jansen R.A., Dopita M.A., 2002, AJ, 124, 3135
Kong X., Charlot S., Brinchmann J., Fall S.M., 2004, MNRAS, 349, 769
Lemoine-Busserolle M., Contini T., Pelló R., Le Borgne J.-F., Kneib J.-P., Lidman C., 2003, A&A, 397, 839
Pérez-González P.G., Zamorano J., Gallego J., Aragón-Salamanca A., Gil de Paz A., 2003, ApJ, 591, 827
Pettini M., Shapley A.E., Steidel C.C., Cuby J.-G., Dickinson M., Moorwood A.F.M., Adelberger K.L., Giavalisco M., 2001, ApJ, 554, 981
Rosa-González D., Terlevich E., Terlevich R. 2002, MNRAS, 332, 283
Stasińka G., Leitherer C., 1996, ApJS, 107, 661
Sullivan M., Mobasher B., Chan B., Cram L., Ellis R., Treyer M., Hopkins A., 2001, ApJ, 558, 72
Teplitz H.I., Collins N.R., Gardner J.P., Hill R.S., Rhodes J., 2003, ApJ, 589, 704
Teplitz H.I., et al., 2000, ApJ, 533, 65
Wang B., Heckman T.M., 1996, ApJ, 457, 645
Wyse R.F.G., Gilmore G., Houdashelt M.L., Feltzing S., Hebb L., Gallagher J.S., Smecker-Hane T.A., 2002, NewA, 7, 395

THE STARBURST PHENOMENON FROM THE OPTICAL/NEAR-IR PERSPECTIVE

Nils Bergvall[1], Thomas Marquart[1], Göran Östlin[2], and Erik Zackrisson[1]
[1] *Uppsala Astronomical Observatory, Box 515, SE-751 20 Uppsala, Sweden*
[2] *Stockholm Observatory, AlbaNova University Center, Roslagstullsbacken 21 SE-106 91 Stockholm, Sweden*
nils.bergvall@astro.uu.se, thomas.marquart@astro.uu.se, ostlin@astro.su.se, ez@astro.uu.se

Abstract The optical/near-IR stellar continuum carries unique information about the stellar population in a galaxy, its mass function and star-formation history. Star-forming regions display rich emission-line spectra, from which we can derive the dust and gas distribution, map velocity fields, metallicities and young massive stars and locate shocks and stellar winds. All this information is very useful in the dissection of the starburst phenomenon. We discuss a few of the advantages and limitations of observations in the optical/near-IR regime, and focus on some results. Special attention is given to the role of interactions and mergers and observations of the relatively dust-free starburst dwarfs. In the future, we expect new and refined diagnostic tools to provide us with more detailed information about the IMF, strength and duration of the burst and its triggering mechanisms.

1. Introduction

Optical/near-IR broad-band photometry of a starburst galaxy gives a first indication of the burst strength, age and distribution of the young and old populations, and of their basic morphological structure parameters. Model-based spectrophotometric tools are provided for more detailed analysis. A rich set of emission lines are used for analysis of the kinematics, chemical abundance, shocks, the stellar upper mass limit and the distribution of dust and molecular gas. Absorption-line indices provide estimates of the age and IMF of the evolved population. Figure 1 shows a synthetic spectrum of a mixture of a young and old population, with a mass ratio of 2:1. A general review of the diagnostic tools and the limitations of the photo-ionization models used in the analysis is discussed by Schaerer (2001).

Heavy dust obscuration, in particular in LIRGs and ULIRGs, has been a problem in the optical/near-IR. Here, we will therefore focus on starbursts in low-extinction regions, notably starburst dwarf galaxies. First, however, we

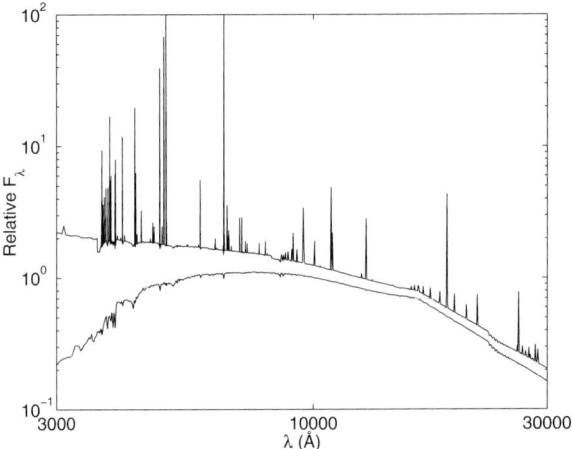

Figure 1. Synthetic spectrum (Zackrisson et al. 2001) of a starburst in an old galaxy. The first burst (lower spectrum) started 10 Gyr ago and had a duration of 100 Myr. The young starburst has a constant SFR and is 10 Myr old. The mass (including stellar remnants) of the old population is twice that of the young population. Both populations have 5% solar metallicity and a Salpeter IMF.

will discuss a widely-debated issue where optical data originally had a strong impact, namely the importance of gravitational interactions as a starburst triggering mechanism.

2. Starbursts and tidal interaction

It is clear from the properties of ULIRGs that mergers are required to trigger major starbursts. But is it a sufficient requirement? How often do mergers and close encounters generate starbursts? To answer this question, it is common to compare two galaxy samples – interacting/merging galaxies (IGs) or pairs, and non-interacting galaxies (NIGs). A problem with the comparison is that NIGs and IGs have evolved in different environments where, e.g., mergers, ram pressure, harassment and gas infall have different influence. Integrated broad-band photometry and Hα emission are the most widely used tools in this context. In the classical paper by Larson & Tinsley (1978) the authors claim, based on UBV data, that interactions frequently trigger a major SF increase involving as much as 5% of the total mass. Many follow-up studies seem to confirm the result but are often influenced by strong selection effects, non-matching morphological type distributions of NIGs/IGs, and are focusing on the most dramatic cases. Studies based on better-constrained samples (Bergvall et al. 2002, Brosch et al. 2004) do not confirm these results but find that tidal interactions have an insignificant influence on the SF history of galaxies in the local Universe. There seems to be agreement, however, on a correlation between interactions and increased SF within the central kpc (first discussed by Keel et al. 1985). *Galaxy pairs* with small separations show similar trends as seen in Hα (Barton et al. 2000, Lambas et al. 2003, Nikolic et al. 2004). The mean increase in both cases is quite moderate, however, and few cases are qualified to be called "nuclear starbursts". Bergvall et al. (2000) and Varela et al. (2004) find that masses of perturbed galaxies are higher than those of NIGs of similar morphology, indicating that they experience mergers more frequently. This may lead to a steady inflow of gas that can explain part of the increased SF in the centre. Varela et al. (2004) also find a *higher frequency of bars in disturbed systems*, in accordance with related studies in the past (see Knapen 2004). Bars are known to generate mass inflows. Thus, it is not clear what is the main triggering mechanism of the central increase in SF. The conclusion must be that there is *no strong support that tidal interactions generate starburst activity that significantly affects the SF history of galaxies in the local Universe*. Estimates leave room for major starbursts among less than a few % of the IGs.

3. Blue compact galaxies

Blue compact galaxies (BCGs) are not a well-defined type, as the galaxies are selected based on either spectroscopic or photometric criteria. The general properties are high surface brightness, low chemical abundance and a high gas mass fraction. They have a wide range of morphologies (Loose & Thuan 1986). Are they bursting? Figure 2 shows $L_B/\mathcal{M}_{HI}$ vs. M_B of different types of gas-rich galaxies. The BCG sample is incomplete but constitutes a representative part of the nearby sample of starburst dwarfs (Mrk, UM, Tololo, etc.). We see that there is a continuous distribution towards high $L_B/\mathcal{M}_{HI}$, but that the properties of most BCGs are similar to dIrr and late-type spirals of similar luminosity, i.e., they are probably not bursting. The high surface brightness of the burst could be due to a high column density (and a small scale length, cf. Papaderos et al. [1996], Salzer et al. [2002]), perhaps caused by low angular momentum. Since their gas mass often constitutes a major fraction of the total mass (Salzer et al. 2002), the diagram shows that starbursts in these galaxies are either short-lived or rare.

Some BCGs have a ∼ tenfold global increase in SFR, i.e., they are true starbursts. What are their specific properties? There is no strong indication of a correlation between SF activity and tidal interactions (Brosch 2004, Hunter & Elmegreen 2004). On the other hand, BCGs appear to be involved in mergers with intense SF more frequently than other dwarfish galaxies (e.g., Gil de Paz 2004). It could indicate that mergers are important triggers and morpholog-

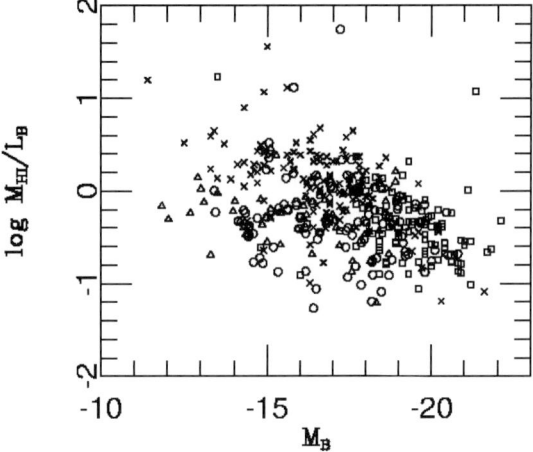

Figure 2. The hydrogen mass per B luminosity as function of absolute B magnitude for four different types of galaxies: BCGs (circles), dIrr (triangles), LSB galaxies (crosses) and late-type spiral galaxies (squares). The data are obtained from different sources in the literature. It is incomplete but estimated to be representative.

ically short-lived. The gas consumption rates are typically shorter than 100 Myr, i.e., similar to the dynamical timescale of a merger.

Ages and masses. Dynamical mass estimates of BCGs are difficult, since the kinematics sometimes are quite chaotic due to the mass motions that cause the burst, and because of the SN winds. To overcome the problem with the stellar winds, it becomes necessary to use stellar absorption features. The only useful lines for this purpose are the CaII triplet lines at about 8500Å. Not until quite recently has this option become accessible (Östlin et al. 2004). The results are very promising and will soon help to solve the question regarding the coupling between gas and stars and facilitate the detailed analysis of velocity fields based on Hα (e.g., Marquart et al., these proceedings). Age and SFR are often estimated from the Hα flux, the Hα equivalent width (EW(Hα)) and broad-band photometry. From this, the "photometric mass" is obtained, assuming that the SFR is constant. The age is, however, difficult to determine, even if we assume that the SFR is constant. In such a case, EW(Hα) is a function of the IMF and age. The IMF slope in starbursts seems to be well constrained in the intermediate stellar mass range (Elmegreen 2004), but not so well for high masses. Figure 3 shows the predicted EW(Hα) for two values of the upper mass limit, 40 and 120 M$_\odot$. It can be seen that the predicted ages differ by a factor of 5–10 over a large age range. There is also an observational problem, in that intense starbursts may have huge Strömgren spheres, from which the

Hα emission may be lost due to a limited aperture size. The uncertainty in the determination of the widely used b parameter (b = SFR/$\langle$SFR$\rangle$) obviously must be quite high, in particular if we consider the poorly-constrained SF history.

For BCGs, there seems to be a simple way to account for the SF history reasonably well. It is based on a two-component model of the galaxy, consisting of a starburst superimposed on a host galaxy with an exponential luminosity profile. If photometric masses are applied to this model we find that there is a fairly tight correlation between mass and central velocity dispersion (Östlin et al. 2001), indicating that this simple model is quite successful.

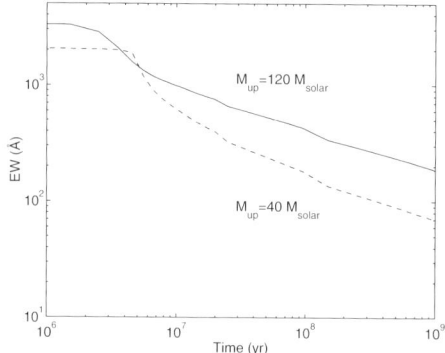

Figure 3. The equivalent width of Hα in emission from a starburst with constant SFR, a Salpeter IMF and 20% solar metallicity. The predicted evolution assuming two different values for the upper mass limit are shown. The model is from Zackrisson et al. (2001).

A very useful method to determine the past starburst activity in a galaxy is based on its rich system of super star clusters and globular clusters (GCs). The GC IMF is Salpeter-like and their stellar content is coeval. This makes them quite reliable as standard clocks and optical/near-IR photometry and spectroscopy can be used to determine their ages and trace the past star-formation history, thereby identifying bursts (e.g. Östlin et al. 2003, de Grijs et al. 2004).

The best method to derive the ages is from colour-magnitude diagrams of the stellar population, but most starburst galaxies are too distant to make this method feasible. The few results available provide no support for strong short-lived bursts separated by quiescent periods (Annibali et al. 2003, Schulte-Ladbeck et al. 2001). A similar conclusion was reached by Westera et al. (2004) in a study of 200 HII galaxies, based on stellar absorption features. Taking the previous discussion into account, these observations indicate that true starbursts are rare rather than short-lived, or that they are short-lived but change morphological type at or after the burst.

The starburst host. It is well established that the luminosity profile of most BCGs can be characterised by a burst superimposed on a host galaxy with (mostly) red colours, typical of an old stellar population, and a morphology resembling an early-type galaxy (e.g., Papaderos et al. 1996, Gil de Paz et al. 2004). An attractive scenario is that a dE is merging with a gas-rich galaxy that triggers the burst. Recently, it was found that the optical/near-IR colours of the host of luminous BCGs at very faint levels has a red excess, difficult to explain with a normal IMF and a low metallicity (Bergvall & Östlin 2002). This problem is discussed in the paper by Zackrisson et al. (these proceedings). It could indicate that a host galaxy of special properties is needed to trigger a true starburst.

References

Annibali F., Greggio L., Tosi M., Aloisi A., Leitherer C., 2003, AJ, 126, 2752
Barton E.J., Geller M.J., Kenyon S.J., 2000, ApJ, 530, 660
Bergvall N., Östlin G., 2002, A&A, 390, 891
Bergvall N., Laurikainen E., Aalto S., 2003, A&A, 405, 31
Brosch N., Almoznino E., Heller A.B., 2004, MNRAS, 349, 357
de Grijs R., Smith L.J., Bunker A., Sharp R.G., Gallagher J.S., Lançon A., O'Connell R.W., Parry I.R., 2004, MNRAS, 352, 263
Elmegreen B.G., 2004, MNRAS, 354, 367
Gil de Paz A., Madore B.F., Pevunova O., 2003, ApJS, 147, 29
Hunter D.A., Elmegreen B.G., 2004, AJ, 128, 2170
Keel W.C., Kennicutt R.C., Hummel E., van der Hulst J.M., 1985, AJ, 90, 708
Knapen J.H., 2004, in: Penetrating Bars Through Masks of Cosmic Dust: The Hubble Tuning Fork Strikes a New Note, Block D.L., Freeman K.C., Puerari I., Groess R., Block E.K., eds., Kluwer: Dordrecht, in press (astro-ph/0407068)
Lambas D.G., Tissera P.B., Alonso M.S., Coldwell G., 2003, MNRAS, 346, 1189
Larson R.B., Tinsley B.M., 1978, ApJ, 219, 46
Loose H.-H., Thuan T.X., 1986, MitAG, 65, 231
Östlin G., Amram P., Bergvall N., Masegosa J., Boulesteix J., Márquez I., 2001, A&A, 374, 800
Östlin G., Cumming R.J., Amram P., Bergvall N., Kunth D., Márquez I., Masegosa J., Zackrisson E., 2004, A&A, 419, L43
Östlin G., Zackrisson E., Bergvall N., Rönnback J., 2003, A&A, 408, 887
Papaderos P., Loose H.-H., Fricke K.J., Thuan T.X., 1996, A&A, 314, 59
Salzer J.J., Rosenberg J.L., Weisstein E.W., Mazzarella J.M., Bothun G.D., 2002, AJ, 124, 191
Schaerer D., 2001, in: Starburst Galaxies: Near and Far, Tacconi L., Lutz D., eds., Springer: New York, p. 197
Schulte-Ladbeck R.E., Hopp U., Greggio L., Crone M.M., Drozdovsky I.O., 2001, ApSSS, 277, 309
Varela J., Moles M., Márquez I., Galletta G., Masegosa J., Bettoni D., 2004, A&A, 420, 873
Westera P., Cuisinier F., Telles E., Kehrig C., 2004, A&A, 423, 133
Zackrisson E., Bergvall N., Olofsson K., Siebert A., 2001, A&A, 375, 814

DISSECTING STARBURST GALAXIES WITH INFRARED OBSERVATIONS

Paul P. van der Werf and Leonie Snijders
Leiden Observatory, P.O. Box 9513, 2300 RA Leiden, The Netherlands
pvdwerf@strw.leidenuniv.nl, snijders@strw.leidenuniv.nl

Abstract The infrared regime contains a number of unique diagnostic features for probing the astrophysics of starburst galaxies. After a brief summary of the most important tracers, we focus in detail on the use of emission lines to probe the upper part of the main sequence in a young super star cluster in the Antennae galaxies, and the compact, dusty starburst in the nucleus of the nearby ultraluminous infrared galaxy Arp 220.

1. Introduction

Since stars form in the cores of dusty molecular clouds, it is not surprising that optical obscuration forms a major stumbling block for studying starburst galaxies. Since dust content and extinction correlate with stellar luminosity (e.g., Kennicutt, these proceedings), this is – in particular – true for the more luminous starbursts. Thus, while optical and ultraviolet (UV) observations of relatively unobscured regions in low to moderate luminosity starburst galaxies still provide excellent astrophysical diagnostics (e.g., Leitherer, these proceedings), more luminous systems require observations at near-infrared (near-IR) and longer wavelengths.

However, reduced extinction ($A_K \sim 0.11 A_V$) is not the only reason for observing starburst galaxies in the infrared. The infrared regime also contains an extensive set of diagnostic features, which are ideal probes of stellar population parameters, as well as of its feedback on the ambient gas:

1 The Lyman continuum flux can be probed using hydrogen recombination lines, principally from the Paschen and Brackett series (as well as by thermal radio continuum emission);

2 The temperature of the ionizing radiation field can be probed using suitable ratios of helium and hydrogen recombination lines, as well as by ratios of suitable combinations of fine-structure lines (e.g., Shields 1993, Lumsden et al. 2003, Rigby & Rieke 2004);

3 A measure of the supernova rate can be obtained from the near-IR [FeII] lines, as well as from non-thermal radio continuum emission (e.g., van der Werf et al. 1993, Alonso-Herrero et al. 2003);

4 The age of the stellar population can be derived from any ratio of tracers that probe different temporal phases of the starburst, such as the Brγ equivalent width, effectively probing the relative importance of O stars and red supergiants (e.g., Leitherer et al. 1999);

5 Warm molecular gas, heated by shocks or by UV radiation, can be probed by the H_2 near-IR ro-vibrational lines and the mid-infrared (mid-IR) rotational lines, as well as low-excitation fine-structure lines such as the [CII] 158μm line;

6 Extinction can be derived from ratios of hydrogen recombination lines, from ratios of H_2 or [FeII] lines arising from the same upper level, and from analysis of ice and silicate absorption features;

7 A possible hidden active galactic nucleus (AGN) can be revealed by high-excitation lines such as [SiVI] 1.96μm;

8 Emission and absorption-line kinematics can be used to derive dynamical masses, and to probe bulk flows;

9 Finally, in luminous and ultraluminous infrared galaxies, the best measure of the total luminosity of the starburst is provided by the integrated far-infrared (far-IR) emission, which typically dominates the bolometric energy output.

However, care should be taken not to apply these diagnostics blindly, since none of them is entirely straightforward, and some are only usable in limited regions of parameter space. A complete discussion of the caveats is beyond the scope of this paper, and the reader is referred to the references above for more details.

A detailed analysis, using a combination of these parameters, has been done for the nearby starburst galaxy M82 (Förster Schreiber et al. 2001, Förster Schreiber et al. 2003), resulting in a detailed view of the spatial and temporal evolution of this starburst. More limited studies, using mostly near-IR data, have been published of other nearby starbursts such as NGC 253 (Forbes et al. 1993, Engelbracht et al. 1998), NGC 1808 (Kotilainen et al. 1996), NGC 7552 (Schinnerer et al. 1997), and IC 342 (Böker et al. 1997).

In this paper, we focus in particular on the use of near-IR hydrogen recombination lines to derive the Lyman continuum flux of obscured starbursts. We first discuss the spectral properties of a powerful young super star cluster in the Antennae galaxies (NGC 4038/4039), and then contrast these with the spectral properties of the nearby ultraluminous infrared galaxy (ULIG) Arp 220. The analysis is based on near-IR H and K-band spectra, obtained with ISAAC at the ESO/Very Large Telescope.

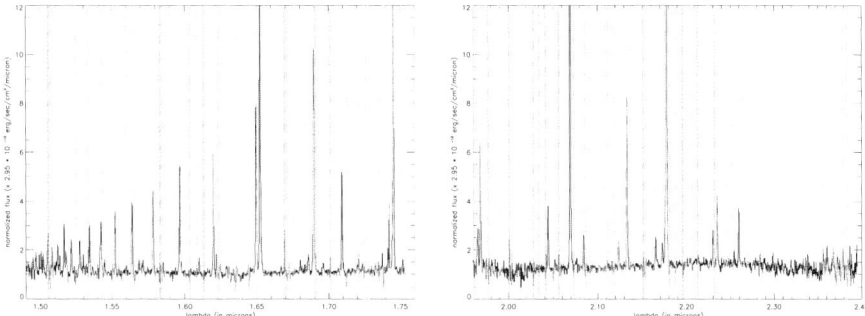

Figure 1. H-band *(left)* and K-band *(right)* spectra of super star cluster 80 in the Antennae. Shaded bands indicate spectral regions affected by sky lines.

2. Near-IR spectra of a super star cluster in the Antennae

In Fig. 1 we present near-IR H and K-band spectra of a young, obscured super star cluster in the Antennae. This cluster (object number 80 in the notation of Whitmore & Schweizer 1995) is a prominent mid-IR source, producing about 15% of the total 15μm flux density of the entire system (Mirabel et al. 1998). The spectra show the typical features expected for obscured star-forming regions: the H band is dominated by the Brackett series, with in addition a bright [FeII] line at 1.64μm, and a prominent HeI line at 1.70μm. The K band is dominated by very strong Brγ and HeI 2.06μm emission (with weaker HeI emission at 2.11μm), and a number of H_2 ro-vibrational lines. CO absorption bands are visible at 2.32 and 2.36μm, but are extremely faint, as expected for this very young cluster ($\sim$ 4 Myr; Gilbert et al. 2000).

The large number of hydrogen recombination lines detected allow an accurate extinction determination. Fitting all lines simultaneously, we obtain an extinction $A_K = 0.5$ mag, located in a foreground screen. This geometry provides a significantly better fit than a model where emitting and absorbing material are co-spatial. After correction for extinction, the derived Lyman continuum flux (under the usual assumptions of case B recombination in ionization bounded, dust-free HII regions) is $Q_0 = 1.5 \times 10^{53}$ s^{-1}, which implies the presence of about 35,000 O-stars within a volume with a half-light radius of only 32 pc. The effective temperature of the radiation field derived from the ratio of helium and hydrogen recombination lines is $T_{\text{eff}} \sim 38{,}500$ K, which corresponds to the most massive stars having a Zero-Age Main Sequence spectral type of approximately O7.5. Given the derived cluster age, stars with spectral types up to O4.5 could have been present, but in that case $T_{\rm eff} \sim 47{,}000$ K would be expected. This value is closer to the $T_{\rm eff} \sim 44{,}000$ K estimated from ratios of mid-IR fine-structure lines (Kunze et al. 1996). The disagreement with the value derived from recombination lines is harmless, since the recombination-line ratios are insensitive to $T_{\rm eff}$ values above about 40,000 K. More detailed analysis is required to see if this procedure can be used to establish the presence of an upper-mass cut-off of the initial mass function.

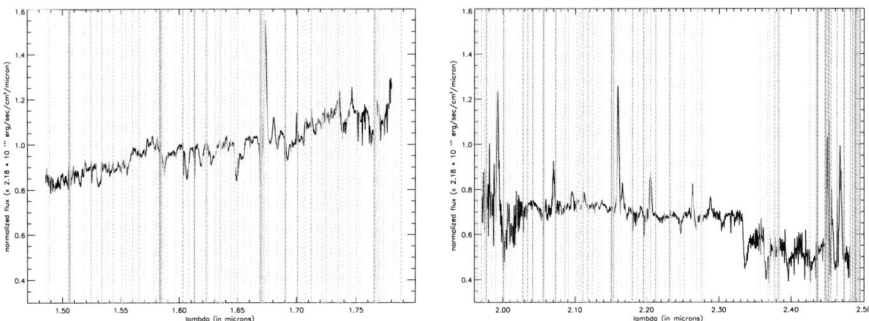

Figure 2. H-band *(left)* and K-band *(right)* spectra of the nuclear starburst in Arp 220. Shaded bands indicate spectral regions affected by sky lines.

3. The compact, dusty starburst in Arp 220

In Fig. 2 we present near-IR H and K-band spectra of the compact, dusty starburst in the nucleus of the ULIG Arp 220. These spectra provide a remarkable contrast with those shown in Fig. 1. In the H band, the only significant emission line is that of [FeII], which probes the supernova remnants created in the starburst. However, the Brackett series is completely absent, except for the Brγ line in the K band, which is, however, dominated by H_2 ro-vibrational lines. Strong photospheric continuum from red supergiants is evident in both the H and K bands. The apparent deficiency in young, hot stars suggested by the faintness of the recombination lines has led to speculations of a hidden AGN in Arp 220, which would provide most of the bolometric luminosity (Armus et al. 1995). Alternatively, extreme foreground obscuration has been invoked to suppress the recombination lines (Sturm et al. 1996). While extinction certainly plays a role, the prominence of red-supergiant features and of the [FeII] lines, as well as several other arguments (van der Werf 2001) argue against extreme foreground obscuration.

The most satisfactory explanation for the faintness of the recombination lines is Lyman continuum absorption by dust within the ionized regions (see also Dopita, these proceedings). If most of the ionizing radiation is absorbed by dust grains rather than by hydrogen atoms, a dust-bounded (rather than hydrogen-bounded) nebula results, and all tracers of ionized gas (recombination lines, fine-structure lines, free-free emission) will be suppressed. If the HII regions in Arp 220 are principally dust-bounded, the observational properties of Arp 220 can be accounted for, even with only moderate extinction. Since the dust also absorbs far-ultraviolet radiation longward of the Lyman limit, the formation of photon-dominated regions is also suppressed, and the thus the same mechanism can account for the faintness of the 158μm [CII] line in Arp 220 and other ULIGs (Fischer et al. 1999).

Is the Arp 220 starburst dominated by dust-bounded HII regions? The *average* molecular gas density in the $\sim 10^{10}$ M$_\odot$ nuclear molecular complex in Arp 220 is $n_{H_2} \sim 2 \times 10^4$ cm^{-3} (Scoville et al. 1997). The strong emission from high-dipole moment molecules such as CS, HCO$^+$ and HCN argues for even higher densities: $\sim 10^{10}$ M$_\odot$ of molecular gas (i.e., *all* of the gas in the nuclear complex) has a density $n_{H_2} \sim 10^5$ cm^{-3} (Solomon et al. 1990). At such densities, the ionized nebulae created by hot stars are *compact* or *ultracompact* HII regions, where 50 to 99% of the Lyman continuum is absorbed by dust (Wood & Churchwell 1989). Observationally, hydrogen-bounded and dust-bounded HII regions can be distinguished by the quantity $R = L_{\text{FIR}}/L_{\text{Br}\gamma}$: for a wide range of parameters, $R < 3570$ implies that the nebula is hydrogen-bounded, while $R > 35700$ implies that the nebula is dust-bounded (Bottorff et al. 1998). For Arp 220, the Brγ luminosity of 1.3×10^6 L$_\odot$ (van der Werf 2001), implies $R = 1.6 \times 10^5$, assuming an obscuring foreground screen with $A_V = 20$ mag (a model consistent with all infrared data). The star formation takes place in (ultra)compact HII regions, where all of the usual tracers of ionized gas (recombination lines, fine-structure lines, free-free emission) are *quenched*, not extincted. While this result significantly complicates the interpretation of diagnostics of massive star formation in ULIGs, it is safe to conclude that the properties of Arp 220 can be accounted for by an intense, and significantly (but not extremely) obscured starburst. There is no reason to invoke the presence of extreme extinction, a strongly aged starburst, or an additional power source in Arp 220.

References

Alonso-Herrero A., Rieke G.H., Rieke M.J., Kelly D.M., 2003, AJ, 125, 1210
Armus L., Shupe D.L., Matthews K., Soifer B.T., Neugebauer G., 1995, ApJ, 440, 200
Böker T., Förster Schreiber N.M., Genzel R., 1997, AJ, 114, 1883
Bottorff M., Lamothe J., Momjian E., Verner E., Vinković D., Ferland G., 1998, PASP, 110, 1040
Engelbracht C.W., Rieke M.J., Rieke G.H., Kelly, D.M., Achtermann J.M., 1998, ApJ, 505, 639
Förster Schreiber N.M., Genzel R., Lutz D., Kunze D., Sternberg A., 2001, ApJ, 552, 544
Förster Schreiber N.M., Genzel R., Lutz D., Sternberg A., 2003, ApJ, 599, 193

Fischer J., et al., 1999, in: The Universe as seen by ISO, Cox P., Kessler M.F., eds., (ESA Publications Division: Noordwijk), ESA-SP, 427, p. 817
Forbes D.A., Ward M.J., Rotaciuc V., Blietz M., Genzel R., Drapatz S., van der Werf P.P., Krabbe A., 1993, ApJ, 406, L11.
Gilbert A.M., et al., 2000, ApJ, 533, L57
Kotilainen J.K., Forbes D.A., Moorwood A.F.M., van der Werf P.P., Ward M.J., 1996, A&A, 313, 771
Kunze D., et al., 1996, A&A, 315, L101
Leitherer C., et al., 1999, ApJS, 123, 3
Lumsden S.L., Puxley P.J., Hoare M.G., Moore T.J.T., Ridge N.A., 2003, MNRAS, 340, 799
Mirabel I.F., et al., 1998, A&A, 333, L1
Rigby J.R., Rieke G.H., 2004, ApJ, 606, 237
Schinnerer E., Eckart A., Quirrenbach A., Böker T., Tacconi-Garman L.E., Krabbe A., Sternberg A., 1997, ApJ, 488, 174
Scoville N.Z., Yun M.S., Bryant P.M., 1997, ApJ, 484, 702
Shields J.C., 1993, ApJ, 419, 181
Solomon P.M., Radford S.J.E., Downes D., 1990, ApJ, 348, L53
Sturm E., et al., 1996, A&A, 315, L133
van der Werf P.P., 2001, in: Starburst galaxies: near and far, Tacconi L., Lutz D., eds., (Springer: New York), p. 151.
van der Werf P.P., Genzel R., Krabbe A., Blietz M., Lutz D., Drapatz S., Ward M.J., Forbes D.A., 1993, ApJ, 405, 522
Whitmore B.C., Schweizer F., 1995, AJ, 109, 960
Wood D.O.S., Churchwell E., 1989, ApJS, 69, 831

WHAT FRACTION OF STARS FORMED IN INFRARED GALAXIES AT HIGH REDSHIFT?

Neil Trentham
Institute of Astronomy, University of Cambridge, Cambridge CB3 0HA, United Kingdom
trentham@ast.cam.ac.uk

Abstract   Star formation happens in two types of environment: ultraviolet-bright starbursts (like 30 Doradus and HII galaxies at low redshift and Lyman-break galaxies at high redshift) and infrared-bright dust-enshrouded regions (which may be moderately star-forming, like Orion in the Galaxy or extreme, like the core of Arp 220). In this work I will estimate how many of the stars in the local Universe formed in each type of environment, using observations of star-forming galaxies at all redshifts at different wavelengths, and of the evolution of the field galaxy population.

1. Introduction

It is now possible to estimate the star-formation history of the Universe. This is performed most directly by summing the contributions from star-forming field galaxies in optical (corresponding to rest-frame ultraviolet at high redshift) surveys. The direct contribution from ultraviolet-bright star-forming galaxies to the co-moving star-formation rate density is about 3×10^{-3} $M_\odot$ yr^{-1} Mpc^{-3} at $z = 0$, rising to 4×10^{-2} $M_\odot$ yr^{-1} Mpc^{-3} at $z = 1$, before slowly declining to 1.5×10^{-2} $M_\odot$ yr^{-1} Mpc^{-3} at $z = 6$ ($h = 0.7$, $\Omega_\Lambda = 0.7$, $\Omega_m = 0.3$, Salpeter IMF; Giavalisco et al. 2004). This is normally presented in the uncorrected form of the "Madau" or "Madau-Lilly" plot.

However, galaxies that are forming stars also experience significant dust extinction – we know this because we see that local star-forming regions like Orion are dusty, and because local spiral galaxies have spectral energy distributions (SEDs) that peak in the far-infrared. Correcting for this, Giavalisco et al. (2004) find that the the total contribution from optically selected galaxies to the co-moving star-formation rate density is about 1.3×10^{-2} $M_\odot$ yr^{-1} Mpc^{-3} at $z = 0$, rising to 0.13 $M_\odot$ yr^{-1} Mpc^{-3} at $z = 1$, where it stays roughly constant out to at least $z = 6$. The corrections used come from the analysis of Adelberger & Steidel (2000), which is based on multi-wavelength studies of a large sample of star-forming galaxies.

An additional contribution may come from extremely dusty galaxies where the dust is optically thick – these may be missing altogether from optical surveys. Local ultraluminous infrared galaxies (ULIGs) are examples of this kind of galaxy; the V extinction to the core of Arp 220 is > 10 mag (Scoville et al. 1998). Another example is the host of GRB 010222 (Frail 2002), which is an optical sub-L^* galaxy, but has a submillimetre star-formation rate of ~ 600 $M_\odot$ yr^{-1}. This kind of galaxy may be similar to the SCUBA galaxies (Blain et al. 2002) seen in submillimetre surveys, which might have a redshift distribution quite different from galaxies in optically selected samples.

In this work I assess the contributions from all three of these modes of star formation in generating the current cosmological density in stars Ω_*. I provide estimates given current observations, and outline how future observations may provide stronger constraints.

2. Definitions

The total density in stars in critical units is $\Omega_* = \Omega_*^{\rm UV} + \Omega_*^{\rm IR/Opt} + \Omega_*^{\rm IR}$.

The density of stars seen forming directly in optical galaxies, $\Omega_*^{\rm UV}$, equals the integral of the uncorrected ultraviolet star formation rate density over redshift. The density of stars that formed in dusty regions within those same galaxies, $\Omega_*^{\rm IR/Opt}$, equals the integral of the ultraviolet star formation rate density multiplied by an extinction correction (which may depend on z) over redshift. Finally, the density of stars $\Omega_*^{\rm IR}$ that formed in highly obscured regions whose presence cannot be inferred from optical observations equals the infrared star formation rate density (as measured by SCUBA or, in the future, ALMA) minus the contributions to the previous integral, integrated over redshift.

The partition between $\Omega_*^{\rm IR/Opt}$ and $\Omega_*^{\rm IR}$ is somewhat arbitrary. Here, we put in $\Omega_*^{\rm IR}$ all star formation in galaxies whose bolometric luminosity is greater than some threshold luminosity corresponding to that at which the optical luminosity no longer tracks the bolometric (mainly far-infrared) luminosity; locally this happens at a 60μm luminosity of 6.3×10^{10} L$_\odot$ (Rieke & Lebofsky 1986). The extinction in these luminous galaxies surely comes from optically thick dust. Optically thin extinction would tend to happen in galaxies contributing to $\Omega_*^{\rm IR/Opt}$.

Star formation that happened in ULIGs would be included in $\Omega_*^{\rm IR}$.

The fraction of star formation that is obscured in any particular galaxy varies from 0% to about 99% (Adelberger & Steidel 2000). Obscured star formation at the low end of this range mostly contributes to $\Omega_*^{\rm IR/Opt}$, and at the high end to $\Omega_*^{\rm IR}$. The host of GRB 010222 would be at the very high end of this range.

3. The current cosmological stellar content: what needs to be produced

The time integral of the cosmic star-formation rate must equal the luminosity integral of the galaxy luminosity function,

$$\Sigma_i \int_L L\,\phi_i(L)\,\Gamma_i\,dL = \int_{t(z)} \dot{\rho}_*\,dt\,, \tag{1}$$

where Γ_i is the mass-to-light ratio of stellar population i (this is derived from stellar population synthesis models). From the combination of the SDSS survey measurements at the bright end (Blanton et al. 2001), and from CCD mosaic surveys (e.g., Trentham & Tully 2002) at the faint end, the galaxy luminosity function appears to be well-described by a Schechter function with $M_R^* = -22.0$ and $\alpha^* = -1.28$ brightward of $M_R = -19$, and a power law with $\alpha = -1.24$ faintward of $M_R = -19$. Performing the sum of integrals on the LHS of this equation,

$$\Omega_* = 0.0036 \pm 0.0020 \tag{2}$$

in units of the critical density. About 3/4 of this is in spheroids and 1/4 in disks. If a Salpeter, not KTG (Kroupa et al. 1993) IMF is used, Ω_* is a factor of two higher.

4. Observational Constraints

Field Galaxy Evolution. Multi-colour photometry of a near-infrared selected sample of galaxies has permitted Dickinson et al. (2003) to measure the evolution of Ω_* with redshift. They found that about 40% of the present-day stars in the Universe formed between $z = 0$ and $z = 1$. Between $z = 1$ and $z = 2$, their best-fitting models suggest that a further fraction > 50% formed and only a few per cent of the stars that we currently see had formed by $z = 2$. However, there is considerable uncertainty in the star-formation histories used in modelling the SEDs of the sample galaxies, and the fraction of stars in place by $z = 2$ may be as high as 25%. An integral of the extinction-corrected Madau Plot (Giavalisco et al. 2004) over redshift from $z = \infty$ up to $z = 2$ gives a fraction of about 25%.

Near-infrared surveys (Glazebrook et al. 2004) have shown that about 30% of the massive early-type galaxies seen today were already in place by $z = 2$ – these are the distant red galaxies (DRGs). Therefore, most of the stars that we believe formed at $z > 2$ are not only in massive galaxies today *but were already in massive galaxies by $z = 2$*. If these formed within the large galaxies, they would have had to form in very extreme bursts.

GRB host galaxies. Long-duration gamma-ray bursts (GRBs) are thought to be linked to the deaths of massive stars, as suggested by the coincidence between SN 2003dh and GRB 030329 (Hjorth et al. 2003). Since massive stars do not live long, this opens up the possibility of using GRB host galaxies as a SFR-selected sample of galaxies.

There are, however, complications. In the context of a collapsar model (MacFadyen & Woosley 1999), GRBs originate preferentially from stars of high mass and low metallicity. This means that there will be more GRBs per unit SFR in galaxies that have a high-mass biased IMF, or a low metallicity. Perhaps these two effects explain the preponderance of GRBs in ULIGs at $z \sim 1$ (Berger et al. 2003) and Lyα emitters (Fynbo et al. 2003; see also Vreeswijk et al. [2004] about the very low metallicity and high gas column density of the host of GRB 030323), neither of which significantly contribute to Ω_*.

Infrared and submillimetre backgrounds and counts. The extragalactic background light (EBL) is high at infrared wavelength: COBE/DIRBE measured it as 32 ± 13 nW m^{-2} sr^{-1} at 140 μm and 17 ± 4 nW m^{-2} sr^{-1} at 240 μm (Schlegel et al. 1998). The submillimetre background is somewhat lower: 0.55 ± 0.15 nW m^{-2} sr^{-1} at 850 μm (Fixsen et al. 1998). The optical EBL light also appears to be high: 12/15/18 nW m^{-2} sr^{-1} at 300/550/800 nm, with an uncertainty of 50% (Bernstein et al. 2002a).

Madau & Pozzetti (2000) estimated the total EBL as 55 ± 20 nW m^{-2} sr^{-1}. This is lower than the value of 100 ± 20 nW m^{-2} sr^{-1} quoted by Bernstein et al. (2002b), the main difference being an additional component from the optical background that was previously undetected (however see Matilla 2003).

Most of this background is generated by stars, not AGN, or else the local density of supermassive black holes would be overproduced (Madau & Pozzetti 2000). Assuming a recycling fraction of 0.4 (much of the material in stars is returned to the ISM via winds and supernovae; Cole et al. 2000), the EBL implies a stellar density of $\Omega_* = 0.003 - 0.006$ (cf. Section 3).

One reason the range here is quite large is that the redshift distribution of infrared star-forming galaxies is unknown, and the contribution of each galaxy to the EBL $\propto (1+z)^{-1}$.

The submillimetre background has been resolved and redshifts determined for a number of bright sources (Chapman et al. 2003). However, there are indications (Chary et al. 2004) that the galaxies that dominate the infrared background are a different population to the galaxies that dominate the submillimetre background. The models described by Chary et al. (2004) suggest that these infrared galaxies have lower bolometric luminosities and lower redshifts than the ULIGs observed by Chapman et al. (2003).

Optical/Infrared observations of star-forming galaxies. The instantaneous cosmic SFR at any redshift equals the infrared and optical contributions.

Making extinction corrections to star-formation rates for high-redshift galaxies is a substitute for direct measurement at infrared wavelengths. Only at low redshift, $z < 1$, can the two be measured directly. The CFRS+ISO survey (Flores et al. 1999) showed that (i) about 30% of star formation at $z < 1$ was visible at optical wavelengths, (ii) most of the remaining 70% that was in infrared galaxies was in disturbed systems with red $(I - K)$ colours, and (iii) about 18% of the star formation happened in ULIGs with SFRs in excess of 100 $M_\odot$ yr^{-1}. An implication of these results is that most of the star formation at $z < 1$ happened in infrared galaxies that are neither optical galaxies with high internal extinction nor ULIGs of the kind seen by SCUBA.

The first Spitzer results (Chary et al. 2004) seem to point towards a similar situation at higher redshifts. The main difference between low and high redshift is that while the redshift distributions of optical and infrared galaxies are similar *within* the $0 < z < 1$ range, they are very different at high redshift.

Another implication of these results is that in optically-selected star-forming galaxies, most of the star formation is being observed directly and these are at the low end of the obscuration range described by Adelberger & Steidel (2000). Many of these may be blue compact emission-line galaxies of the type described at this conference by Lowenthal and Bershady.

5. The IMF and density in infrared galaxies

The local star-forming region 30 Doradus has a stellar IMF that is Salpeter ($\propto m^{-2.35}$) above 3 $M_\odot$ (Selman et al. 1999). Lyman-break galaxies at high redshift also have Salpeter IMFs at high masses, an inference based on optical spectroscopy of cB58 at $z = 2.7$ (Pettini et al. 2000). This IMF is attractive in that it evolves into the KTG IMF seen locally (Kroupa & Weidner 2003). Unfortunately, no equivalent analysis can be made for infrared galaxies and it remains a possibility that they have a different IMF – if it is high-mass biased, then they will generate more energy per unit mass of stars formed than will optical galaxies.

Stars that form in infrared galaxies probably form in dense star clusters which need to dissipate in order to produce local galaxies. Dissipation timescales are long for dense clusters but are shorter if the cluster is embedded in a gaseous medium. Simulations of this physical process (Lamers et al. 2005), along with inferences about the star-formation history of the Universe from stellar populations of nearby galaxies (Heavens et al. 2004) will provide additional constraints on the total amount of cosmic star formation that occurred in infrared galaxies and when it happened.

6. Concluding thoughts

My current thinking is that (i) roughly equal amounts of stars formed in each of the following four types of environments: optically visible regions, dust-enshrouded regions within optical galaxies, heavily obscured galaxies with $L_{\mathrm{IR}} < 10^{12}$ $L_\odot$ and ULIGs with $L_{\mathrm{IR}} < 10^{12}$ $L_\odot$, and (ii) most star formation in optically visible regions was in small galaxies, while most star formation in ULIGs happened at high redshift, perhaps producing the DRGs. But these are not strongly held convictions and I share the optimism felt at this conference that the puzzle of star formation in galaxies will be solved over the next few years.

References

Adelberger K.L., Steidel C.C., 2000, ApJ, 544, 218
Berger E., Cowie L.L., Kulkarni S.R., Frail D.A., Aussel H., Barger A.J., 2003, ApJ, 588, 99
Bernstein R.A., Freedman W.L., Madore B.F., 2002a, ApJ, 571, 56
Bernstein R.A., Freedman W.L., Madore B.F., 2002b, ApJ, 571, 107
Blain A.W., Smail I., Ivison R.J., Kneib J.-P., Frayer D.T., 2002, Phys. Rep., 369, 111
Blanton M.R., et al., 2001, AJ, 121, 2358
Chapman S.C., Blain A.W., Ivison R.J., Smail I.R., 2003, Nature, 422, 695
Chary R., et al., 2004, ApJS, 154, 80
Cole S., Lacey C.G., Baugh C.M., Frenk C.S., 2000, MNRAS, 319, 168
Dickinson M., Papovich C., Ferguson H.C., Budavari T., 2003, ApJ, 587, 25
Fixsen D.J., Dwek E., Mather J.C., Bennett C.L., Shafter R.A., 1998, ApJ, 508, 123
Flores H., et al., 1999, ApJ, 517, 148
Frail D.A., 2002, ApJ, 565, 829
Fynbo J.P.U., et al., 2003, A&A, 406, L63
Giavalisco M., et al., 2004, ApJ, 600, L103
Glazebrook K., et al., 2004, Nature, 430, 181
Heavens A., Panter B., Jimenez R., Dunlop J., 2004, Nature, 428, 625
Hjorth J., et al., 2003, Nature, 423, 847
Kroupa P., Tout C.A., Gilmore G., 1993, MNRAS, 262, 545
Kroupa P., Weidner C., 2003, ApJ, 598, 1076
Lamers H.J.G.L.M., Gieles M., Portegies Zwart S.F., 2005, A&A, 429, 173
MacFadyen A.I., Woosley S.E., 1999, ApJ, 524, 262
Madau P., Pozzetti L., 2000, 312, L9
Matilla K., 2003, ApJ, 591, 119
Pettini M., Steidel C.C., Adelberger K.L., Dickinson M., Giavalisco M., 2000, ApJ, 528, 96
Rieke G.H., Lebofsky M.J., 1986, ApJ, 304, 326
Schlegel D.J., Finkbeiner D.P., Davis M., 1998, ApJ, 500, 525
Scoville N.Z., et al., 1998, ApJ, 492, L107
Selman F., Melnick J., Bosch G., Terlevich R., 1999, A&A, 347, 532
Trentham N., Tully R.B., 2002, MNRAS, 335, 7
Vreeswijk P.M., et al., 2004, A&A, 419, 927

SHADES: THE SCUBA HALF DEGREE EXTRAGALACTIC SURVEY

James S. Dunlop
Institute for Astronomy, University of Edinburgh, UK
jsd@roe.ac.uk

Abstract   SHADES is a new, major, extragalactic sub-mm survey currently being undertaken with SCUBA on the JCMT. The aim of this survey is to map 0.5 square degrees of sky at a depth sufficient to provide the first, major ($\simeq 300$ sources), unbiased sample of bright ($S_{850} \simeq 8$ mJy) sub-mm sources. Combined with extensive multi-frequency supporting observations already in hand, we aim to measure the redshift distribution, clustering and AGN content of the sub-mm population. Currently $\simeq 40\%$ complete, the survey is due to run until early 2006. Here I provide some early example results which demonstrate the potential power of our combined data set, and highlight a series of forthcoming papers which will present results based on the current interim sample of $\simeq 130$ 850μm sources detected within the Lockman Hole and SXDF SHADES survey fields.

1. Survey Rationale

The sub-mm galaxy population continues to present a major challenge to theories of galaxy formation (e.g., Somerville 2004, Baugh et al. 2005), as current semi-analytic models cannot naturally explain the existence of a substantial population of dust-enshrouded starburst galaxies at high redshift. However, while now regarded as of key importance by theorists, the basic properties of sub-mm galaxies are not, in fact, well defined. Several redshifts have been measured (Chapman et al. 2003), some masses have been determined from CO observations (Genzel et al. 2004), and several individual SCUBA-selected galaxies have been studied in detail (e.g., Smail et al. 2003). However, these follow-up studies have had to rely on small, inhomogeneous, and often deliberately biased (e.g., lensed or radio pre-selected) samples of sub-mm sources, and until now no robust, complete, unbiased and statistically significant (i.e., > 100 sources) sample of sub-mm sources has been constructed.

SHADES (http://www.roe.ac.uk/ifa/shades), the Scuba Half Degree Extragalactic Survey, was designed to remedy this situation. It aims to map 0.5 square degrees with SCUBA to an r.m.s. noise level of $\simeq 2$ mJy at

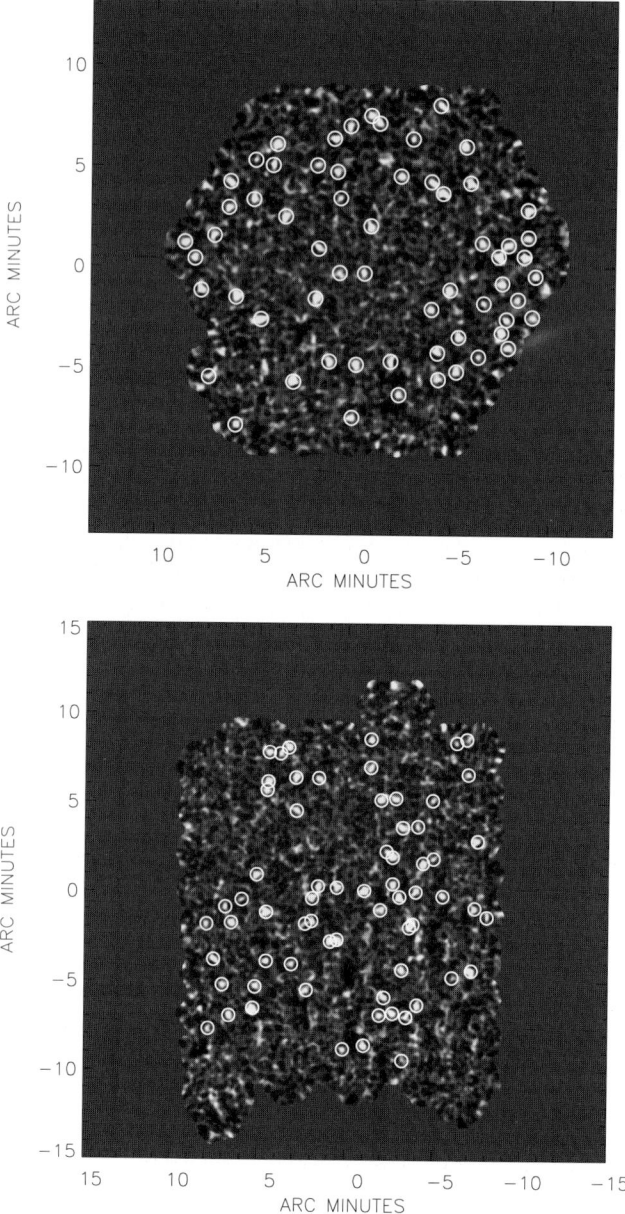

Figure 1. Signal-to-Noise images of the current 850μm SHADES images of the SXDF (upper plot), and Lockman Hole (lower plot). Sources confirmed by the multiple reductions are circled.

850μm. The SHADES consortium consists of nearly the entire UK sub-mm cosmology community, coupled with a subset of the BLAST balloon-borne observatory consortium.

The survey has many goals (see Mortier et al. 2004), but the primary objective is to provide a complete and consistent sample of a few hundred sources with $S_{850} > 8$ mJy, with sufficient supporting multi-frequency information to establish the redshift distribution, clustering properties, and AGN fraction of the bright sub-mm population. The aim is to provide this information, within the 3-year lifetime of the survey, by co-ordinating the SCUBA mapping observations with (i) deep VLA and Merlin radio imaging, (ii) Spitzer mid-infrared imaging, (iii) far-infrared maps of the same fields made with BLAST, (iv) optical and near-infrared imaging with UKIRT and the Subaru telescope, and (v) deep optical spectroscopy with Gemini, Keck and the VLT.

2. SCUBA mapping

SHADES is split between two fields – the Subaru SXDF field at RA $02^h18^m00^s$, Dec $-05°00'00''$ (J2000), and the Lockman Hole centred on RA $10^h52^m51^s$, Dec $57°27'40''$ (J2000), with the goal being to map 0.25 square degrees in each. These two fields were chosen both to provide a spread in RA (to allow observation with the JCMT throughout the majority of the year), and because each field offered unique advantages in terms of low Galactic cirrus emission (crucial for BLAST and Spitzer observations) and existing/guaranteed supporting data at other wavelengths.

SHADES SCUBA observations commenced at the JCMT in December 2002. Full details on the observing technique can be found in Mortier et al. (2004).

SCUBA Signal-to-Noise maps for the SXDF and Lockman SHADES fields obtained by Spring 2004 are shown in Fig. 1. The total area covered by these two maps is 700 square arcmin (402 square arcmin in the Lockman Hole, 294 square arcmin in the SXDF), of which an effective area of $\simeq 650$ square arcmin has complete coverage.

The survey is therefore $\simeq 40\%$ complete, and to date has yielded a total of 130 sources, whose reality is confirmed by 4 independent reductions undertaken within the consortium. (61 sources in the SXDF image [24 at $> 4\sigma$, 53 at $> 3.5\sigma$] and 69 sources in the Lockman Hole image [22 at $> 4\sigma$, 47 at $> 3.5\sigma$]). Based on this interim reduction, we predict a final source list of $\simeq 300$ sources.

3. VLA and Spitzer identifications

In Fig. 2 we show 20 example 20×20 arcsec^2 postage stamps extracted from our deep VLA 1.4 GHz images of the SHADES fields, centred on the positions of the SCUBA sources. Contours from the radio images are shown

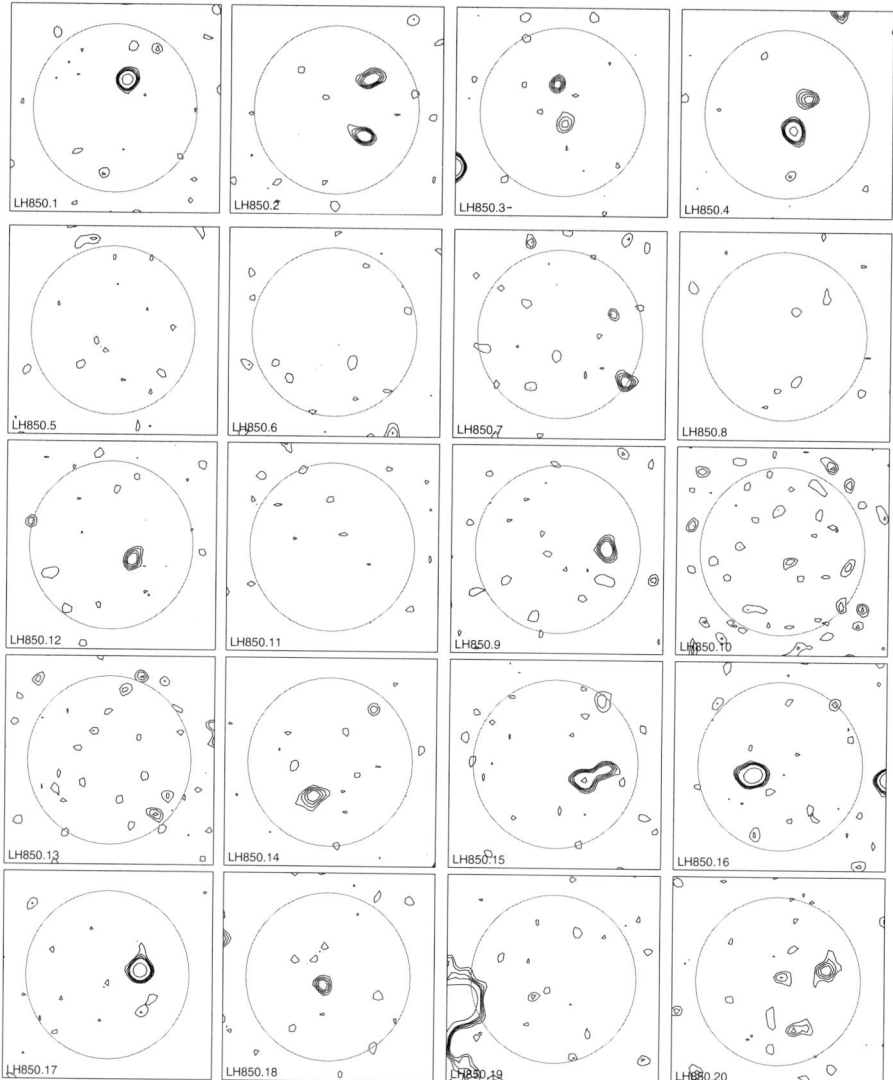

Figure 2. VLA postage stamps centred on the positions of the top 20 SHADES sources in the Lockman Hole field. 70% of these 20 sources have a VLA counterpart within the illustrated search radius of 8 arcsec, and consideration of the lower-resolution B+C array data alone raises this figure to 80% (see Fig. 3)

at 2, 3, 4, 5 and 10σ. The circles have a radius of 8 arcsec, which represents an appropriate search radius for radio counterparts, given the uncertainties in the SCUBA positions. What is striking about this plot is that $\simeq 15$ (i.e., 75%) of these sources have radio counterparts. This figure is significantly higher than

SHADES: The Scuba HAlf Degree Extragalactic Survey 125

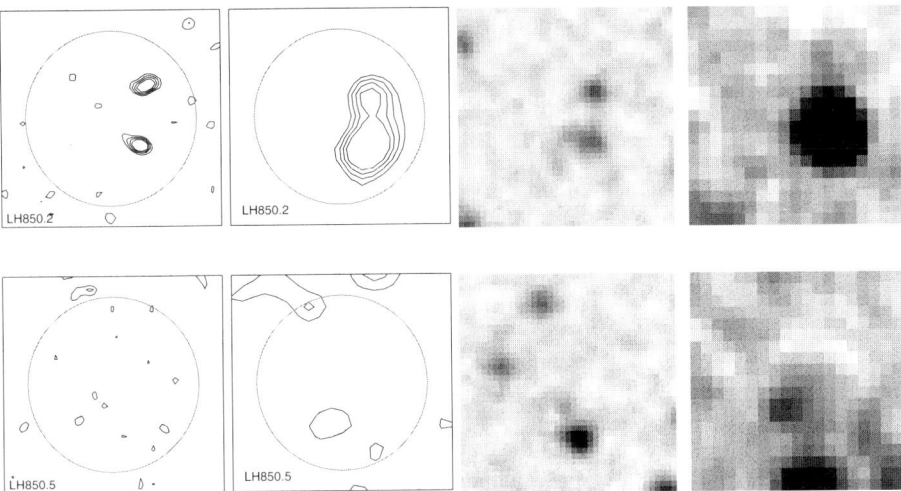

Figure 3. Two examples of the power of combining high-resolution and low-resolution radio imaging (with the VLA at 1.4 GHz) with Spitzer IRAC 4.5μm and MIPS 24μm imaging to determine the galaxy counterparts of SHADES sub-mm sources. The upper row of panels are 20 × 20 arcsec^2 postage stamp images extracted around the position of SHADES source Lockman 850.2 from the VLA A+B+C array map, the VLA B+C array map, the Spitzer IRAC 4.5μm map, and the Spitzer MIPS 24μm map. The circle on the radio images has a radius of 8 arcsec, which is the typical search radius for identifying the counterpart of a SHADES source with 95% confidence. For this source, the VLA imaging reveals two potential counterparts, but the Spitzer imaging reveals that it is the southerly candidate which is dusty. The second row of images shows the same information for SHADES source Lockman 850.5. In this example there is no clear radio counterpart in the high-resolution radio map, but the tentative radio counterpart provided by the lower-resolution radio imaging is confirmed as real by the Spitzer data. The radio data will be published in full by Ivison et al. (2005). The Spitzer data for the Lockman Hole sources have been provided to the SHADES consortium by Eichii Egami and George Rieke.

found from the radio follow-up of any previous SCUBA survey. Such a high radio-identification rate confirms the reality of the vast majority of the 850μm sources. The fact that the figure is so high will also be in part due to the depth and quality of the radio data, but nevertheless it already seems clear that few of the SHADES sources can lie at redshifts significantly in excess of $z \simeq 3$. A corollary to this is that we can now realistically expect to obtain an accurate ($\simeq 1$ arcsec) position for the vast majority of the SHADES sources, providing excellent prospects for subsequent unambiguous optical/IR identification, and optical/infrared spectroscopy. Moreover, for those sources which escape optical spectroscopy, the existence of a radio detection will still assist enormously in the derivation of accurate redshift estimates, especially in combination with BLAST and Spitzer data.

Even the high success rate of radio identification shown in Fig. 3 is not the whole story. The production of lower-resolution B+C array VLA maps (i.e., deliberately leaving out the A-array data) has revealed that even when, at first sight, a source appears to lack a radio identification, often an extended radio counterpart is found to exist in the lower-resolution map (resolved out in the A-array data). An example of this is shown in Fig. 3.

Also illustrated in Fig. 3 is the added value of combining the radio data with Spitzer imaging. Spitzer data for SHADES is being provided for the Lockman Hole field in collaboration with the Spitzer GTO consortium, and for the SXDF as part of the SWIRE survey. Figure 3 shows how these data can help both to differentiate between alternative radio counterparts, and to confirm the reality and dusty nature of tentative radio identifications.

4. Forthcoming papers

Detailed predictions of the extent to which SHADES can constrain the redshift distribution and clustering of submm sources are presented by van Kampen et al. (2004), while an overview of the survey strategy and design is provided in Mortier et al. (2004). In addition, three journal papers based on the current interim data set are currently in preparation: Scott et al. (2005) will present the current sub-mm maps, source list and number counts, Ivison et al. (2005) will present radio and Spitzer identifications, and Aretxaga et al. (2005) will report the estimated redshift distribution of the current SHADES sample. This first set of data papers will be followed by publication of the full multi-frequency study of SHADES sources, and by a detailed comparison of our results with the predictions of current models of galaxy formation.

Acknowledgments

I thank the members of the SHADES consortium, and also the staff of the JCMT, for their role in turning SHADES from a simulation into a real data set.

References

Baugh C.M., Lacey C.G., Frenk C.S., Granato G.L., Silva L., Bressan A., Benson A.J., Cole S., 2005, MNRAS, 356, 1191
Chapman S.C., Blain A.W., Ivison R.J., Smail I., 2003, Nature, 422, 695
Genzel R., et al., 2004, in: Multiwavelength mapping of galaxy formation and evolution, Bender R., Renzini A., eds., ESO Astroph. Symp., (Springer: New York), in press (astro-ph/0403183)
van Kampen E., et al., 2004, MNRAS, subm. (astro-ph/0408552)
Mortier A., 2004, MNRAS, subm.
Smail I., Ivison R.J., Gilbank D.G., Dunlop J.S., Keel W.C., Motohara K., Stevens J.A., 2003, ApJ, 583, 551
Somerville R.S., 2004, in: Multiwavelength mapping of galaxy formation and evolution, Bender R., Renzini A., eds., ESO Astroph. Symp., (Springer: New York), in press (astro-ph/0401570)

COMPACT EXTRAGALACTIC STAR FORMATION: PEERING THROUGH THE DUST AT CENTIMETER WAVELENGTHS

James S. Ulvestad
National Radio Astronomy Observatory, Socorro, New Mexico, USA
julvesta@nrao.edu

Abstract We summarize results of arcsecond and milliarcsecond-resolution radio imaging of starbursts. Thermal radio emission from young super star clusters (SSCs) indicates that they are powered by the equivalent of hundreds to thousands of O7 stars. Milliarcsecond imaging of Arp 299 reveals individual young supernovae that are factors of ~ 100 more powerful than Cas A, and that may show the locations of older SSCs in which the most massive stars already have gone supernova.

1. Introduction

Extragalactic star formation is studied at a variety of wavelengths, with many different instruments. The unique capability of centimeter-wavelength interferometry is that this technique is able to achieve high-resolution imaging while peering through the dust that obscures the youngest regions of intense star formation. Questions that can be influenced by results of sub-arcsecond radio imaging include the following:

- What are the properties of the youngest massive star clusters? How do they evolve to become globular clusters today? What is the luminosity function of super star clusters (SSCs)?

- Is optical/infrared modeling of star formation in SSCs consistent with radio observations?

- How do supernovae evolve in dense environments? What is the luminosity function of radio supernovae in starbursts?

In this contribution, we review a few results from high-resolution radio imaging that bear on the questions listed above.

2. Radio Emission From Extragalactic Starbursts

The youngest starburst regions contain large quantities of dust, and optical radiation from these regions may be obscured by more than 20 magnitudes of visual extinction. Radio emission at centimeter wavelengths originates within the starbursts from two principal causes. (i) The youngest starbursts contain complexes of dense HII regions that are energized by SSCs. Thermal radio emission from these complexes can be used to estimate the total ionizing flux of the embedded stars, and hence to assess the massive star population. (ii) Individual supernova remnants or young supernovae generate synchrotron radio emission. Imaging of this radiation over time can be used to study the evolution of individual objects, which in turn may be used to infer an approximate supernova rate.

The Very Large Array (VLA) has been the instrument of choice for imaging extragalactic starbursts. Its present sensitivity and sub-arcsecond resolution enable imaging of SSCs or SSC complexes out to distances of a few tens of Megaparsecs, with sensitivity to embedded stellar populations ranging from a few dozen O7-equivalent stars up to tens of thousands of such stars.

The stellar content of the thermal sources can be estimated by computing the number of ionizing photons necessary to account for the radio emission:

$$N(\mathrm{UV})\,\mathrm{s}^{-1} = 10^{51} \left(\frac{D}{2.5\,\mathrm{Mpc}}\right)^2 \left(\frac{S_{5\mathrm{GHz}}}{1\,\mathrm{mJy}}\right) \quad , \qquad (1)$$

where 10^{49} UV photons per second is the number expected from a single O7 star (e.g., Vacca 1994).

3. Stellar Content of SSCs

The nearby starbursts M82 and NGC 253, at distances near 3 Mpc, were first imaged using the VLA in the 1980s (Kronberg & Sramek 1985, Antonucci & Ulvestad 1987). At resolutions of a few parsecs and less, these starbursts contained numerous thermal radio sources (flat or inverted radio spectra) and young supernova remnants (steep radio spectra). Most of these compact radio sources have shown little or no variability over time-scales of 10–20 years. In NGC 253, the strongest compact thermal radio source emits ~ 8 mJy at 5 GHz within a region just a few parsecs in diameter (Ulvestad & Antonucci 1997). For a distance of 2.5 Mpc, Eq. (1) implies that this source is powered by ionizing photons from the equivalent of 800 O7 stars. Of course, an estimate of the total stellar mass depends critically on the stellar mass function, which is unknown in most individual SSCs.

Nearby dwarf galaxies often have all their radio and mid-infrared emission dominated by SSCs; examples are He 2-10 (Johnson & Kobulnicky 2003) and NGC 5253 (Turner & Beck 2004). The SSCs sometimes are optically thick even at long centimeter wavelengths, and observations at frequencies of 8 GHz and higher may be necessary to reveal the intrinsic radio powers. Simple modeling indicates that the SSCs have radii of a few parsecs, total masses of $10^5 M_\odot$ or more, and central densities above 10^5 cm^{-3}. Pressures are in the range $nT = P/k_B \sim 10^8$ cm^{-3}K or greater, much higher than in the typical galactic interstellar medium (Johnson & Kobulnicky 2003, Turner & Beck 2004). The compact ionized regions may be in pressure equilibrium with their parent dusty molecular clouds or may be confined by the gravity of the embedded massive stars.

Merger galaxies typically are at larger distances of tens of Megaparsecs. The available sensitivity and resolution permits only complexes of multiple SSCs to be imaged; examples are NGC 4038/9 (Neff & Ulvestad 2000) and Arp 299 (Neff, Ulvestad & Teng 2004). In NGC 4038/9 (the "Antennae"), the most powerful ionized complex requires $\sim 3 \times 10^{53}$ UV photons to energize the radio emission, corresponding to 30,000 O7-equivalent stars in a region roughly 100 pc in diameter.

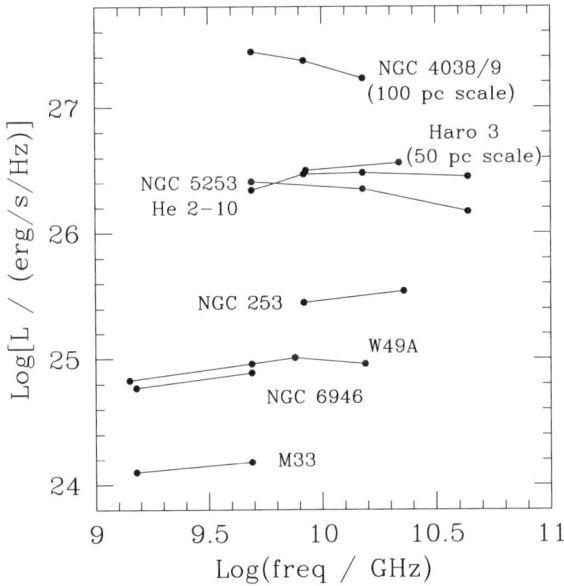

Figure 1. Radio spectra of the strongest thermal radio sources in a selection of SSCs and star-formation regions.

Figure 1 shows the radio luminosity of the most luminous thermal radio sources in a sample of starburst galaxies, along with the nearby galaxy M33 and the galactic source W49A. For two galaxies, Haro 3 and NGC 4038/9, the size scales of 50–100 pc may include multiple SSCs, while the rest of the points plotted generally refer to a single SSC or ionized region with diameter no larger than a few parsecs. The range of a factor of more than 1000 in radio luminosity corresponds to a similar range in numbers of O7-equivalent stars.

4. Young Radio Supernovae

Dominance by thermal radio emission in many SSCs indicates that less than a few million years have passed since the bulk of their star formation occurred. The ionizing luminosity in an SSC is dominated by the highest-mass stars, which have main-sequence lifetimes in the range of 3–4 Myr. For older SSCs, the ionizing luminosity will drop dramatically as the most massive stars go supernova. Then, the radio emission will be dominated by synchrotron radiation from the young supernovae and supernova remnants, and these objects may be used to distinguish the locations of the evolving SSCs.

Arp 299, at a distance of 41 Mpc, serves as an interesting illustration of an older starburst. It is a merger of two disk galaxies, containing galaxy nuclei ("A" and "B1") that apparently underwent their peak star-formation episodes 6–8 Myr in the past (Alonso-Herrero et al. 2000). VLA imaging has shown that the dominant radio nuclei are 1 arcsec (200 pc) or smaller at centimeter wavelengths, and appear to be dominated by optically thin synchrotron radiation. The Very Long Baseline Array (VLBA), sometimes augmented by the Green Bank Telescope (GBT), has been used to search for even more compact radio sources within these heavily obscured arcsecond nuclei. Because of its long baselines, the VLBA acts as a spatial filter, resolving away all the ionized regions with brightness temperatures of 10^4 K or less. Any detected radio sources will have brightness temperatures in excess of 10^6 K, and must be young supernovae, supernova remnants, or (perhaps) active galactic nuclei.

Neff et al. (2004) detected at least five milliarcsecond radio sources within nucleus A of Arp 299, in an area roughly 0.5 arcsec (100 pc) in diameter. Three of these sources were found within a diameter of ~ 10 pc, and two were separated by only 3 pc. It is likely that all of the compact radio sources are young supernovae, and trace the locations of ageing SSCs in Arp 299. In fact, one object has a strongly inverted radio spectrum, and is likely to be a very young supernova that has not yet broken out of its circumstellar shell.

More recently, we have made repeated images of Arp 299 at intervals of six months, in order to search for new supernovae and for evolution of the older objects. By stacking multiple images, we have been able to detect more young supernovae. Figure 2 shows a two-epoch VLBA+GBT image of nucleus A, revealing at least 11 young supernovae, now spread over a diameter of nearly 200 pc. Four additional objects have been detected so far at 8.4 GHz, and another four young supernovae now have been detected in nucleus B1. The ratios of the numbers of supernovae in the nuclei A and B1 are plausibly consistent with respective supernova rates estimated at ~ 0.6 yr^{-1} and ~ 0.1 yr^{-1} (Alonso-Herrero et al. 2000).

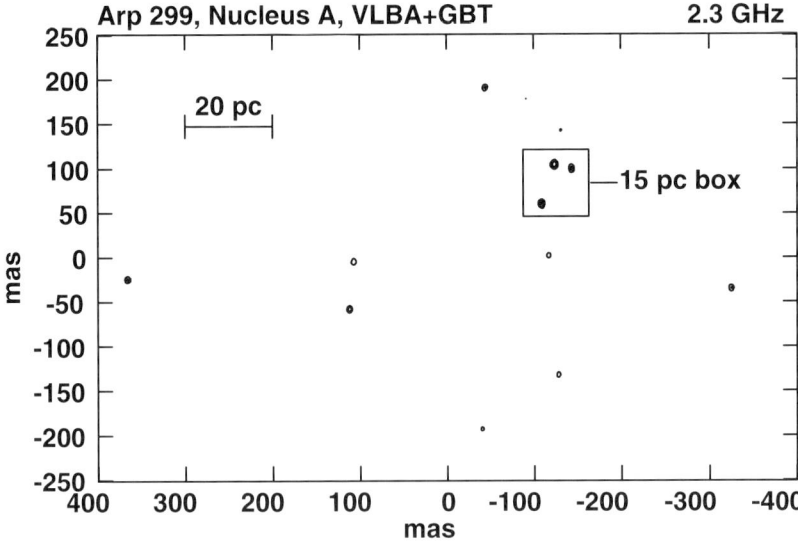

Figure 2. VLBA+GBT image of nucleus A in Arp 299. Radio contours are separated by factors of 2, starting at 5 times the rms noise of 31 μJy beam^{-1}.

Enough young supernovae have been detected in nucleus A of Arp 299 that we can begin to assess the top end of the radio luminosity function of these supernovae. Figure 3 shows a histogram of the radio luminosities at 2.3 GHz, compared to the radio luminosities of young supernova remnants imaged in M82 (Muxlow et al. 1994). The radio counts are roughly consistent with a ratio of a factor of 6 in the relative supernova rates, although the small numbers of objects mean that the VLBI observations cannot be used by themselves to estimate this ratio reliably. From Fig. 3, we see that the Arp 299 supernovae have radio powers as high as 10^{20} W Hz^{-1}, reaching the range of weak active galactic nuclei, a factor of 100 or more above the Galactic supernova remnant Cas A. Clearly, we are detecting only the youngest and most powerful supernovae or supernova remnants in Arp 299, consistent with the fact that the detected milliarcsecond sources contain only a few per cent of the total centimeter radio emission observed by the VLA.

Acknowledgments

The National Radio Astronomy Observatory is a facility of the U.S. National Science Foundation, operated under cooperative agreement by Associated Uni-

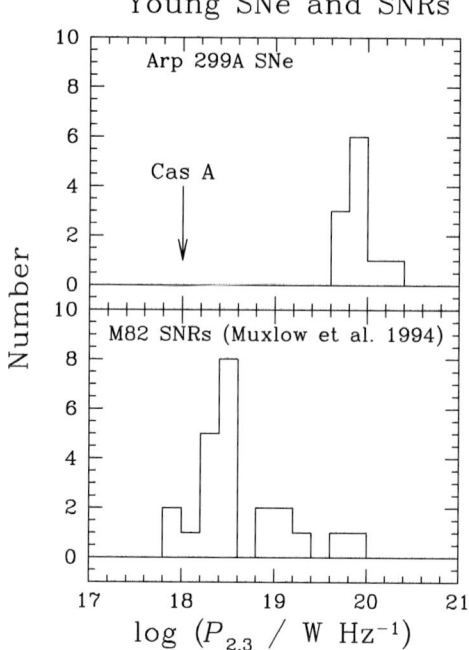

Figure 3. Histograms of 2.3 GHz radio powers in Arp 299 (nucleus A) and M82. The M82 powers have been converted from the 5 GHz values (Muxlow et al. 1994) by using a spectral index $\alpha = -0.7$, where $S_\nu \propto \nu^{+\alpha}$.

versities, Inc. The author thanks Kelsey Johnson for helpful comments and for supplying material in advance of publication.

References

Alonso-Herrero A., Rieke G.H., Rieke M.J., Scoville N.Z., 2000, ApJ, 532, 845
Antonucci R.R.J., Ulvestad J.S., 1987, ApJ, 330, L97
Johnson K.E., Kobulnicky H.A., 2003, ApJ, 597, 923
Kronberg P.P., Sramek R.A., 1985, Science, 227, 28
Muxlow T.W.B., Pedlar A., Wilkinson P.N., Axon D.J., Sanders E.M., de Bruyn A.G., 1994, MNRAS, 266, 455
Neff S.G., Ulvestad J.S., 2000, AJ, 120, 670
Neff S.G., Ulvestad J.S., Teng S.H., 2004, ApJ, 611, 186
Turner J.L., Beck S.C., 2004, ApJ, 602, L85
Ulvestad J.S., Antonucci R.R.J., 1997, ApJ, 488, 621
Vacca W.D., 1994, ApJ, 421, 140

DUST ATTENUATION AND STAR FORMATION IN THE NEARBY UNIVERSE: THE ULTRAVIOLET AND FAR-INFRARED POINTS OF VIEW

Véronique Buat and Jorge Iglesias-Páramo
Laboratoire d'Astrophysique de Marseille, France
veronique.buat@oamp.fr, jorge.iglesias@oamp.fr

Abstract   We make use of the on-going All-sky Imaging Survey of the UV GALEX satellite, cross-correlated with the IRAS all-sky survey to build samples of galaxies truly selected in the far-infrared or in the ultraviolet. We discuss the amount of dust attenuation and the star-formation rates for these galaxies and compare the properties of the galaxies selected in the FIR or in the UV.

Introduction

The measure of the star-formation rate (SFR) in galaxies is based on the analysis of the emission from young stars that escapes the galaxies or is absorbed and re-emitted by the gas or the dust. In this paper, we will focus on UV and IR emission. Emission at these wavelengths is closely related to the energy budget at work in star-forming galaxies: the UV light that does not escape the galaxy is absorbed by the interstellar dust and re-emitted in the far-IR. Therefore, both types of emission originate from the same stellar populations and their comparison is a powerful tracer of the dust attenuation (Buat & Xu 1996, Gordon et al. 2000). They are also closely related to the recent star-formation rate over similar time-scales (Buat & Xu 1996, Kennicutt 1998). In this paper we will combine the new GALEX data with the existing far-IR data from IRAS.

1. The data

We have worked on 600 deg^2 observed by GALEX in NUV ($\lambda = 231$ nm and FUV ($\lambda = 153$ nm) to build two samples of galaxies. The first one, called the *NUV-selected sample*, includes all galaxies brighter than m(NUV) = 16 mag (AB scale); among the 88 selected galaxies (excluding ellipticals and active galaxies), only 3 are not detected by IRAS at 60 μm. The second sample, called the *FIR-selected sample*, is based on the IRAS PSCz (Saunders et al. 2000): 118 galaxies from this catalog lie within our GALEX fields, only 1 is not detected in the NUV.

2. Dust attenuation in the nearby universe

Mean values of the dust attenuation. For both samples we measure the dust attenuation using the $F(\text{dust})/F(\text{NUV})$ ratio. This ratio is a quantitative measure of the dust attenuation at UV wavelengths (e.g., Buat & Xu 1996, Gordon et al. 2000). The formulae used here are adapted to the GALEX bands (Buat et al. 2005). The total $(8 - 1000\mu m)$ dust emission is calculated from the fluxes at 60 and 100 μm using the Dale et al. (2001) recipe.

Moderate attenuation is found in the NUV-selected sample, with $0.8^{+0.4}_{-0.3}$ mag in the NUV and $1.1^{+0.5}_{-0.4}$ mag in the FUV. As expected, the dust attenuation is greater in the FIR-selected sample with $2.1^{+1.2}_{-0.8}$ mag in the NUV and $2.9^{+1.2}_{-1.2}$ mag in the FUV.

A comparison of the FIR and UV luminosity densities from IRAS (Saunders et al. 1990) and GALEX (Thilker et al. 2005) leads to a mean dust attenuation at $z \sim 0$ of 1.1 mag in the NUV and 1.6 mag in the FUV: the nearby Universe is not very obscured.

The "IRX-β" relation. From their study of starburst galaxies, Meurer and colleagues (e.g., Meurer et al. 1999) have found a strong correlation between the slope, β, of the UV continuum (fitted as a power law $F(\lambda) \propto \lambda^\beta$ between 1200 and 2500 Å) and $F(\text{dust})/F(\text{FUV})$. They concluded that β is a reliable tracer of the dust attenuation in galaxies. However, further studies showed that this law is not universal and that ultra-luminous infrared galaxies (Goldader et al. 2002) and normal spiral galaxies (Bell 2002) do not follow the relation for starbursts.

The FUV–NUV color from the GALEX observations is directly linked to β (Kong et al. 2004). We plot this color versus $F(\text{dust})/F(\text{FUV})$ for our two samples of galaxies in Fig. 1, and compared it to the predictions for starburst galaxies. Obviously, β is not a reliable tracer of the dust attenuation in our NUV and FIR-selected samples. Moreover, different types of behavior are found within both samples: whereas the FIR-selected galaxies spread over a large area of the diagram, most of the NUV-selected galaxies lie below the starburst relation. This may be due to age effects in the stellar populations: assuming a b parameter (defined as the ratio of the recent to the average past SFR) equal to 0.25 leads to a reasonable fit using the models of Kong et al. (2004).

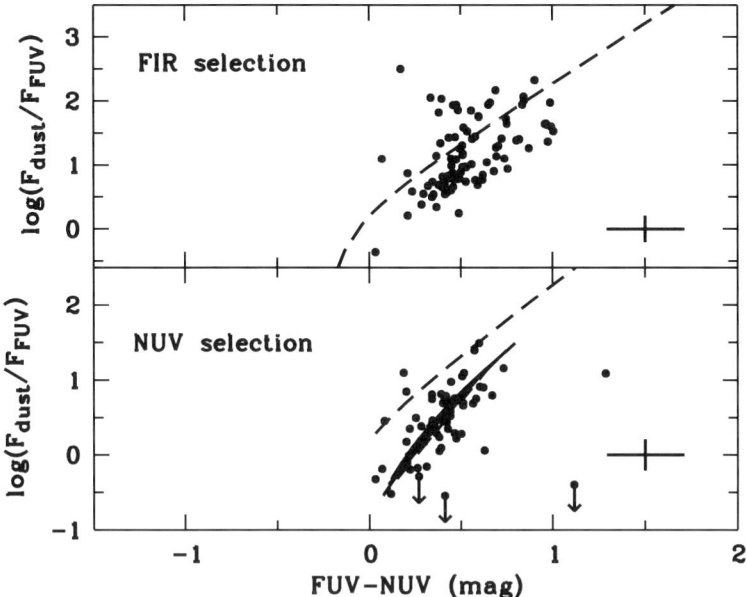

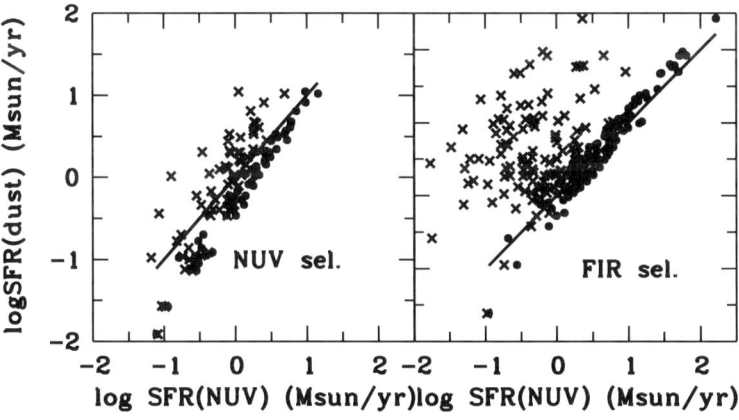

Figure 1. Top: $\log(F(\text{dust})/F(\text{FUV}))$ versus the FUV−NUV color for the NUV and FIR-selected samples. The dashed line is the mean relation expected for starburst galaxies, the solid line is the locus of Kong et al. (2004) models with $b = 0.25$ (see text). Bottom: SFR deduced from the dust luminosity versus the SFR deduced from the NUV luminosity directly observed (crosses) and corrected for dust attenuation (points) for the NUV-selected (left panel) and the FIR-selected samples (right panel). The lines correspond to equality on both axes.

3. The measure of the star-formation rate

UV and total dust emission can be calibrated in terms of a recent star-formation rate assuming a recent (over $\sim 10^8$ yr) star-formation history and an initial mass function (Buat & Xu 1996, Kennicutt 1998). When using the dust luminosity one must add an additional assumption regarding the absorption of the stellar light by the dust. The classical hypothesis (also adopted here) is that all of the stellar light from the young stars is absorbed by the dust (Kennicutt 1998). In this paper, we also assume a constant SFR over 10^8 yr, and a Salpeter IMF. Using STARBURST99 synthesis models we obtain:

$$\log(L_{\text{NUV}})(\text{L}_\odot) = 9.73 + \log(\text{SFR})(\text{M}_\odot\text{yr}^{-1}) \qquad (1)$$

and

$$\log(L_{\text{dust}})(\text{L}_\odot) = 10.168 + \log(\text{SFR})(\text{M}_\odot\text{yr}^{-1}) \quad . \qquad (2)$$

In Fig. 1 we plot the SFR estimated from the NUV luminosity versus the SFR from the dust luminosity. In both samples, the *observed* NUV luminosity strongly underestimates the SFR, the effect being worse for the FIR-selected sample. When the NUV-selected sample is corrected for dust attenuation the agreement between both estimates of the SFR in the FIR-selected sample is very good – as is expected, since we are dominated in each case by the dust emission. Conversely, in the NUV-selected sample the SFR estimated from the dust luminosity alone is found to underestimate the SFR, compared to the NUV-corrected luminosity. The discrepancy increases toward the low SFRs, to reach a factor 3 for SFR of ~ 0.3 $\text{M}_\odot$ yr^{-1}. Therefore, using the dust emission alone to measure the total SFR in all galaxies can be misleading. The best solution is to combine UV and IR emission to estimate reliable SFRs.

References

Bell E., 2002, ApJ, 577, 150
Buat V., Xu C., 1996, A&A, 306, 61
Buat V., et al., 2005, ApJ, 619, L51
Dale D., Helou G., Contursi A., Silbermann N., Kolhatkar S., 2001, ApJ, 549, 215
Goldader J.D., Meurer G., Heckman T.M., Seibert M., Sanders D.B., Calzetti D., Steidel C.C., 2002, ApJ, 568, 651
Gordon K, Clayton G., Witt A., Misselt K., 2000, ApJ, 533, 236
Kennicutt R., 1998, ARA&A, 36, 189
Kong X., Charlot S., Brinchmann J., Fall M., 2004, MNRAS, 349, 769
Meurer G.R., Heckman T.M., Calzetti D., 1999, ApJ, 521, 64
Saunders W., et al., 2000, MNRAS, 317, 55
Saunders W., Rowan-Robinson M., Lawrence A., Efstathiou G., Kaiser N., Ellis R.S., Frenk C.S., 1990, MNRAS, 242, 318
Thilker D.A., 2005, ApJ, 619, L67

THEORETICAL PAN-SPECTRAL ENERGY DISTRIBUTIONS OF STARBURST GALAXIES

Michael A. Dopita

Research School of Astronomy & Astrophysics, The Australian National University, Cotter Road, Weston Creek, ACT2611, Australia
Michael.Dopita@anu.edu.au

Abstract We have combined the STARBURST99 code with the MAPPINGS IIIq nebular modelling code, and a self-consistent 1D dynamical evolution model for H II regions, to produce theoretical spectral energy distributions (SEDs) extending from the Lyman Limit up to beyond 1 mm. We show that two parameters control the form of the SED, the mean pressure in the ISM of the starburst, and the time-scale required to dissipate the molecular clouds surrounding individual clusters.

1. Introduction

A knowledge of the star-formation rate (SFR) is fundamental if we are to understand the formation and evolution of galaxies. In a much-cited seminal paper, Madau et al. (1996) connected the star formation in the distant Universe to that estimated from low-redshift surveys, by plotting the estimated star-formation rate per unit co-moving volume against redshift. However, at the present time an unacceptable degree of uncertainty still attaches to the Madau plot, and hence to our overall understanding of the evolution of star formation in the Universe, so it is very important to understand the source of these uncertainties, and how best they can be eliminated.

Part of the problem is that a wide variety of emission line or continuum techniques contribute towards populating this diagram with observational points, many of which have been discussed at this conference, and each of which has its advantages and its problems. These techniques range from the very direct to the quite indirect. As an example of a direct technique, we may use the continuum produced by the stars, such as the UV flux (Calzetti 2001). A somewhat less direct approach is to use the flux reprocessed into the far infrared by dust. Both depend on the geometry of the dust with respect to the ionizing sources. For example, the UV must give a lower limit on the SFR, and the far-IR methods will only give the correct SFR if the star-forming region is completely surrounded by a thick layer of dust. There are also systematic effects due to both metallicity (low metallicity will allow escape of UV photons) and due to the SFR itself, since a variety of studies show that the SFR correlates with dustiness over a very wide range of SFR (Buat et al. 1999, Adelberger & Steidel 2000, Dopita et al. 2002, Kewley et al. 2002). Inoue (2002), and also Buat (this conference), has discussed a number of these issues.

Amongst emission-line techniques, both recombination lines and [OII] lines are frequently used. These are only sensitive to the SFR within the last 10 Myr. However, ionising photons may be lost by dust absorption within the HII regions (Dopita et al. 2003), and the [OII] technique is, in addition, sensitive to metallicity and ionisation parameters. With careful calibration, some of these problems can be corrected for (Kewley et al. 2004).

Other techniques have also been proposed, including the use of the PAH emission features (Roussel et al. 2001), which depend on the C abundance, the X-ray emission, which depends on the X-ray binary population (Ghosh & White, 2001), and the enigmatic non-thermal radio continuum. This depends on Fermi acceleration of electrons in supernova remnants and on the loss of energy of these electrons in regions of high magnetic field (Bressan et al. 2002), but which is nonetheless correlated with the far-IR emission over ~ 5 decades of magnitude and with less than 0.3 dex dispersion (Yun et al. 2001).

In order to calibrate many of these SFR indicators and ultimately to reduce the scatter in the Madau plots, we have generated a series of theoretical pan-spectral SED models which, for the first time take into account the physical evolution of HII regions in the multi-phase interstellar medium of starburst galaxies. These, and the main results, are briefly described in this paper.

2. The Model

In Dopita et al. (2005) we have used the most recent version of the STARBURST99 code (Leitherer et al. 1999) to generate the theoretical stellar SEDs for solar metallicity, a standard Salpeter initial mass function, and upper and lower mass cut-offs of 120 $M_\odot$ and 1 $M_\odot$, respectively on the zero-age main sequence. We assume that the starburst region occupies a roughly spherical region filled with individual HII regions surrounding clusters, of mean mass 10^4 $M_\odot$ and having all possible ages up to 10 Myr. The number of HII regions is normalised to supply a mean SFR of 1.0 $M_\odot$ yr^{-1}. These HII regions expand as a consequence of the over-pressure produced by the energy deposition within them (given by the STARBURST99 models). They therefore have a radius determined by both their age and by the pressure in the ISM. In high pressure environments, the HII regions are smaller and denser, and the dust in the surrounding photo-dissociation regions (PDRs) is therefore hotter.

The radiative transfer through the HII regions and the surrounding PDRs is computed using the MAPPINGS IIIq code. This includes all relevant gas and dust physics needed to compute the gas ionisation and temperature structure, the strengths of all emission lines and of the atomic continua. We also compute the grain charge and the temperature distribution function of each grain size and species, allowing an accurate computation of the shape of the far-IR spectrum. The dust grain model largely follows the formulation of Weingartner & Draine (2001a,b). Our models also include PAH physics, including absorption, photoelectric charging, photoelectric destruction, heating, and re-emission in the PAH features in the IR. The assumptions relevant to the HII region SEDs are given in Dopita et al. (2005), and the complete details of the computational methods are given in Groves (2004).

3. Results

The fact that the HII regions are smaller in the high-pressure environments means that the dust temperatures are, in general, higher in higher-pressure environments. This is clearly seen in Fig. 1. Thus warm IRAS sources can be identified with regions of high specific star formation, since high pressure is needed to confine them.

A second controlling factor on the SED is the time required to clear away the molecular clouds. Alternatively, this could be thought of as the time taken for the OB stars to diffuse away from the dense regions where they are born. Both of these would have a similar effect on the SED. When the clearing timescale is long, the dust captures and reprocesses the UV photons for longer. This depresses the UV flux, and means that the stars that are seen in the UV have a greater average age. It also means that the far-IR flux is larger, and that the peak of the far-IR occurs at longer wavelengths, because the dust shells have a larger diameter, on average. These effects are evident in Fig. 2.

The SEDs shown in Figs. 1 and 2 do not include the absorption of any foreground dust. Models of such attenuation in starburst galaxies show that it can be well-represented by the effect of a (distant) foreground screen (e.g., Meurer 1999). In an earlier paper (Fischera et al. 2003) we showed that many features of the Calzetti (2001) attenuation law for starbursts could be reproduced by a turbulent screen having a log-normal local density distribution. In a later paper (Fischera & Dopita 2005) we investigated mathematical properties of such a screen and showed that it can be represented to high accuracy by a log-normal column density distribution. This allows us to compute the absorption characteristics over a wide variety of physical conditions and so model the SEDs of starburst galaxies over the whole range from 912Å up to 1000μm.

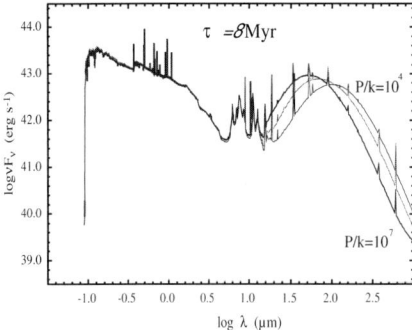

Figure 1. The variation of the starburst SED with pressure in the ISM. In high-pressure environments, the HII regions are smaller in size, the dust temperatures are hotter, and consequently the peak of the SED in the far-IR is found at shorter wavelengths.

Figure 2. The variation of the starburst SED with the time-scale over which the molecular clouds are destroyed. When this time-scale is long, less UV radiation escapes, the mean age of the stars seen in the UV is greater, and the far-IR peak in the SED occurs at longer wavelengths.

Acknowledgments

M. Dopita acknowledges the support of the Australian National University and the Australian Research Council (ARC) through his Federation Fellowship, and of the ARC Discovery project grant DP0208445.

References

Adelberger K.L., Steidel C.C., 2000, ApJ, 544, 218
Bressan A., Silva L., Granato G.L., 2002, A&A, 392, 377
Buat V., Donas J., Milliard B., Xu C., 1999, A&A, 352, 371
Calzetti D., 2001, PASP, 113, 1449
Dopita M.A., Pereira M., Kewley L.J., Capaccioli M., 2002, ApJS, 143, 47
Dopita M.A., Groves B.A., Sutherland R.S., Kewley L.J., 2003, ApJ, 583, 727
Dopita M.A., et al. 2005, ApJ, in press (astro-ph/0407008)
Fischera J., Dopita M.A., Sutherland R.S., 2003, ApJ, 599, L21
Fischera J., Dopita M.A. 2005, ApJ, 619, 340
Ghosh P., White N.E., 2001, ApJ, 559, L97
Groves B., 2004, Ph.D. thesis, Australian National University
Inoue A.K., 2002, ApJ, 570, L97
Kewley L.J., Geller M.J., Jansen R.A., 2004, AJ, 127, 2002
Leitherer C., et al., 1999, ApJS, 123, 3
Madau P., Ferguson H.C., Dickinson M.E., Giavalisco M., Steidel C.C., Fruchter A., 1996, MNRAS, 283, 1388
Meurer G.R., Heckman T.M., Calzetti D., 1999, ApJ, 521, 64
Roussel H., Sauvage M., Vigroux L., Bosma A., 2001, A&A, 372, 427
Weingartner J.C., Draine B.T., 2001a, ApJ, 548, 296
Weingartner J.C., Draine B.T., 2001b, ApJS, 134, 263
Yun M.S., Reddy N.A., Condon J.J., 2001, ApJ, 554, 803

Session IV

Triggering and Quenching of Starbursts and the Effects of Galactic Interactions

MERGER-INDUCED STARBURSTS

François Schweizer
Carnegie Observatories, Pasadena, CA 91101, USA
schweizer@ociw.edu

Abstract Extragalactic starbursts induced by gravitational interactions can now be studied from $z \approx 0$ to $\gtrsim 2$. The evidence that mergers of gas-rich galaxies tend to trigger galaxy-wide starbursts is strong, both statistically and in individual cases of major disk–disk mergers. Star-formation rates appear enhanced by factors of a few to $\sim 10^3$ above normal. Detailed studies of nearby mergers and ULIRGs suggest that the main trigger for starbursts is the rapidly mounting pressure of the ISM in extended shock regions, rather than high-velocity, 50–100 km s^{-1} cloud–cloud collisions. Numerical simulations demonstrate that in colliding galaxies the star-formation rate depends not only on the gas density, but crucially also on energy dissipation in shocks. An often overlooked characteristic of merger-induced starbursts is that the spatial distribution of the enhanced star formation extends over large scales ($\sim 10 - 20$ kpc). Thus, although most such starbursts do peak near the galactic centers, young stellar populations pervade merger remnants and explain why (i) age gradients in descendent galaxies are mild and (ii) resultant cluster systems are far-flung. This review presents an overview of interesting phenomena observed in galaxy-wide starbursts and emphasizes that such events continue to accompany the birth of elliptical galaxies to the present epoch.

1. Introduction

This brief review concentrates on three items. First, I report on recent progress in our understanding of the dynamical triggers at work in merger-induced starbursts. Then, I address two issues that are often ignored or misunderstood, yet are of fundamental importance to the subject: the large spatial extent of merger-induced starbursts, and the implications of such starbursts for the formation of elliptical galaxies at low and high redshifts.

The basic reason why tidal interactions and mergers help fuel bursts of star formation has been understood for over three decades. Under the headline *Stoking the Furnace?* Toomre & Toomre (1972) wrote: "Would not the violent mechanical agitation of a close tidal encounter – let alone an actual merger – already tend to bring *deep* into a galaxy a fairly *sudden* supply of fresh fuel in the form of interstellar material?" Subsequent numerical simulations that included gas have fully corroborated this notion (e.g., Negroponte & White 1983, Noguchi 1988, Barnes & Hernquist 1991).

Similarly, many observational studies have established beyond any doubt that mergers invigorate star formation well beyond the levels observed in quiescent disk galaxies. As early as 1970, Shakhbazian pointed out the presence of extraordinary stellar *"super associations"* with luminosities of up to $M_V \approx -15.5$ in the Antennae. In a landmark paper, Larson & Tinsley (1978) showed that the UBV colors of Arp's peculiar galaxies can best be explained if tidal interactions engender short, but intense bursts of star formation involving up to $\sim 5\%$ of the total mass. Infrared observations confirmed the notion of *super* starbursts in major disk–disk mergers (Joseph & Wright 1985). Since then, a steady stream of papers from surveys (e.g., 2dF: Lambas et al. 2003; SDSS: Nikolic et al. 2004) has continued to support and refine this picture. A nice twist was the discovery – fostered by *HST*'s high resolution – that star clusters and, specifically, globular clusters form in large numbers during galactic mergers (Schweizer 1987, Holtzman et al. 1992, Whitmore et al. 1993).

2. Gas Supply and Dynamical Triggers

Even after ~ 13 Gyr of evolution, many present-day galaxies still have significant gas supplies available for star formation during interactions and mergers. The median gas fraction of neutral hydrogen alone, expressed relative to the total baryonic mass, is 15%, 10%, and 4% for dIrr, Sc, and Sa galaxies, respectively (Roberts & Haynes 1994). Even more impressive is the median fraction of *all* gas relative to the dynamical mass, $M_{\text{HI}+\text{H}_2}/M_{\text{dyn}} \approx 25\%$, 15%, and 3% for the same three types of galaxies (Young & Scoville 1991). Since, even at high redshifts, no galaxy can be more than 100% gaseous, the relatively high gas fractions of local Sc and later-type galaxies tell us that there is less than one order-of-magnitude difference between the fractional gas contents of many local disk galaxies and their high-z counterparts. Hence, the often-heard objection that mergers at $z \approx 2-5$ were completely different from local mergers is weak, and studying local mergers can, in fact, help us understand high-z mergers.

Molecular gas masses observed in local mergers and distant quasars support this point of view. Locally, M_{H_2} ranges from 0.6×10^{10} $M_\odot$ for an aging merger remnant, like NGC 7252, through 1.5×10^{10} $M_\odot$ for an ongoing merger, like the Antennae, to $\sim 3 \times 10^{10}$ $M_\odot$ for extreme ULIRGs, while $M_{\text{H}_2} \approx 2 \times 10^{10}$ $M_\odot$ in a QSO at $z = 2.56$ (Solomon et al. 2003) and also in one at $z = 6.42$ (Walter et al. 2003).

With gas amply available for star formation during mergers at both low and high redshifts, what are the *dynamical triggers* for merger-induced starbursts?

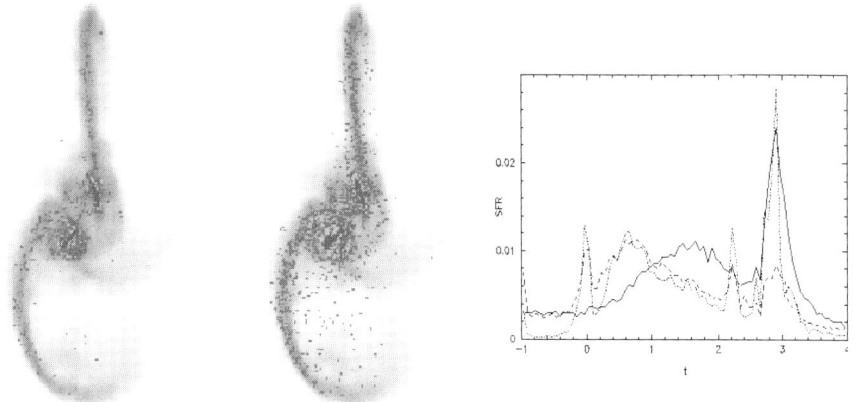

Figure 1. Simulations of star formation in the Mice for *(left)* density-dependent and *(middle)* shock-induced star-formation recipes; halftones mark old stars, points mark star formation. *(Right)* Star-formation rate vs. time for density-dependent (solid line) and various shock-induced (dashed and dotted) star-formation recipes (from Barnes 2004).

During disk galaxy interactions, gravitational torques arise between the bars induced in the gas and among the stars. Because the gaseous bar leads, the gas experiences braking, which – in turn – leads to its infall. The resulting pressure increase in the gas has long been understood to be the root cause of interaction-induced starbursts (Noguchi 1988, Hernquist 1989, Barnes & Hernquist 1991).

Although the vehemence of this pressure increase clearly depends on such factors as the presence or absence of a central bulge (Mihos & Hernquist 1996) and the encounter geometry (Barnes & Hernquist 1996), the small-scale details of the dynamical triggers have been less clear, until recently.

It now appears that *shocks* play a very major role, both in affecting the spatial distribution of star formation (Barnes 2004) and in squeezing giant molecular clouds (hereafter GMCs) into rapid star and cluster formation (Jog & Solomon 1992, Elmegreen & Efremov 1997). Questions such as whether cloud–cloud collisions are important and what role magnetic fields play are just beginning to be addressed by observers, as detailed below.

Merger-induced shocks can be fierce. In a simulation of two merging equal-mass disks (Barnes & Hernquist 1996), massive rings of dense gas form around the center of each galaxy and collide, during the third passage, head-on with a relative velocity of ~ 500 km s^{-1}! This extreme final smash is made possible by the rapid $\sim 90\%$ loss of orbital angular momentum that the two gas rings experience within $\sim 1/4$ disk rotation period.

Even during milder encounters, shock-induced star formation may dominate. Because of the ubiquity of shocks in mergers involving gas, Barnes (2004) proposes a new star-formation recipe that, in addition to the local gas density, ρ_{gas}, includes the local rate, $\dot{u}$, of mechanical heating due to shocks and PdV work:

$$\dot{\rho}_* = C_* \times \rho_{\text{gas}}^n \times \max(\dot{u}, 0)^m . \quad (1)$$

Assuming that energy dissipation balances the heating rate, setting $m > 0$ and $n = 1$ yields purely *shock-induced* star formation, while setting $m = 0$ and $n > 1$ yields *density-dependent* star formation (with Schmidt's law as a special case). Barnes compares simulations run for these two limit cases of star formation with observations of the Mice (see Fig. 1 here, and color figs. 3 and 4 in his paper) and shows convincingly that shock-induced star formation is *spatially more extended* and *occurs earlier* during the merger, which is in significantly better accord with the observations.

One long-standing question has been whether high-velocity cloud–cloud collisions ($50 - 100$ km s^{-1}) contribute significantly to the triggering of starbursts (Kumai, Basu & Fujimoto 1993) or not. To address this issue, Whitmore et al. (in prep.) measured Hα velocities of the gas associated with young massive clusters in the Antennae, using *HST*/STIS and positioning the 52 arcsec long slit of STIS along different groups of clusters. From many clusters in 7 regions, the measured cluster-to-cluster velocity dispersion is $< 10 - 12$ km s^{-1}, which argues *against* high-velocity cloud–cloud collisions as a major trigger of starbursts. Instead, the squeezing of GMCs by the general pressure increase in the ISM (Jog & Solomon 1992) appears favored.

The role of magnetic fields in triggering starbursts in mergers remains unclear at present, but is beginning to be studied observationally. Chyży & Beck (2004) have used the VLA to produce detailed maps of radio total power and polarization in NGC 4038/39 (see Chyży, these proceedings). The derived mean total magnetic field of $\sim 20\mu$G is twice as strong as in normal spirals and appears *tangled* in regions of enhanced star formation. The field peaks at $\sim 30\mu$G in the southern part of the Overlap Region, suggesting strong compression where the star-formation rate is highest. The crucial question to address over the coming years is whether the enhanced magnetic field observed in mergers merely traces compression, or whether it contributes to the triggering of starbursts.

3. Spatially Extended Starbursts

Interaction-induced starbursts tend to be spatially extended ($\sim 10 - 20$ kpc) for most of their duration. Only relatively late in a merger do they become strongly concentrated.

For example, any good *HST*, Spitzer, or Chandra image of NGC 4038/39 shows that enhanced star formation extends over a projected area of $\sim 8 \times 11$ kpc (Fig. 2). In the optical, Hα and blue images are best at showing the extended nature of the starburst. In the infrared, a Spitzer/IRAC image at 8μm

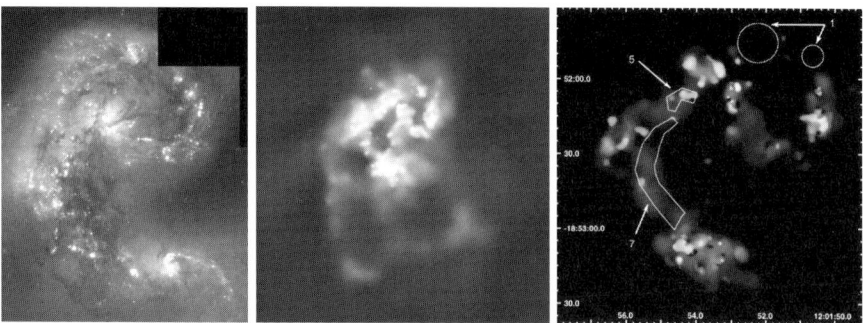

Figure 2. Extended starburst in NGC 4038/39, as imaged by *(left)* HST ($\sim 8 \times 11$ kpc field of view) and *(middle)* Chandra. *Right:* Metallicity map for hot ISM, showing spotty chemical enrichment (from Whitmore et al. 1999, Fabbiano et al. 2004, and Baldi et al. 2005).

emphasizes the warm dust associated with star formation throughout the two disks, glowing especially bright in the optically obscured disk contact ("Overlap") region (Wang et al. 2004). And in X-rays, a deep Chandra image displays not only two disks filled with superbubbles of hot gas (typical diameter ~ 1.5 kpc, $T \approx 5 \times 10^6$ K, $M \approx 10^{5-6}$ M$_\odot$), but also two giant, 10 kpc-size loops extending to the south (Fig. 2, middle panel). Their nature remains unclear (wind blown, or tidal ejecta?). These images illustrate that early in a merger the extended starburst heats the ISM in a chaotic manner, rather than leading to well-directed bipolar superwinds.

An interesting consequence of the extended starburst in NGC 4038/39 is the *spotty chemical enrichment* of the hot ISM, observed for the first time in any merger galaxy (Fabbiano et al. 2004, Baldi et al. 2005). The high signal-to-noise (S/N) ratio of the Chandra emission-line spectra permits the determination of individual Fe, Ne, Mg, and Si abundances in ~ 20 regions. Figure 2 (3rd panel, also color fig. 3 in Fabbiano et al. 2004) shows a metallicity map, with various shades of gray marking individual elements. The α elements are enhanced by up to $20 - 25\times$ solar and follow an enhancement pattern distinctly different from Fe, as one would expect if they were recently produced by SNe II. A question for future study is how such spotty chemical enrichment may affect stars still to form.

Another important consequence of the large spatial extent of merger-induced starbursts is that newly-formed stars decouple from the inward-trending gas continuously and at many different radii. This process differs sharply from the widely held misconception that such starbursts occur mainly in the central kiloparsec, where they are being fueled by infalling gas. As a result of this extended star formation, radial age gradients in merger remnants are weak (e.g., Schweizer 1998). This is also the likely reason why in ellipticals age gradients

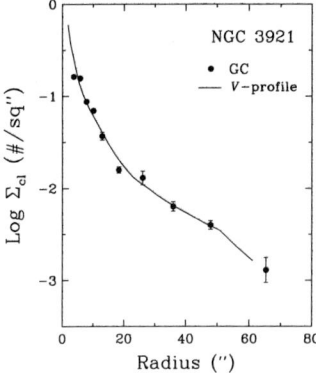 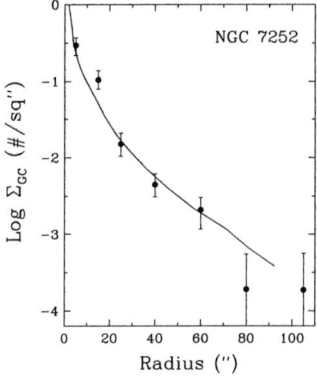

Figure 3. Radial distributions of second-generation globular clusters (data points) and background V-band light (lines) in the merger remnants NGC 3921 (Schweizer et al. 1996) and NGC 7252 (Miller et al. 1997). The far-flung cluster distributions are remnant signatures of extended starbursts.

are near zero, and mean metallicity gradients are only $\sim 40\%$ per decade in radius (Davies et al. 1993, Trager et al. 2000, Mehlert et al. 2003).

The strongest evidence linking extended starbursts to merger remnants and ellipticals is the wide radial distribution of the resultant star clusters. In both remnants (Fig. 3) and Es, second-generation metal-rich globular clusters track the underlying light distribution of their host galaxies with surprising accuracy. It is true that they tend to be more centrally distributed than metal-poor globulars, but only by little. Typically, half of them lie within $R_{\rm eff} \approx 3 - 5$ kpc from the center. This is consistent with some additional gaseous dissipation, but completely inconsistent with nuclear-only ($\lesssim 1$ kpc) starbursts. Hence, wide-flung globular-cluster systems are signatures of ancient extended starbursts.

4. Cosmological Implications

In 1972, Toomre & Toomre put forth the bold hypothesis that most giant ellipticals might be the remnants of major disk–disk mergers. Toomre (1977) elaborated on this idea, proposing a sequence of 11 increasingly merged disk pairs and refining the argument that from the current merger rate one could expect between 1/3 and all local ellipticals to be remnants of ancient mergers. Much evidence has since been accumulated to support this hypothesis.

Yet, beginning with the 1996 release of the Hubble Deep Field data, a new generation of astronomers has begun to study galaxy formation directly at high redshifts, often with remarkable success, but too often also making claims about elliptical formation that run afoul of the merger hypothesis and its strong supporting evidence in the local Universe. For example, claims about (i) an

"E formation epoch" ending around $z \approx 2$ and (ii) constant co-moving space densities of ellipticals since then abound, but are clearly mistaken.

Few astronomers would contest that disk–disk mergers are occurring locally ($z \approx 0$), and form remnants remarkably similar to young ellipticals. Evidence that some field Es contain intermediate-age stellar populations is also increasingly being accepted. What remains controversial is how most older ellipticals formed, say the majority that formed during the first half of the Hubble time and now appear uniformly old, crammed as they are into a small, 0.3 dex logarithmic age interval. Did they form by major disk mergers as well, or did they form by a process more akin to "monolithic collapse"?

First, the similarities between recent, $\lesssim 1$ Gyr-old merger remnants like NGC 3921 or NGC 7252 and giant Es (e.g., Toomre 1977, Schweizer 1982, 1996, Barnes 1998) are worth re-emphasizing. The above two remnants currently have luminosities of $\sim 2.8 L_V^*$ and will still shine with $\sim 1.0 L_V^*$ after 10–12 Gyr of evolution. They feature $r^{1/4}$-type light distributions, power-law cores, $UBVI$ color gradients, and velocity dispersions typical of Es. Both also possess many young, metal-rich halo globular clusters. They show integrated "E + A" spectra indicative of $b \gtrsim 10\%$ starbursts (Fritze-v. Alvensleben & Gerhardt 1994), as do many other similar young merger remnants in the local Universe (Zabludoff et al. 1996). Hence, claiming that E formation ceased around $z \approx 2$ is as mistaken as would be any claim that star formation ceased then. Local starbursts and merger remnants tell a different story.

Second, although the age distribution of local E and S0 galaxies is clearly weighted toward old ages, it does show a tail of youngish galaxies, especially in the field, with luminosity-weighted mean population ages of $\sim 1.5 - 5$ Gyr (González 1993, Trager et al. 2000, Kuntschner et al. 2002). Hence, in the field E + S0 formation has clearly not ceased yet.

Third and to astronomers' surprise, massive disk galaxies not unlike the Milky Way have been discovered at $1.4 \lesssim z \lesssim 3.0$ (Labbé et al. 2003) and thus *were available for major mergers at the epoch of peak QSO formation.* These galaxies seem to represent $\sim$ half of all galaxies with $L_V \geq 3L_V^*$ at those redshifts. Complementing such IR–optical observations, Genzel et al. (2003) have found a large disk galaxy at $z = 2.8$, whose rapidly rotating CO disk indicates a dynamical mass of $\gtrsim 3 \times 10^{11}$ M$_\odot$. Even more surprising is a massive *old* disk galaxy at $z = 2.48$ that shows a pure exponential disk ($\alpha \approx 1.7$ kpc) and no bulge, has a luminosity of $\sim 2L_V^*$, and has not formed stars for the past ~ 2 Gyr (Stockton et al. 2004). This galaxy indicates that massive Milky Way-size disks were available for E formation through major mergers even at $z > 3$.

With this high-z availability of disks and the above evidence that disk mergers continue to form E-like remnants to the present epoch, it is instructive to revisit Toomre's (1977) argument that most ellipticals may be merger rem-

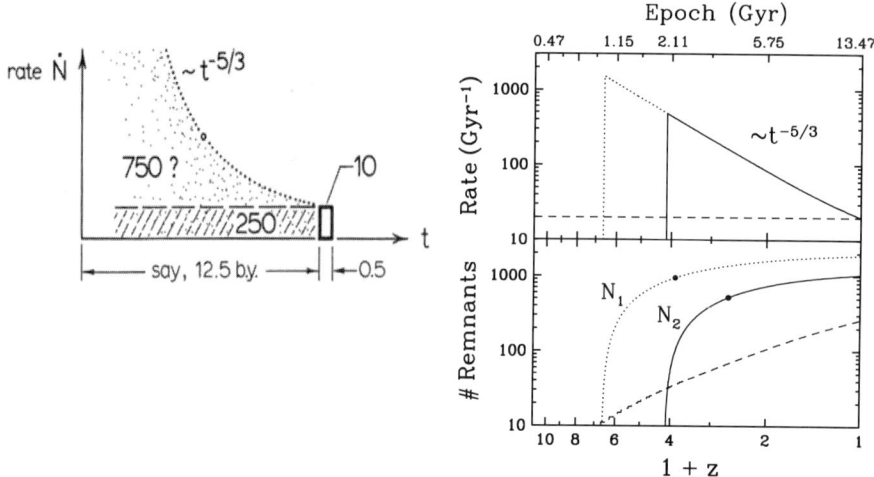

Figure 4. Merger rates and numbers of merger remnants as functions of cosmic epoch. *Left:* Toomre's (1977) original sketch; *Right:* a modern version of it. For details, see text.

nants. Figure 4 shows, to the left, his original sketch of the merger rate, $\dot{N}(t)$, as a function of time, t, and, to the right, a modern version of it, in which I have plotted the rate and computed number of remnants vs. $(1+z)$. From the ~ 10 ongoing disk–disk mergers among $\sim 4000+$ NGC galaxies and their mean age of ~ 0.5 Gyr, Toomre argued that there should be *at least* 250 remnants among these NGC galaxies, had the rate stayed constant, and more likely ~ 750 remnants if the merger rate declined as $t^{-5/3}$ (assuming a flat distribution of binding energy for binary galaxies). The latter number being close to the number of Es in the catalog, he suggested that *most* such galaxies may be old merger remnants.

The top panel of the modern diagram shows the same rate, $\dot{N} \propto t^{-5/3}$, plotted vs. $(1+z)$ assuming that major disk merging began 1 Gyr (*dotted lines*) or 2 Gyr (*solid*) after the Big Bang, with the corresponding numbers of remnants labeled N_1 and N_2 in the bottom panel. Dashed lines mark the case of constant $\dot{N}(t)$ for comparison, and epochs for the standard ΛCDM cosmology are given at the top. Notice that major disk mergers beginning at 1 Gyr ($z \approx 5.6$) would produce more remnants than needed to explain all Es among present-day NGC galaxies, while such mergers beginning at 2 Gyr ($z = 3.15$) would produce just about the right number. Interestingly, half of the N_2 remnants would already have formed at $z = 1.64$ (*dot* on N_2 curve), which may explain why observers are having a hard time deciding whether the co-moving space density of Es changes from $z \approx 1.5$ to 0 or not.

In summary, with massive disk galaxies already present at $z \gtrsim 3$, major disk mergers must have contributed to a growing population of elliptical galaxies ever since. Like star formation, E formation through major mergers is an ongoing process in which gaseous dissipation and starbursts play crucial roles.

Acknowledgments. I thank A. Baldi, J.E. Barnes, G. Fabbiano, A. Toomre, and B.C. Whitmore for permission to reproduce figures, and gratefully acknowledge support from the NSF through Grant AST–02 05994.

References

Baldi A., Raymond J.C., Fabbiano G., Zezas A., Rots A.H., Schweizer F., King A.R., Ponman T.J., 2005, ApJ, in press (astro-ph/0410192)
Barnes J.E., 1998, in: Galaxies: Interactions and Induced Star Formation, Friedli D., Martinet L., Pfenniger D., eds., (Springer: Berlin), p. 275
Barnes J.E., 2004, MNRAS, 350, 798
Barnes J.E., Hernquist L.E., 1991, ApJ, 370, L65
Barnes J.E., Hernquist L., 1996, ApJ, 471, 115
Chyży K.T., Beck R., 2004, A&A, 417, 541
Davies R.L., Sadler E.M., Peletier R.F., 1993, MNRAS, 262, 650
Elmegreen B.G., Efremov Y.N., 1997, ApJ, 480, 235
Fabbiano G., et al., 2004, ApJ, 605, L21
Fritze-v. Alvensleben U., Gerhard O.E., 1994, A&A, 285, 775
Genzel, R., Baker A.J., Tacconi L.J., Lutz D., Cox P., Guilloteau S., Omont A., 2003, ApJ, 584, 633
González J.J., 1993, Ph.D. thesis, UC Santa Cruz, USA
Hernquist L., 1989, Nature, 340, 687
Holtzman J.A., et al., 1992, AJ, 103, 691
Jog C.J., Solomon P.M., 1992, ApJ, 387, 152
Joseph R.D., Wright G.S., 1985, MNRAS, 214, 87
Kumai Y., Basu B., Fujimoto M., 1993, ApJ, 404, 144
Kuntschner H., Smith R.J., Colless M., Davies R.L., Kaldare R., Vazdekis A., 2002, MNRAS, 337, 172
Labbé I., et al., 2003, ApJ, 591, L95
Lambas D.G., Tissera P.B., Alonso M.S., Coldwell G., 2003, MNRAS, 346, 1189
Larson R.B., Tinsley B.M., 1978, ApJ, 219, 46
Mehlert D., Thomas D., Saglia R.P., Bender R., Wegner G., 2003, A&A, 407, 423
Mihos J.C., Hernquist L., 1996, ApJ, 464, 641
Miller B.W., Whitmore B.C., Schweizer F., Fall S.M., 1997, AJ, 114, 2381
Negroponte J., White S.D.M., 1983, MNRAS, 205, 1009
Nikolic B., Cullen H., Alexander P., 2004, MNRAS, 355, 874
Noguchi M., 1988, A&A, 203, 259
Roberts M.S., Haynes M.P., 1994, ARA&A, 32, 115
Schweizer F., 1982, ApJ, 252, 455
Schweizer F., 1987, in: Nearly Normal Galaxies, Faber S.M., ed., (Springer: New York), p. 18
Schweizer F., 1996, AJ, 111, 109
Schweizer F., 1998, in: Galaxies: Interactions and Induced Star Formation, Friedli D., Martinet L., Pfenniger D., eds., (Springer: Berlin), p. 105

Schweizer F., Miller B.W., Whitmore B.C., Fall S.M., 1996, AJ, 112, 1839
Shakhbazian R.K., 1970, Afz, 6, 367 (English transl.: Astroph., 6, 195)
Solomon P., Vanden Bout P., Carilli C., Guelin M., 2003, Nature, 426, 636
Stockton A., Canalizo G., Maihara T., 2004, ApJ, 605, 37
Toomre A., Toomre J., 1972, ApJ, 178, 623
Toomre A., 1977, in: The Evolution of Galaxies and Stellar Populations, Tinsley B.M., Larson R.B., eds., (Yale Univ. Obs.: New Haven), p. 401
Trager S.C., Faber S.M., Worthey G., González J.J., 2000, AJ, 119, 1645
Walter F., et al., 2003, Nature, 424, 406
Wang Z., et al., 2004, ApJS, 154, 193
Whitmore B.C., Schweizer F., Leitherer C., Borne K., Robert C., 1993, AJ, 106, 1354
Whitmore B.C., Zhang Q., Leitherer C., Fall S.M., Schweizer F., Miller B.W., 1999, AJ, 118, 1551
Young J.S., Scoville N.Z., 1991, ARA&A, 29, 581
Zabludoff A., Zaritsky D., Lin H., Tucker D., Hashimoto Y., Shectman S.A., Oemler A., Kirshner R.P., 1996, ApJ, 466, 104

GALAXY COLLISIONS: MODELING STAR FORMATION IN DIFFERENT ENVIRONMENTS

Chris Mihos
Department of Astronomy, Case Western Reserve University, USA
mihos@case.edu

Abstract   Starbursts in interacting galaxies are mediated by the dynamical response of the host galaxy to the collisional perturbation. Environmental effects may lead to systematic differences in the nature of the encounter such that the properties of induced starbursts are likely to be different in cluster and field environments. I first discuss these differences in the context of numerical simulations of interacting galaxies, then describe the role of cluster tides in influencing interaction-induced star formation.

1. Introduction

Collisions and mergers of galaxies provide an efficient mechanism for driving starburst activity in galaxies. Interacting galaxies show evidence for elevated star-formation rates in all star-formation tracers, including Hα luminosities, infrared emission, and radio continuum. Samples of ultraluminous infrared galaxies are dominated by interacting systems, with a strong tendency toward late-stage mergers (see, e.g., the review by Kennicutt 1998). However, the *detailed* star-forming properties of interacting systems depend on a variety of factors, and can be very different between the cluster and field environments.

2. Interactions: The Dynamical Trigger

Computational modeling of interacting galaxies has illustrated the dynamical mechanism by which interactions drive starburst activity in disk galaxies. After the initial encounter, the self-gravity of the disk amplifies the $m = 2$ perturbation of the companion into spiral arms or a central bar, which mediates an inflow of gas to the central regions of the disk. If the interaction is strong enough that the galaxies ultimately merge, a second phase of inflow can occur during the coalescence phase. Strong starbursts are not confined to this late phase, however; depending on the strength of the initial inflow, strong central starbursts can result well before the galaxies merge.

Because of the connection between the self-gravity of the disk and the triggering of inflows, the detailed pattern of star formation in interacting systems depends strongly on the structure of the host galaxies (Mihos & Hernquist 1996, Mihos et al. 1997). Suppressing the self-gravity of the disk, either through the presence of a dominant central bulge or reduced disk surface density, results in milder inflows. This shifts the mode of star formation from central starbursts to a milder, disk-wide increase in star-forming activity.

Figure 1 illustrates this behavior, showing the star-forming response of a pair of equal-mass galaxies on a prograde orbit (disks inclined 45° to the orbital plane) with closest approach of $r_{\rm peri} = 5$ disk scale lengths. In this encounter, one galaxy is a pure disk + dark halo model, while the other galaxy possesses a central bulge (with a 1:3 bulge:disk ratio). The curves marked "Slow" show the evolution of the star-formation rate during a slow, parabolic encounter representative of collisions in field galaxies. In this encounter, the disk-dominated model (left) develops an intense central starburst shortly after perigalacticon (at $t = 24$), while the disk/bulge model shows a more modest (but still significant) increase in the global SFR.

Certainly, the detailed star-formation evolution in models such as these are sensitive to the parameterization of the star-formation law and treatment of feedback from supernovae and stellar winds into the ISM. Nonetheless, the overall picture of the timing and relative intensity of starbursts being mediated by disk instabilities has been proven fairly robust in a variety of computational studies with different approaches towards star formation and feedback (Mihos & Hernquist 1996, Bekki 1998, Gerritsen 1998, Springel 2000; although see interesting differences noted by Barnes 2004).

3. Cluster vs. Field Interactions

While the internal dynamics of the interacting galaxies play a crucial role in the star-forming response, so too do the parameters of the interaction. Of particular interest is the relative encounter velocity; the encounter velocities in galaxy clusters can be much higher than in the field, making the perturbation more impulsive in nature. To investigate the resulting behavior of the star-formation rate, the simulations described above were repeated with a hyperbolic encounter velocity twice that of the slow, parabolic encounter. The curves marked "Fast" in Fig. 1 show the star-formation evolution in this simulation. In the pure disk galaxy, the disk still amplifies the perturbation into a strong bar (although it takes longer to develop), driving a central starburst. The model with a central bulge, however, proves quite impervious to the interaction: the combination of a fast encounter and the stabilizing effect of a central bulge means that the disk/bulge galaxy experiences little evolution in its SFR at all.

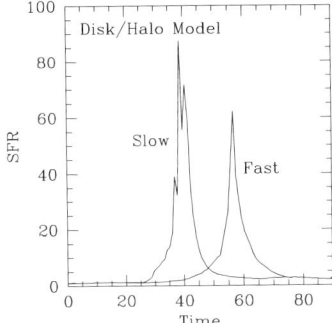 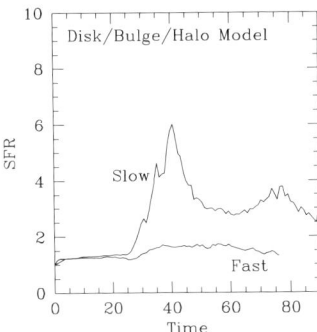

Figure 1. Evolution of the global star-formation rate (in relative units) vs. time for slow and fast collisions of equal mass disk galaxies with and without central bulges. Unit time is roughly 13 Myr, and closest encounter occurs at $t = 24$. The encounters are distant enough that the galaxies do not merge over the timescales shown here. Note the scale change in the vertical axis between the two figures.

These simulations suggest that the evolution of interacting galaxies depends strongly on both the environment and the nature of the progenitor. Large, early-type cluster spirals may be relatively impervious to collisionally-induced starbursts, while late-type spirals will react more strongly. If late-type low surface brightness spirals are preferentially dark matter dominated (de Blok & Mc-Gaugh 1998), the suppression of disk instabilities may reduce their response over that shown in the pure disk/halo models in Fig. 1 (Mihos et al. 1997), but in this case *repeated* encounters may ultimately drive strong evolution (Moore et al. 1996).

Other caveats remain. It is not clear at all that the *typical* encounter velocity of infalling cluster galaxies should be so high; galaxies are accreted in infalling groups, where the relative encounter velocities may be more like those in the field. Also, ram-pressure stripping by the hot intracluster gas may deprive galaxies of gas to fuel a starburst response; however, many blue and post-starburst galaxies in clusters appear to be interacting systems (e.g., Lavery & Henry 1994, Caldwell et al. 1999), which argues that ram-pressure stripping cannot completely inhibit interaction-induced starburst activity.

4. The Role of Tidal Debris

Aside from star formation induced in the main body of the galaxies, the tidal debris may also participate in the star-forming response of the system. Star-forming knots have been observed in many tidal tails, and have been suggested to be the progenitors of dwarf galaxies (e.g., Duc et al. 2000). Additionally, gas expelled in tidal tails during a merger will fall back into the remnant over long timescales (Hibbard & Mihos 1995), leading to late star formation and perhaps the reformation of low-mass disks (Hibbard et al. 1994, Schweizer 1998, Barnes 2002).

These scenarios have been well-demonstrated in field mergers through numerical simulations showing compression and collapse of tidal debris (Barnes & Hernquist 1992, Elmegreen et al. 1993) as well as late infall and disk formation (Mihos & Hernquist 1996, Barnes 2002), the efficacy of these mechanisms in the cluster environment will be strongly suppressed. The tidal field of the cluster can efficiently strip the loosely-bound tidal debris from interacting systems, even well outside the cluster core. Factor in ram-pressure stripping of the low density, patchy gas in the tidal debris, and the possibility for star formation in the tidal debris is small indeed – the efficacy of building tidal dwarfs in this environment is questionable. Similarly, late infall of stripped material is strongly curtailed — disk rebuilding (like that which may be occurring in Centaurus A) will be impossible in cluster galaxies. This presents significant problems for models for the production of cluster S0's which rely on the re-accretion or reformation of disk gas during strong encounters with the cluster environment.

References

Barnes J.E., 2002, MNRAS, 333, 481
Barnes J.E., 2004, MNRAS, 350, 798
Barnes J.E., Hernquist L.E., 1992, Nature, 360, 715
Bekki K., 1998, ApJ, 502, L133
Caldwell N., Rose J.A., Dendy K., 1999, AJ, 117, 140
de Blok W.J.G., McGaugh S.S., 1998, ApJ, 499, 41
Duc P.-A., Brinks E., Springel V., Pichardo B., Weilbacher P., Mirabel I.F., 2000, AJ, 120, 1238
Elmegreen B.C., Kaufman M., Thomasson M., 1993, ApJ, 412, 90
Gerritsen J., 1998, Ph.D. thesis, University of Groningen, the Netherlands
Hibbard J., Guhathakurta P., van Gorkom J.H., Schweizer F., 1994, AJ, 107, 67
Hibbard J., Mihos J.C., 1995, AJ, 110, 140
Kennicutt R.C., 1998, in: Saas-Fee Advanced Course 26, Galaxies: Interactions and Induced Star Formation, (Geneva Obs. Press: Geneva), p. 1
Lavery R.J., Henry J.P., 1994, ApJ, 426, 524
Mihos J.C., Hernquist L., 1996, ApJ, 464, 641
Mihos, J.C., McGaugh S.S., de Blok W.J.G., 1997, ApJ, 477, L79
Moore B., Katz N., Lake G., Dressler A., Oemler Jr. A., 1996, Nature, 379, 613
Schweizer F., 1998, in: Saas-Fee Advanced Course 26, Galaxies: Interactions and Induced Star Formation, (Geneva Obs. Press: Geneva), p. 105
Springel V., 2000, MNRAS, 312, 859

STAR AND CLUSTER FORMATION IN EXTREME ENVIRONMENTS

Richard de Grijs
Department of Physics & Astronomy, University of Sheffield, Hicks Building, Hounsfield Road, Sheffield S3 7RH, UK
R.deGrijs@sheffield.ac.uk

Abstract   Current empirical evidence on the star-formation processes in the extreme, high-pressure environments induced by galaxy encounters (mostly based on high-resolution *Hubble Space Telescope* observations) strongly suggests that star *cluster* formation is an important and perhaps even the dominant mode of star formation in such starburst events. The sizes, luminosities, and mass estimates of the young massive star clusters (YMCs) are entirely consistent with what is expected for young Milky Way-type globular clusters (GCs). Recent evidence lends support to the scenario that GCs, which were once thought to be the oldest building blocks of galaxies, are still forming today. Here, I present a novel empirical approach to assess the shape of the initial-to-current YMC mass functions, and hence their possible survival chances for a Hubble time.

1. Star clusters as starburst tracers

The production of luminous, massive yet compact star clusters seems to be a hallmark of the most intense star-forming episodes in galaxies. Young massive star clusters (YMCs; with masses often significantly exceeding $M_{cl} = 10^5 M_\odot$) are generally found in intense starburst regions, often in galaxies involved in gravitational interactions of some sort (e.g., de Grijs et al. 2001, 2003a,b,c,d,e, and references therein).

An increasingly large body of observational evidence suggests that a large fraction of the star formation in starbursts actually takes place in the form of such concentrated clusters, rather than in small-scale star-forming "pockets". YMCs are therefore important as benchmarks of cluster formation and evolution. They are also important as tracers of the history of star formation of their host galaxies, their chemical evolution, the initial mass function (IMF), and other physical characteristics in starbursts.

Using optical observations of the "Mice" and "Tadpole" interacting galaxies (NGC 4676 and UGC 10214, respectively) – based on a subset of the Early Release Observations obtained with the Advanced Camera for Surveys on board the *Hubble Space Telescope (HST)* – and the novel technique of pixel-by-pixel analysis of their colour-colour and colour-magnitude diagrams, we deduced the systems' star and star cluster formation histories (de Grijs et al. 2003e). In both of these interacting systems we find several dozen YMCs (or, alternatively, compact star-forming regions), which overlap spatially with regions of active star formation in the galaxies' tidal tails and spiral arms (from a comparison with Hα observations that trace active star formation; Hibbard & van Gorkom 1996). The tidal tail of the Tadpole system is dominated by star-forming regions, which contribute $\sim 70\%$ of the total flux in the *HST* F814W filter (decreasing to $\sim 40\%$ in the F439W filter). If the encounter occurs between unevenly matched, gas-rich galaxies then, as expected, the effects of the gravitational interaction are much more pronounced in the smaller galaxy. For instance, when we compare the impact of the interaction as evidenced by star cluster formation between M82 (de Grijs et al. 2001, 2003b,c) and M81 (Chandar et al. 2001), or the star cluster formation history in M51 (Bik et al. 2003), which is currently in the process of merging with the smaller spiral galaxy NGC 5194, the evidence for enhanced cluster formation in the larger galaxy is minimal if at all detectable.

Nevertheless, we have shown that star cluster formation is a major mode of *newly-induced* star formation in galactic interactions, with $\geq 35\%$ of the active star formation in encounters occurring in YMCs (de Grijs et al. 2003e).

The question remains, however, whether or not at least a fraction of the numerous compact YMCs seen in extragalactic starbursts, may be the progenitors of GC-type objects. If we could settle this issue convincingly, one way or the other, the implications of such a result would have profound and far-reaching implications for a wide range of astrophysical questions, including (but not limited to) our understanding of the process of galaxy formation and assembly, and the process and conditions required for star (cluster) formation. Because of the lack of a statistically significant sample of similar nearby objects, however, we need to resort to either statistical arguments or to the painstaking approach of one-by-one studies of individual objects in more distant galaxies.

2. From YMC to old globular cluster?

The present state-of-the-art teaches us that the sizes, luminosities, and – in several cases – spectroscopic mass estimates of most (young) extragalactic star cluster systems are fully consistent with the expected properties of young Milky Way-type GC progenitors.

However, the postulated evolutionary connection between the newly formed YMCs in intensely star-forming areas, and old GCs similar to those in the Galaxy is still a contentious issue. The evolution and survivability of YMCs depend crucially on the stellar IMF of their constituent stars (cf. Smith &

Gallagher 2001): if the IMF is too shallow, i.e., if the clusters are significantly depleted in low-mass stars compared to (for instance) the solar neighbourhood, they will disperse within a few (galactic) orbital periods, and likely within about a billion years of their formation (e.g., Smith & Gallagher 2001, Mengel et al. 2002). Ideally, one would need to obtain (i) high-resolution spectroscopy of all clusters in a given cluster sample in order to obtain dynamical mass estimates (we will assume, for the purpose of the present discussion, that our YMCs are fully virialised) and (ii) high-resolution imaging (e.g., with the *HST*) to measure their luminosities and sizes. However, individual YMC spectroscopy, while feasible today with 8m-class telescopes for the nearest systems, is very time-consuming, since observations of large numbers of clusters are required to obtain statistically significant results. Instead, one of the most important and most widely used diagnostics, both to infer the star (cluster) formation history of a given galaxy, and to constrain scenarios for its expected future evolution, is the distribution of cluster luminosities, or – alternatively – their associated masses, commonly referred to as the cluster luminosity and mass functions (CLF, CMF), respectively.

Starting with the seminal work by Elson & Fall (1985) on the young cluster system in the Large Magellanic Cloud (LMC; with ages $\leq 2 \times 10^9$ yr), an ever increasing body of evidence, mostly obtained with the *HST*, seems to imply that the CLF of YMCs is well described by a power law. On the other hand, for the old GC systems in the local Universe, with ages ≥ 10 Gyr, the CLF shape is well established to be roughly lognormal (Whitmore et al. 1993, Harris 1996, 2001, Harris et al. 1998).

This type of observational evidence has led to the popular – but thus far mostly speculative – theoretical prediction that not only a power-law, but *any* initial CLF (and CMF) will be rapidly transformed into a lognormal distribution (e.g., Elmegreen & Efremov 1997, Gnedin & Ostriker 1997, Ostriker & Gnedin 1997, Fall & Zhang 2001). We recently reported the first discovery of an approximately lognormal CLF (and CMF) for the star clusters in M82's fossil starburst region "B", formed roughly simultaneously in a pronounced burst of cluster formation (de Grijs et al. 2003b; see also Goudfrooij et al. 2004). This provides the very first sufficiently deep CLF (and CMF) for a star cluster population at intermediate age (of ~ 1 Gyr), which thus serves as an important benchmark for theories of the evolution of star cluster systems.

The CLF shape and characteristic luminosity of the M82 B cluster system is nearly identical to that of the apparently universal CLFs of the old GC systems in the local Universe. This is likely to remain virtually unchanged for a Hubble time, if the currently most popular cluster disruption models hold. With the very short characteristic cluster disruption time-scale governing M82 B (de Grijs et al. 2003c), its cluster mass distribution will evolve toward a higher characteristic mass scale than that of the Galactic GC system by the time it reaches a similar age. Thus, this evidence, combined with the similar cluster sizes (de Grijs et al. 2001), lends strong support to a scenario in which the current M82 B cluster population will eventually evolve into a significantly depleted old Milky Way-type GC system dominated by a small number of high-mass clusters (de Grijs et al. 2003b). This implies that (metal-rich) GCs, which were once thought to be the oldest building blocks of galaxies, are still forming today in galaxy interactions and mergers.

3. The $L_V - \sigma_0$ relation as a diagnostic tool

We have recently started to explore a new, empirical approach to assess the long-term survival chances of YMCs formed profusely in intense starburst environments (de Grijs, Wilkinson & Tadhunter 2005). The method hinges on the empirical relationship for old Galactic and M31 GCs, which occupy tightly constrained loci in the plane defined by their V-band luminosities, L_V (or, equivalently, absolute magnitudes, M_V) and central velocity dispersions, σ_0 (Djorgovski et al. 1997 and references therein; see Fig. 1).

Encouraged by the tightness of the GC relationship, we also added the available data points for the YMCs in the local Universe, including nuclear star clusters (objects 1–5), for which velocity dispersion information was readily available. In order to be able to compare them to the ubiquitous old Local Group GCs, we evolved their luminosities to a common age of 12 Gyr, adopting the "standard" Salpeter IMF covering masses from 0.1 to 100 M$_\odot$, and assuming stellar evolution as described by the GALEV SSPs (cf. Anders & Fritze–v. Alvensleben 2003). Based on a careful assessment of the uncertainties associated with this luminosity evolution, we conclude that the most important factor affecting the robustness of our conclusions is the adopted form of the stellar IMF.

We find that if we adopt the universal solar neighbourhood IMF as the basis for the YMCs' luminosity evolution, the large majority will evolve to loci within twice the observational scatter around the best-fitting GC relationship. In the absence of significant external disturbances, this implies that these objects may potentially survive to become old GC-type objects by the time they reach a similar age. Thus, these results provide additional support to the suggestion that the formation of proto-GCs appears to be continuing until the present. Detailed one-to-one comparisons between our results based on this new method with those obtained previously and independently based on dynamical mass estimates and mass-to-light (M/L) ratio considerations lend strong support to the feasibility and robustness of our new method. The key characteristic and main advantage of this method compared to the more complex analysis involved in using dynamical mass estimates for this purpose is its simplicity and empirical basis. Where dynamical mass estimates require

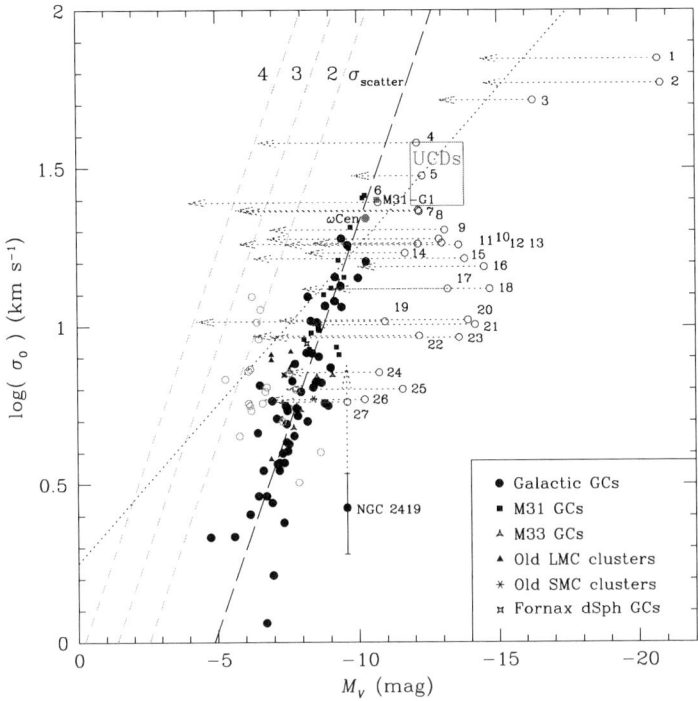

Figure 1. Diagnostic figure used to predict the chances of survival to old GC-type ages for YMCs with (central) velocity dispersion measurements available in the literature. The filled symbols correspond to the old GCs in the Local Group (see legend); the best-fitting relation for these old clusters is shown by the long-dashed line. The short-dashed lines are displaced from this best-fitting relationship by, respectively, 2, 3, and 4 times the scatter in the data points around the best-fitting line, $\sigma_{\rm scatter}$. The dotted line is the Faber-Jackson relationship for elliptical galaxies (see de Grijs et al. 2005), which bisects the locus of the ultracompact dwarf galaxies (UCDs). The numbered open circles are the locations of the YMCs with measured velocity dispersions, which we have evolved to a common age of 12 Gyr (represented by the dotted arrows) using the GALEV SSP models for the appropriate metallicity and age of these objects. The remaining open circles are the young compact clusters in the LMC and SMC. The most massive GCs in both the Galaxy and M31 (ω Cen and G1, respectively) are also indicated.

one to obtain accurate size estimates and to make assumptions regarding a system's virialised state and M/L ratio, these complications can now be avoided by using the empirically determined GC relationship as reference. The only observables required are the system's (central or line-of-sight) velocity dispersion and photometric properties.

Careful analysis of those YMCs that would overshoot the GC relationship significantly if they were to survive for a Hubble time show that their unusually high ambient density likely has already had a devastating effect on their stellar content, despite their young ages, thus altering their present-day mass function (PDMF) in a such a way that they have become unstable to survive for any significant length of time. This is, again, supported by independent analyses, thus further strengthening the robustness of our new approach. The expected loci in the $L_V - \sigma_0$ plane that these objects would evolve to over a Hubble time are well beyond any GC luminosities for a given velocity dispersion, thus leading us to conclude that they will either dissolve long before reaching GC-type ages, or that they must be characterised by a PDMF that is significantly depleted in low-mass stars (or highly mass segregated). This, therefore, allows us to place moderate limits on the functionality of their PDMFs.

References

Anders P., Fritze–v. Alvensleben U., 2003, A&A, 401, 1063
Bik A., Lamers H.J.G.L.M., Bastian N., Panagia N., Romaniello M., 2003, A&A, 397, 473
Chandar R., Ford H.C., Tsvetanov Z., 2001, AJ, 122, 1330
de Grijs R., O'Connell R.W., Gallagher J.S., 2001, AJ, 121, 768
de Grijs R., Anders P., Lynds R., Bastian N., Lamers H.J.G.L.M., Fritze–v. Alvensleben U., 2003a, MNRAS, 343, 1285
de Grijs R., Bastian N., Lamers, H.J.G.L.M., 2003b, ApJ, 583, L17
de Grijs R., Bastian N., Lamers H.J.G.L.M., 2003c, MNRAS, 340, 197
de Grijs R., Fritze–v. Alvensleben U., Anders P., Gallagher J.S., Bastian N., Taylor V.A., Windhorst R.A., 2003d, MNRAS, 342, 259
de Grijs R., Lee J.T., Mora Herrera M.C., Fritze–v. Alvensleben U., Anders P., 2003e, New Astron., 8, 155
de Grijs R., Wilkinson M.I., Tadhunter C.N., 2005, MNRAS, subm.
Djorgovski S.G., Gal R.R., McCarthy J.K., Cohen J.G., de Carvalho R.R., Meylan G., Bendinelli O., Parmeggiani G., 1997, ApJ, 474, L19
Elmegreen B.G., Efremov Y.N., 1997, ApJ, 480, 235
Elson R.A.W., Fall, S.M., 1985, ApJ, 299, 211
Fall S.M., Zhang Q., 2001, ApJ, 561, 751
Gnedin O.Y., Ostriker J.P., 1997, ApJ, 474, 223
Goudfrooij P., Gilmore D., Whitmore B.C., Schweizer F., 2004, ApJ, 613, L121
Harris W.E., 1996, AJ, 112, 1487
Harris W.E., 2001, in: Star Clusters, Saas-Fee Advanced Course 28, (Springer: New York), p. 223
Harris W.E., Harris G.L.H., McLaughlin D.E., 1998, AJ, 115, 1801
Hibbard J.E., van Gorkom J.H., 1996, AJ, 111, 655
Mengel S., Lehnert M.D., Thatte N., Genzel R., 2002, A&A, 383, 137
Ostriker J.P., Gnedin O.Y., 1997, ApJ, 487, 667
Smith L.J., Gallagher J.S., 2001, MNRAS, 326, 1027
Whitmore B.C., Schweizer F., Leitherer C., Borne K., Robert C., 1993, AJ, 106, 1354

THE RECURRENT NATURE OF CENTRAL STARBURSTS

Curtis Struck
Department of Physics & Astronomy Iowa State University, Ames, IA, USA
curt@iastate.edu

Abstract New hydrodynamic models with feedback show that feedback-driven turbulence and subsequent relaxation can drive recurrent starbursts, although most of these bursts fizzle due to premature, asymmetric ignition. Strong bursts are terminated when the turbulence inflates the multiphase central disk. The period between bursts is about twice a free-fall time onto the central disk. Transient spirals and bars are common through the burst cycle.

There has been much discussion at this meeting on triggering mechanisms for starbursts. This topic is especially interesting when galaxy interactions, bar-driven inflows and other obvious sources of triggering are absent. Two other interesting questions are, what turns off starbursts, and are starbursts naturally recurrent given an adequate gas supply? And finally, if they are recurrent, is the process deterministic or stochastic?

There are many candidate mechanisms for terminating bursts (see review of Leitherer 2001), including (i) gas consumption, (ii) gas loss to the wind, (iii) conversion of cold gas to warm/hot phases, and (iv) inflation of the central disk without total conversion to hot phases (as in dwarf galaxy models). The latter two allow recurrence, the first two do not. Since there are many complex processes involved, with incomplete sampling in any one observational waveband, it is hard to assemble a complete picture. Many models have been made of particular parts of the coupled starburst plus wind phenomena.

Although models for all of the thermo-hydrodynamic aspects of the starburst/wind phenomena are not possible, exploratory models of some of the important couplings can shed light on the questions above. As a step towards that goal, I present here some preliminary results of numerical hydrodynamical models. These are N-body/SPH models carried out with the HYDRA3.0 code of Couchman, Thomas & Pearce (1995), which includes optically thin radiative cooling (Sutherland & Dopita 1993) for temperatures above 10^4K. The model galaxies consist of three components: a dark matter halo (10,000 collisionless particles), a stellar disk (9,550 collisionless particles), and a gas disk (9,550 SPH particles). The mass per particle is 6×10^6 M$_\odot$, and the total galaxy mass is 1.75×10^{11} M$_\odot$.

A feedback prescription was also added to the HYDRA code. A gas particle is marked as star forming in this prescription when its temperature is less than a threshold value of about 10^4K, its density is above a threshold value of about 0.14 cm^{-3}, and these thresholds are not re-crossed during a time of about 10^7 yr. The latter condition is a computationally cheap way of insuring that the cold, dense region is likely bound. The small value of the density threshold is a symptom of the modest particle resolution. Regions that are much denser are rarely resolved, but we assume that they exist within the densest regions.

Once the heating is initiated it is maintained for 10^7 yr, and the particle internal energy is increased by 10% in each time-step (typically about 10^5 yr) until a maximum temperature of about 10^6K is reached. This corresponds to an energy input of 1.1×10^{54} erg per star-forming particle. (This feedback formulation is discussed in more detail, and compared to others, in Smith [2001].)

If we assume that a typical supernova (SN) injects about 10^{51} erg into the gas, then about 1100 SNe are needed to generate this feedback energy. If we assume that the mass of a typical SN is about 10 M$_\odot$, and adopt a Salpeter mass function over a range of $0.2 - 100$ M$_\odot$, then about 10% of the stars are SN progenitors, and we need to form a star cluster of mass $\simeq 10^5$ M$_\odot$ to obtain the needed energy. This is about 2% of the gas particle mass. Gas consumption is not included in these models, or equivalently we assume instant replenishment of the (small) losses.

These energy estimates do not include losses due to cooling (until the heating is terminated). However, feedback heating should not be directly equated to thermal energy. It is also an algorithm for inducing increased kinetic energy and mass motions via local pressure effects, which begin at a "sub-grid" scale. Once the heating is turned off, the affected particles can cool rapidly due to both adiabatic expansion and radiative cooling. The net pressure generated by the feedback, and its effects on surrounding particles seem generally realistic, even if the thermal details are unresolved. In the future, these details must be resolved to determine quantities like the mass fractions and scale heights of different thermal phases.

The first result is that with sufficient gas supply, and reasonable feedback parameter values, the models are intrinsically bursty at a moderate level, see Fig. 1 (and the accompanying CD-ROM for supplementary material). The initial model was allowed to relax with an adiabatic equation of state plus cooling, but without feedback. The feedback was turned on at the start of these runs, which resulted in some large amplitude bursts. After that the models settled down to a more "steady, bursty" character with the following characteristic properties. (i) The largest bursts are roughly periodic, with a period of a bit less than 100 Myr. This period is much longer than the time delays of the feedback model. (ii) The amplitude of the largest bursts varies substantially. The horizontal line in the figure serves as a useful guide. The case shown by the dashed curve has about an equal number of burst peaks above this line and just below it. The case shown by the solid curve has bursts that peak significantly below the line for most of a Gyr, then several bursts that peak above it, including the double burst between the vertical lines. (iii) The burst durations are typically about 10 Myr (like the feedback time delay), but double or multiple echo bursts are not uncommon.

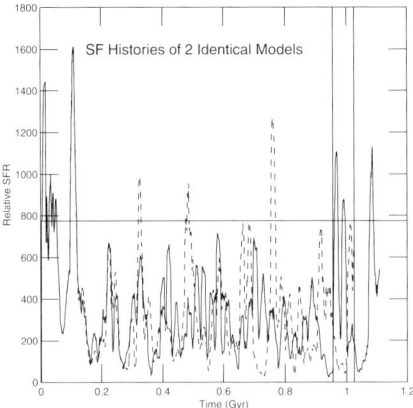

Figure 1. The number of actively heating (feedback) particles, assumed proportional to the SFR, as a function of time in two models with essentially identical initial conditions. The horizontal line in the figure is drawn at the level of the mean plus 2 standard deviations, see text.

I have analyzed the bursts that peak above the 2σ horizontal line, but excluding the initial transients, in more detail. Adopting the star-formation (SF) mass per particle and IMF described above, the stellar mass produced in these bursts ranges from $(4-50) \times 10^6$ $M_\odot$, with a mean of 30×10^6 $M_\odot$. The corresponding star-formation rate (SFR) ranges from $1.6 - 3.6$ $M_\odot$ yr^{-1}. These seem to be fairly reasonable values for an isolated core starburst in a late type galaxy, and so support the parameter values used in the feedback algorithm. However, a number of questions remain.

What determines the burst period? The burst period is slightly less than twice the free-fall time of a particle at $2-3$ kpc above the disk plane, or the time for a boosted gas particle to travel to the top of its trajectory and return to the plane, as observed in the models.

Why do the burst peaks vary so much? Viewed from above, the nuclear disk gas is usually found to be concentrated in flocculent spirals, and transient, rather irregular bars. After a starburst large bubbles and voids often appear, and the spirals and bars are disrupted or rearranged. A great deal of turbulence is generated. The next burst is more likely to be of large amplitude if these waves reform symmetrically and transfer a relatively large amount of gas to the center. More frequently, these waves develop asymmetrically, and an off-center gas concentration triggers a SF hot spot prematurely. This often leads to some propagating SF, but destroys the chance for a large burst. In the case, shown by the solid curve in Fig. 1, successive "fizzles" extend the time between large bursts to nearly 1.0 Gyr.

Where is the wind? Observations indicate that the wind mass is roughly equal to the mass of stars produced in the burst (see Strickland 2004). The hot wind is not resolved in these models. The models do suggest that the SF that generates it occurs in patchy concentrations. Thus, the wind is probably a sum of local gusts. This seems in accord with recent observations of M82 and Arp 284 presented at this meeting and observations of other starburst galaxies.

What about all the gas boosted out of the plane? First, there is evidence indicating that substantial masses of gas are kicked out over kpc distances by bursts. This evidence includes the large HI scale heights in M82 and NGC 2403 (Fraternali et al. 2004), extended dust distributions in edge-on galaxies (e.g., NGC 891; Howk & Savage 2000), evidence that molecular clouds are broken down but not destroyed in starburst regions (Gao & Solomon 2004), and the small filling factor of the hot gas in wind galaxies (Strickland 2004). The high-latitude gas in the models has a wide range of thermal phases.

In summary, the simulations suggest a mosaic model for core starbursts, with the following properties. (i) Star clusters form in the densest regions, create hot spots, which may eventually break out as wind gusts. (ii) Hot spots also drive turbulence over a wider area, and can propagate the SF. (iii) Eventually, central regions become so turbulently stirred, shredded and puffed up that SF crashes. (iv) If not too much gas is consumed or blown out, clouds reform and generate recurrent bursts. The models suggest that starbursts are naturally recurrent. We have not yet undertaken models of bursts driven by rapid mass transfer or merging, but it seems likely that they can overcome the fizzle effect and drive large burst amplitudes.

References

Couchman H., Thomas P., Pearce F., 1995, ApJ, 452, 797
Fraternali F., Oosterloo T., Sancisi R., 2004, A&A, 424, 485
Gao Y., Solomon P.M., 2004, ApJ, 606, 271
Howk J.C., Savage B.D., 2000, AJ, 119, 644
Leitherer C., 2001, in: Astrophysical Ages and Timescales, von Hippel T., Simpson C., Manset N., eds., ASP Conf. Ser., (ASP: San Francisco), vol. 245, p. 390
Smith D.C., 2001, Ph.D. thesis, Iowa State University
Strickland D.K., 2004, in: The Interplay among Black Holes, Stars, and the ISM in Galactic Nuclei, IAU Symp. 222, Storchi Bergmann L.C., Ho L.C., Schmitt H.R., (ASP: San Francisco), in press (astro-ph/0404316)
Sutherland R.S., Dopita M.A., 1993, ApJS, 88, 253

EFFICIENCY OF THE DYNAMICAL MECHANISM

F. Combes
LERMA, Observatoire de Paris, 61 Av. de l'Observatoire, F-75014 Paris, France
francoise.combes@obspm.fr

Abstract The most extreme starbursts occur in galaxy mergers, and it is now acknowledged that dynamical triggering has a primary importance in star formation. This triggering is due partly to the enhanced velocity dispersion provided by gravitational instabilities, such as density waves and bars, but mainly to the radial gas flows they drive, allowing large amounts of gas to condense towards nuclear regions on a short time scale. Numerical simulations including several gas phases, taking into account the feedback to regulate star formation, have explored the various processes, using recipes like the Schmidt law, moderated by the gas instability criterion. Perhaps the most fundamental parameter in starbursts is the availability of gas: this sheds light onto the amount of external gas accretion in galaxy evolution. The detailed mechanisms governing gas infall in the inner parts of galaxy disks are discussed.

1. Introduction

The most spectacular evidence for dynamical triggering of starbursts is that ULIRGs are all mergers of galaxies (e.g., Sanders & Mirabel 1996). They have much more gas, dust and young stars than normal spiral galaxies, but they are quite rare objects in the nearby Universe. However, interacting galaxies do not exhibit intense starbursts (e.g., Bergvall et al. 2003, but see Barton et al. 2000, Nikolic et al. 2004), or only in their centers. From many observational studies, it appears that galaxy interactions are a necessary condition, but not a sufficient condition to trigger a starburst. Another necessary condition, of course, is the presence of large amounts of gas.

For small systems, interactions are even not necessary, since spontaneous star formation can occur intermittently in bursts. Starbursting dwarf galaxies have no excess of companions (Telles & Maddox 2000, Brosch et al. 2004, except Blue Compact Dwarfs according to Hunter & Elmegreen 2004). Thus, for dwarf galaxies tides are not very important. It is possible instead that the gas in these objects, having been accreted recently, is not yet in dynamical equilibrium: observed asymmetries could be due to gas sloshing inside dark haloes.

It is possible today to trace the star formation and chemical enrichment history of nearby galaxies, by studying their stellar populations and their metallicity in detail. The star-formation history in the Small Magellanic Cloud reveals some bursts corresponding to pericenters with the Milky Way (Zaritsky & Harris 2004). The tidally-induced fraction of star formation could be between 10 and 70%. A good fit is impossible, however, without large amounts of gas infall, at least 50%. For two local dwarfs, Skillman et al. (2003) conclude also that the bulk of star formation is recent, unlike the predictions of an exponentially decreasing star-formation history, if the system had acquired most of its mass at early times (Fig. 1).

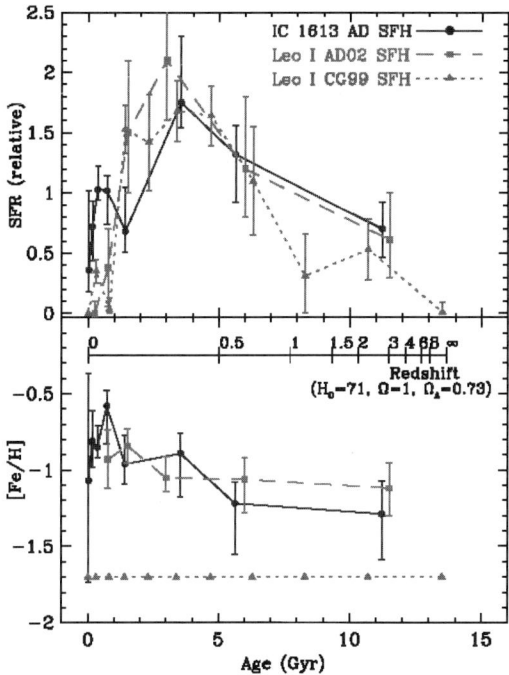

Figure 1. Star formation and metal enrichment histories derived for IC 1613 and the Leo I dwarf by Skillman et al. (2003). It is remarkable that the bulk of the star formation and metal enrichment has occurred since $z = 1$.

2. Dynamical Processes

Empirically, star formation is observed to obey a global Schmidt law, where the rate of SF per unit surface area is a power $n = 1.5$ of the average gas surface density in a galaxy (e.g., Kennicutt 1998). It is remarkable that this law holds with the same slope and is continuous, for interacting and non-interacting objects, pointing to the gas supply as the main factor.

This empirical law can be interpreted through several processes: Jeans instability, since the SFR is then proportional to the density, ρ, and inversely proportional to the dynamical time in $\rho^{-1/2}$, or cloud-cloud collisions (Elmegreen 1998), contagious star formation, associated with feedback (generating chaotic conditions), etc. These processes are able, without dynamical trigger, to yield episodic bursts of star formation, and this is well suited to dwarf galaxies, see Fig. 2 (Köppen et al. 1995, Pelupessy et al. 2004)

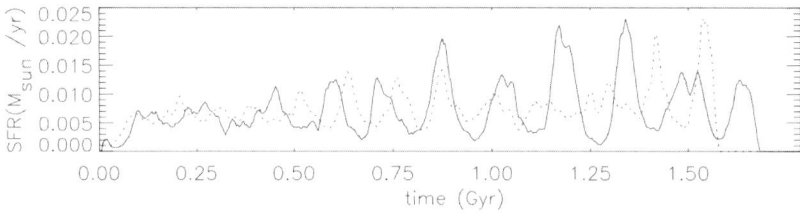

Figure 2. Star-formation history of a simulated dwarf galaxy. The dotted line indicates the SFR for a run with 50% reduced feedback strength (from Pelupessy et al. 2004).

For larger systems, large-scale dynamical instabilities must be invoked. Density waves, in creating shocks and concentrations of mass in spiral arms, can favor star formation, but starbursts require gathering large amounts of gas in a small area. Radial gas flows due to bars, or spiral torques are then at work, leading to molecular gas concentrations, and circumnuclear starbursts (e.g., Buta & Combes 1996, Sakamoto et al. 1999, Knapen 2005). In galaxy clusters, star formation could be induced by shocks due to interactions with the intra-cluster medium (Bekki & Couch 2003).

ULIRGs

Ultraluminous infrared galaxies have not only more gas and star formation, but also an enhanced star-formation efficiency (SFE), defined as the ratio of SFR traced by the far-infrared luminosity to the available fuel, traced by the CO emission (for the H_2 gas). More generally, in interacting galaxies, the CO emission relative to blue luminosity is multiplied by 5 and more concentrated (Braine & Combes 1993). This certainly means that the H_2 content is larger; the interpretation in terms of a lower CO-to-H_2 conversion factor would lead to an excessive star-formation efficiency SFE = $L(\text{FIR})/M(H_2)$.

This enhanced gas amount and concentration can be explained by the gravitational torques of the interactions driving gas very quickly to the centers. Gas in ULIRGs is concentrated in central nuclear disks or rings (Downes &

Solomon 1998). The condition to have a starburst is to accumulate gas in a time short enough that feedback mechanisms have no time to regulate. Also, the tidal forces are generally compressive in the centers, which favors cloud collapse.

Compressive tidal forces

For a spherical density profile, modelled as a power-law $\rho(r) \sim r^{-\alpha}$, the corresponding acceleration is in $r^{1-\alpha}$, so the gravitational attraction can increase with distance from center, if $0 < \alpha < 1$. Therefore the tidal force is then compressive: $F_{\text{tid}} \sim (1-\alpha)r^{-\alpha}$, in particular for a core with constant density ($\alpha = 0$). The rotation curve V_{rot} in $r^{1-\alpha/2}$, would then be almost rigid rotation. Molecular clouds inside the core are then compressed, and star formation can be triggered.

This phenomenon can also explain the formation of nuclear starbursts and young nuclear stellar disks in some barred galaxies. Decoupled stellar nuclear disks are frequently observed in double-barred Seyfert galaxies (Emsellem et al. 2001). The observed velocity dispersion reveals a characteristic drop in the center. The proposed interpretation invokes star formation in a decoupled nuclear gas disk (Wozniak et al. 2003).

Star-formation recipes

Numerical simulations use recipes for star formation and feedback phenomena, since this is sub-grid physics (Katz 1992, Mihos & Hernquist 1994, 1996). These recipes include the Schmidt law with exponent $n = 1.5$, together with a gas density threshold. The star-formation rate is, however, generally decreasing exponentially with time in isolated systems, even taking into account stellar mass loss (see Fig. 3). When comparing the star-formation history in an isolated galaxy with respect to a merger, the exponential law dominates, unless the SFR is normalised to the isolated case.

According to the detailed geometry, mass ratios or dynamical state of the merging galaxies, star formation can be delayed until the final merger, but the availability of gas is the main issue.

Importance of gas accretion

Galaxies in the middle of the Hubble sequence have experienced about constant SFR during their lives (Kennicutt et al. 1994, Brinchmann et al. 2004 [SDSS]). The study of stellar populations in the large SDSS sample has shown that only massive galaxies have formed most of their stars at early times, while dwarfs are still forming now (Heavens et al. 2004). Only intermediate masses have, on average, maintained their star-formation rate over a Hubble time.

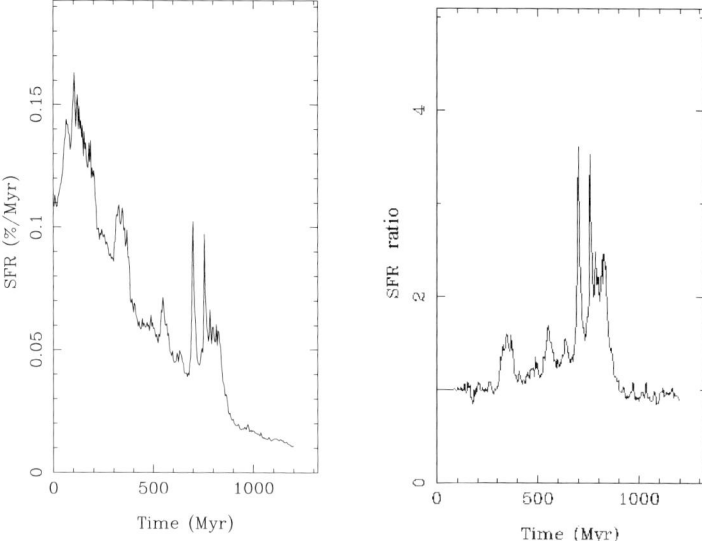

Figure 3. Star-formation history during a major merger of two Sb spiral galaxies: *Left:* the star-formation rate versus time, showing the global exponential decline; *Right:* the ratio between SFR in the merger run and the corresponding control run with the two galaxies isolated.

Even taking into account stellar mass loss, an isolated galaxy should have an exponentially decreasing star-formation history. Galaxies must therefore accrete large amounts of gas mass during their lives to fuel star formation.

Large amounts of gas accretion are also required to explain the observed present-day bar frequency (Block et al. 2002). Numerical simulations reveal that bars in gaseous spirals are quickly destroyed, and only gas accretion can trigger their reformation (Bournaud & Combes 2002). To have the right frequency of bars at the present time, gas accretion must double the galaxy mass in 10 Gyr. This gas cannot come from dwarf companions: they can provide at most 10% of the required gas and their dynamical interactions heat the disk. What is required is continuous cold gas accretion, which could come from the cosmological filaments in the close proximity of galaxies.

3. Conclusion

Star formation depends essentially on the gas supply. External gas accretion is essential for the efficiency of dynamical triggering. Galaxy interactions, and the accompanying bars and spirals, help to drive the accreted gas radially inwards and trigger central starbursts.

Gas accretion regulates not only the star-formation history in galaxies, but also their dynamics (bars, spirals, warps, $m = 1$, ...), since bars require gas to reform, in a self-regulating process.

Depending on the environment, hierarchical merging or secular evolution prevail. In the field, accretion is dominant, and explains bars and spirals, and the constant star-formation rate for intermediate types. In rich environments, galaxy evolution is faster, interactions and mergers are much more important; secular evolution of galaxies is halted at $z \sim 1$, since galaxies are stripped from their gas reservoirs.

References

Barton E., Geller M., Kenyon S., 2000, ApJ, 530, 660
Bekki K., Couch W., 2003, ApJ, 596, L13
Bergvall N., Laurikainen E., Aalto S., 2003, A&A, 405, 31
Block D., Bournaud F., Combes F., Puerari I., Buta R., 2002, A&A, 394, L35
Bournaud F., Combes F., 2002, A&A, 392, 83
Braine J., Combes F., 1993, A&A, 269, 7
Brinchmann J., Charlot S., White S.D.M., Tremonti C., Kauffmann G., Heckman T., Brinkmann J., 2004, MNRAS, 351, 1151
Brosch N., Almoznino E., Heller A.B., 2004, MNRAS, 349, 357
Buta R., Combes F., 1996, Fund. Cosmic Phys., 17, 95
Downes D., Solomon P.M., 1998, ApJ, 507, 615
Elmegreen B.G., 1998, in: Origins of Galaxies, Stars, Planets and Life, ASP Conf. Ser., 148, Woodward C.E., Shull J.M., Thronson H.A. Jr., eds., (ASP: San Francisco), p. 150
Emsellem E., Greusard D., Combes F., Friedli D., Leon S., Pécontal E., Wozniak H., 2001, A&A, 368, 52
Heavens A., Panter B., Jimenez R., Dunlop J., 2004, Nature, 428, 625
Hunter D.A., Elmegreen B.G., 2004, AJ, 128, 2170
Katz N., 1992, ApJ, 391, 502
Kennicutt R.C., Tamblyn P., Congdon C.E., 1994, ApJ, 435, 22
Kennicutt R.C., 1998, ARA&A, 36, 189
Knapen J., 2005, A&A, 429, 141
Köppen J., Theis C., Hensler G., 1995, A&A, 296, 99
Mihos J.C., Hernquist L., 1994, ApJ, 437, 611
Mihos J.C., Hernquist L., 1996, ApJ, 464, 641
Nikolic B., Cullen H., Alexander P., 2004, MNRAS, 355, 874
Pelupessy F.I., van der Werf P.P., Icke V., 2004, A&A, 422, 55
Sakamoto K., Okumura S.K., Ishizuki S., Scoville N.Z., 1999, ApJ, 525, 691
Sanders D., Mirabel I.F., 1996, ARA&A, 34, 749
Skillman E.D., Tolstoy E., Cole A.A., Dolphin A.E., Saha A., Gallagher J.S., Dohm-Palmer R.C., Mateo M., 2003, ApJ 596, 253
Telles E., Maddox S., 2000, MNRAS, 311, 307
Wozniak H., Combes F., Emsellem E., Friedli D., 2003, A&A, 409, 469
Zaritsky D., Harris J., 2004, ApJ, 604, 167

2D KINEMATICS AND MASS DERIVATIONS IN ULIRGS

M. García-Marín[1], L. Colina[1], S. Arribas[2,3], and A. Monreal[4,5]
[1] Departamento de Astrofísica Molecular e Infrarroja, IEM, CSIC, Spain
[2] Space Telescope Science Institute, USA
[3] On leave from the Instituto de Astrofísica de Canarias–CSIC, Spain
[4] Astrophysikalisches Institut Potsdam, Germany
[5] Instituto de Astrofísica de Canarias, Spain

Abstract Integral field optical fiber spectroscopy obtained with the INTEGRAL system, together with archival *HST* (WFPC2 and NICMOS) images, is used to carry out a program aimed at studying the internal physical structure and kinematics of a representative sample of Ultraluminous Infrared Galaxies (ULIRGs). Our goals are to characterize each individual system in order to infer generic properties of ULIRGs, and to perform simulations of high-redshift galaxy observations with JWST instruments. This work will be useful for stablishing connections between local ULIRGs and high-z galaxies, such as sub-mm galaxies, Lyman break galaxies and Spitzer galaxies. Here, we present results related to the kinematic properties of these systems, and the implications when deriving dynamical masses of high-redshift galaxies. Some results for IRAS 16007+3743 are shown too.

1. Introduction

ULIRGs ($L_{\rm bol} \approx L_{\rm IR} \geq 10^{12}$ L$_\odot$) are the most luminous objects in the local Universe. With large amounts of gas and dust, the vast majority of the local ULIRGs show signs of interactions and mergers (Bushouse et al. 2002), which trigger the intense starburst activity; this activity is believed to be their major energy source (Genzel et al. 1998). On the other hand, ULIRGs seem to be transforming gas-rich disk galaxies into moderate mass ($< M^*$) ellipticals through merging processes (Genzel et al. 2001). This could be the physical mechanism for forming the massive old galaxies recently detected at high z (Cimatti et al. 2004).

At high redshift, sub-mm galaxies (SMGs) have large quantities of internal gas and dust, and intense starburst activity triggered by mergers and/or AGN activity (Blain et al. 2002); for these reasons the sub-mm sources are believed to be ULIRGs at high redshift. One of the fundamental parameters of the galaxies is the dynamical mass, which has not yet been studied in detail at high z; this parameter will be derived by measuring the kinematics of the warm, ionized gas with optical emission lines, through studying the CO emission lines of the cold molecular gas, or the absorption lines of the near-IR CO band. Thus, it is important to investigate the agreement between the different methods used to trace the mass, and to establish how reliable they are in a sample of low-z ULIRGs.

2. The sample of ULIRGs

Our program is based on Integral Field Spectroscopy (IFS) data of low-z ULIRGs using the INTEGRAL system, complemented with *HST* high-resolution images. The sample consists of 20 galaxies, and covers different phases of interaction processes and all types of activity, from starbursts to QSO-like galaxies. Warm ($f_{25\mu m}/f_{60\mu m} \geq 0.2$) and cool ($f_{25\mu m}/f_{60\mu m} < 0.2$) galaxies are included in the sample too; all objects have $z < 0.2$.

The IFS data were obtained with INTEGRAL, a fiber system on the William Herschel Telescope (WHT; Arribas et al. 1998). The current configuration provides a 12×16 arcsec^2 FOV, a fiber diameter of 0.9 arcsec, and is capable of obtaining ~ 200 spectra simultaneously, covering the range $5000 - 9000$ Å at ~ 4.5 Å resolution.

3. IRAS 16007+3743

As an example, the 2D maps obtained from INTEGRAL spectra are shown for the galaxy IRAS 16007+3743 (García-Marín 2005), a ULIRG with $L_{IR} = 10^{12.11}$ L$_\odot$, located at $z \sim 0.18$; for this redshift 1 arcsec = 3.0 kpc ($H_0 = 70$ km s^{-1} Mpc^{-1}, $\Omega_m = 0.3$ and $\Omega_\Lambda = 0.7$).

The *HST* I-band image compares well with the IFS red continuum, when the different spatial resolutions are taken into account. The system is composed of two galaxies in the process of merging, with two well-differentiated nuclei, identified as A and B in Fig. 1. Morphologically, this ULIRG shows a complex structure including well-developed tidal tails and plumes, as observed in many other ULIRGs.

The brightest Hα, Hβ and [OIII]λ5007 emission (identified as C) shows the location of star-forming regions along the tidal tail, whereas the peak of the red continuum (nucleus B) is located about 9 kpc to the south. On the other hand, the [SII] emission-line peak coincides spatially with that of the red continuum.

As expected, dust is concentrated in the nuclear zones, and produces a visual extinction ranging from $A_V = 2.2$ (region C) to $A_V = 5.6$ (nucleus B) magnitudes. Although region C has an apparent Hα flux $\simeq 2$ times larger than that of nucleus B, the extinction-corrected Hα luminosities correspond to 2.4×10^{41} erg s^{-1} (region C) and 5.9×10^{42} erg s^{-1} (nucleus B). The Hα-derived star-

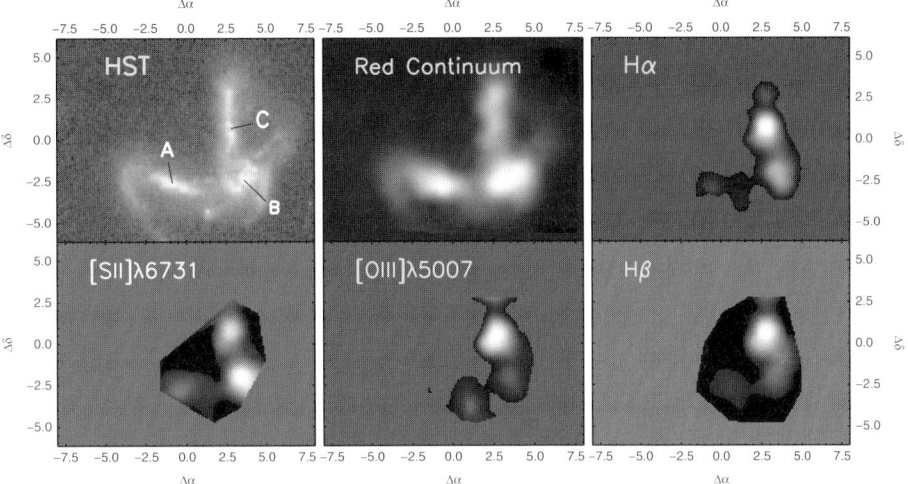

Figure 1. This panel summarizes the results for IRAS 16007+3743 from INTEGRAL IFS data, and an *HST*/WFPC2 *I*-band (F814W) image. The stellar component is traced by the red (0.7μm) continuum, while the ionized gas is traced by different emission lines (Hα, Hβ, [SII] and [OIII]).

formation rate (Kennicutt 1998) values are ~ 1.9 M$_\odot$ yr^{-1} (region C), and ~ 46.6 M$_\odot$ yr^{-1} (nucleus B).

The IFS data allow us to classify different regions of the system using the standard relations between different emission-line ratios. We find that region C and nucleus B have line ratios that correspond to those of HII regions. We measure $\log([OIII]/H\beta) \simeq 0.01$, $\log([NII]/H\alpha) \simeq -0.56$ and $\log([SII]/H\alpha) \simeq -1.04$ for region C, and $\log([OIII]/H\beta) \simeq 0.08$, $\log([NII]/H\alpha) \simeq -0.27$ and $\log([SII]/H\alpha) \simeq -0.77$ for nucleus B.

4. 2D kinematics results

A 2D kinematics study of a small sample of ULIRGs has recently been done to determine the best indicator of the dynamical mass of these systems. In this study the kinematic properties of the ionized gas from IFS data have been compared to those of the CO molecular gas and the stellar near-IR CO bands. In this section, a few results will be discussed (for more details see Colina et al. 2005).

The central velocity amplitude for the ionized gas is defined as half the peak-to-peak velocity difference at radii of $1 - 2$ arcsec on either side of the galaxy. The stellar and ionized gas velocities agree (with some discrepancies), while the cold molecular gas shows, on average, a central velocity amplitude a factor two larger than the stars and ionized gas (Fig 2, left panel). These results suggest that, in general, the stars and ionized gas share the same velocity field

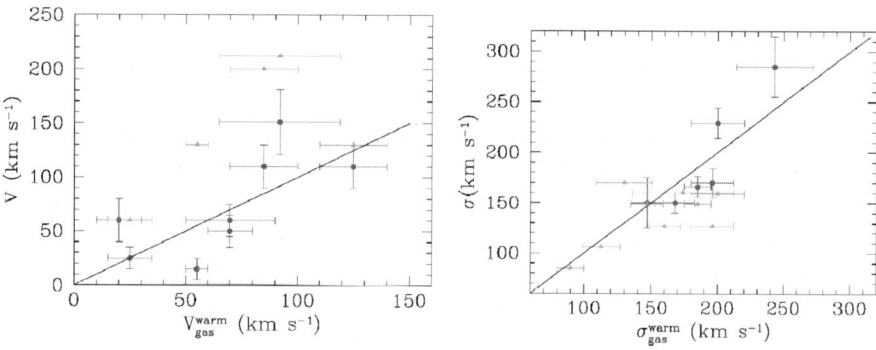

Figure 2. The stellar (filled circle) and cold molecular gas (filled triangle) kinematics versus the warm ionized gas values are presented for the central velocity (left panel) and the velocity dispersion (right panel).

in the central regions of the galaxy, which is not necessarily that governing the molecular gas.

The central velocity dispersion shows that the stellar and cold gas velocity dispersions correlate well with the value obtained for the warm ionized gas (Fig 2, right panel), supporting the idea that the central velocity dispersion of the ionized gas is dominated by the gravitational potential of the system (i.e., its mass) as traced by the stellar velocity dispersion. Based on these results, the central velocity amplitude is not a reliable observable to trace the dynamical mass, while the velocity dispersion is a homogeneous tracer of the internal kinematics, and therefore a more robust indicator for the dynamical mass of the system, independent of the observational method used.

If high-z galaxies, such as sub-mm galaxies or Lyman break galaxies, show merger morphologies similar to those observed in local ULIRGs, the central velocity dispersion will be the preferred method to determine the dynamical mass, while in general the velocity gradients should be treated with caution.

References

Arribas S., et al., 1998, SPIE, 3355, 821
Blain A.W., Smail I., Ivison R.J., Kneib J.-P., Frayer D.T., 2002, Phys. Rep., 369, 111
Bushouse H.A., et al., 2002, ApJS, 138, 1
Cimatti A., et al., 2004, Nature, 430, 184
Colina L., Arribas S., Monreal A., 2005, ApJ, in press
García-Marín M., 2005, Master's thesis, Univ. Autónoma de Madrid, Spain, in prep.
Genzel R., et al., 1998, ApJ, 498, 579
Genzel R., Tacconi L.J., Rigopoulou D., Lutz D., Tecza M., 2001, ApJ, 563, 527
Kennicutt R.C. Jr., 1998, ARA&A, 36, 189

INTERNAL KINEMATICS OF LUMINOUS COMPACT BLUE GALAXIES

Matthew A. Bershady[1], M. Vils[1], C. Hoyos[2], R. Guzmán[3], and D.C. Koo[4]
[1]*University of Wisconsin, USA;* [2]*Universidad Autónoma de Madrid, Spain;* [3]*University of Florida, USA;* [4]*University of California, Santa Cruz, USA*

Abstract   We describe the dynamical properties which may be inferred from *HST*/STIS spectroscopic observations of luminous compact blue galaxies (LCBGs) between $0.1 < z < 0.7$. While the sample is homogeneous in blue rest-frame color, small size and line width, and high surface brightness, their detailed morphology is eclectic. Here, we determine the amplitude of rotation versus random, or disturbed motions of the ionized gas. This information affirms the accuracy of dynamical mass and M/L estimates from Keck integrated line widths, and hence also the predictions of the photometric fading of these unusual galaxies. The resolved kinematics indicates this small subset of LCBGs are dynamically hot, and unlikely to be embedded in disk systems.

1. Introduction

The evolution of LCBGs is a matter of debate. These galaxies are unusual in their blue colors, small sizes and line widths, yet large luminosities. We have suggested that at least a subset of these sources are the progenitors of dEs such as NGC 205 (Koo et al. 1995, Guzmán et al. 1998), while others counter these are bulges in formation (Hammer et al. 2001, Barton & van Zee 2001). Surveys at intermediate redshift are not uniformly defined, and each contains heterogeneous samples – objects span a range in size, color, luminosity, surface brightness, and image concentration. The broad "LCBG" class contributes as much as 45% of the co-moving SFR between $0.4 < z < 1$ (Guzmán et al. 1997); the proposed dE progenitors are a fraction of this class. Here, we focus on an extreme LCBG subclass that are among the smallest, bluest and highest surface brightness (Koo et al. 1995): $M_B \sim -21$ ($H_0 = 70$ km s^{-1} Mpc^{-1}, Ω=1, Ω_Λ=0.7), rest-frame $(B-V) \sim 0.25$, half-light radii of $R_e \sim 2$ kpc, mean surface brightness within R_e of ~ 19 mag arcsec^{-2} (rest-frame B band), and integrated line widths of $\sigma_{\rm gas} \sim 65$ km s^{-1}. Many of these are good dE progenitor candidates. If so, their internal kinematics should reveal they are dynamically hot, while deep imaging should show they lack outer disks.

2. STIS Spectra: Are LCBGs Dynamically Hot or Cold?

We have derived ionized-gas position–velocity and position–line width diagrams from STIS long-slit measurements along what appears to be the photometric major axes of 6 LCBGs between $0.2 < z < 0.7$, and one other source at $z \sim 0.1$ which is 2–3 mag lower luminosity than the others. One example is given in Fig. 1. With 0.2 arcsec slits, STIS delivers instrumental resolutions (σ) of 13–19 km s^{-1}. Line emission is not always centered on the continuum (Hoyos et al. 2004); the continuum centroid is adopted as the kinematic center.

We find Keck HIRES integrated line widths (Koo et al 1995) agree in the mean with the resolved velocity dispersions from STIS spectroscopy: integrated dispersions are not due to large-scale, bulk, motion. This secures our previous dynamical estimates of M/L and their use as constraints on photometric fading (e.g., Guzmán et al. 1998). Only the low-L, low-z system shows clear rotation and substantially different integrated versus resolved line-widths.

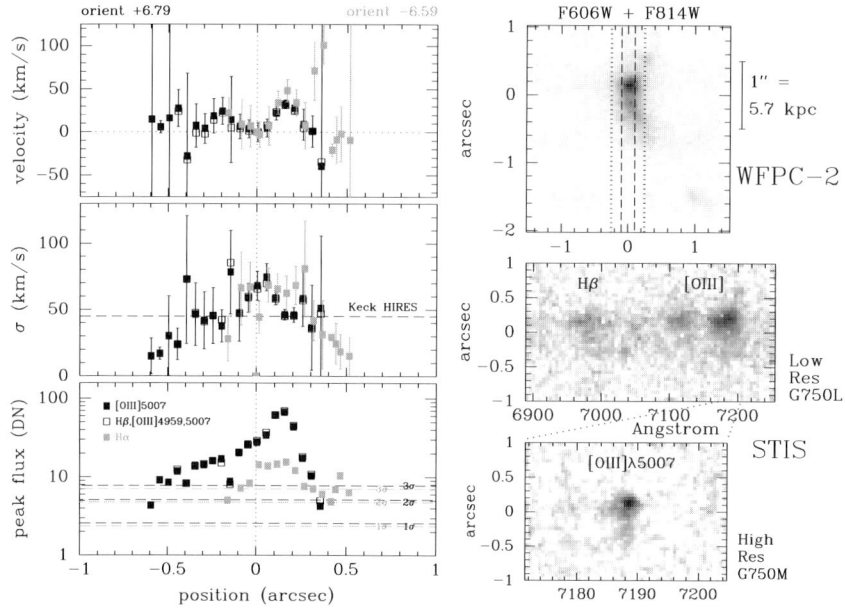

Figure 1. Morphology and kinematics of LCBG 313088 at $z = 0.44$. *Left:* HST/WFPC2 image showing distorted, tail-like source morphology and mean STIS slit positions (0.2 and 0.5 arcsec). STIS spectra for low and high-resolution gratings are at bottom. *Right:* Position vs. velocity, line width, and line flux for two sets of high-resolution data taken at two position angles varying by $\sim 13°$. (Hα spectrum not shown). Spectral data consistently show extended line emission with little velocity gradient, no evidence for rotation, and dispersions that agree in the mean with Keck HIRES integrated measurements (dashed line, middle panel).

The lack of rotation coupled with their ellipticity squarely places these sources in the "spheroidal" region of the V/σ – ellipticity plane, illustrated in Fig. 2.

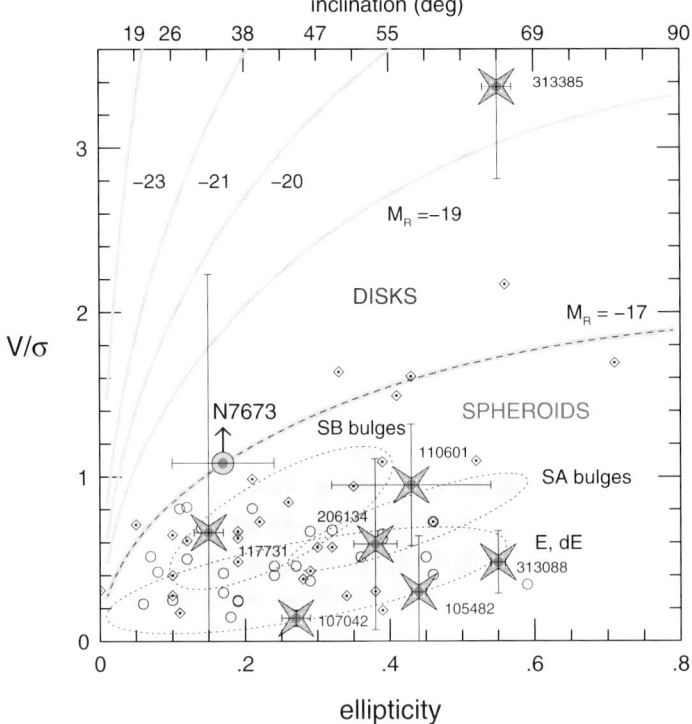

Figure 2. V/σ versus ellipticity or inclination for LCBGs (dotted stars), NGC 7673 (circle/lower limit) and early-types galaxies from Simien & Prugniel (2002; circles, dotted diamonds). Lines represent trajectories as a function of inclination for disk galaxies of different luminosity, assuming they lie on the Tully-Fisher relation and $\sigma_{\rm gas} = 25$ km s^{-1} (Andersen et al., in prep.). Shaded regions are adopted from Kormendy & Kennicutt (2004).

For fair comparison to local samples, ellipticities are measured at the half-light radius near rest-frame V band from *HST* images; rotation velocities are set to half the difference between minimum and maximum velocities; and σ is the observed central velocity dispersion. While there is a range of observed V/σ, LCBGs lie well below the region inhabited by disk systems, with values comparable to local dEs and other spheroidals, particularly if recent observations of dEs with larger rotational components are considered (Pedraz et al. 2002, van Zee et al. 2004): **LCBGs are dynamically HOT**.

3. Progenitors and Descendants

Do we know that we are not just sampling a bulge, or a face-on nuclear starburst? NGC 7673 has been suggested by Homeier et al. (2002) as a nearby example. Indeed, this source has the right color, surface brightness, size, luminosity and integrated line width, and has a faint, extended outer disk. However, Homeier & Gallagher's (1999) Hα velocity map shows clear evidence for rotation in the inner, starburst region. Our re-analysis confirms this: we would see similar structure *if it existed* in our STIS spectra. Independently, deep CFHT imaging reveals no strong evidence for extended, normal disks around the types of (and specific) sources presented here (Barton et al., in prep.).

In summary, the preponderance of evidence is against bulge formation in disk systems and in favor of a dE-like descendant scenario *for all of the specific sample presented here* with $M_B < -20$ (6 out of 7). However, the disturbed morphology and kinematics makes clean interpretations difficult. What is the gas really telling us about dynamics? Are these systems in dynamical equilibrium? While their morphology and resolved kinematics would argue otherwise, the agreement between integrated velocity dispersions and resolved profiles indicates the systems cannot be too far out of equilibrium. Stellar velocity and dispersion profiles would provide a much clearer dynamical picture.

Finally we comment on issues raised at the conference about environment: is NGC 205 a good example of a faded, LCBG descendant? If so, where are the M31-like neighbors? Are there field dEs? If dSphs are the low-mass cousins of dEs, then the presence of isolated dSphs such as Tucana, Cetus, and the recently discovered Apples 1 (Pasquali et al. 2005) should give us pause about accepting assertions that dEs do not exist outside of rich environments or far from massive galaxies. The space density of LCBGs presented here is $(1.25 \pm 0.15) \times 10^{-5}$ Mpc^{-3} for $M_B < -20$ between $0.3 < z < 0.7$. Even allowing a factor of ~ 20 higher relic density (given the time interval in this redshift slice and assuming the LCBG phase is a few $\times 10^8$ yr) to find even one descendant requires an all-sky local survey volume reaching out to ~ 10 Mpc. At this distance, the half-light radius of NGC 205 is ~ 12 arcsec. Are our local surveys this complete?

References

Barton E., van Zee L., 2001, ApJ, 550, L35
Guzmán R., Gallego J., Koo D.C., Phillips A.C., Lowenthal J.D., Faber S.M., Illingworth G.D., Vogt N.P., 1997, ApJ, 489, 559
Guzmán R. Jangren A., Koo D.C., Bershady M.A., Simard L., 1998, ApJ, 495, L13
Hammer F., Gruel N., Thuan T.X., Flores H., Infante L., 2001, ApJ, 550, 570
Homeier N.L., Gallagher J.S., 1999, ApJ, 522, 199
Homeier N.L., Gallagher J.S., Pasquali A., 2002, A&A, 391, 857
Hoyos C., Guzmán R., Bershady M.A., Koo D.C., Díaz A.I., 2004, AJ, 128, 1541
Koo D.C., Guzmán R., Faber S.M., Illingworth G.D., Bershady M.A., Kron R.G., Takamiya M., 1995, ApJ, 440, L49
Kormendy J., Kennicutt R.C., 2004, ARA&A, 42, 603
Pasquali A., Larsen S., Ferreras I., Gnedin O.Y., Malhotra S., Rhoads J.E., Pirzkal N., Walsh J., 2005, AJ, 129, 148
Pedraz S., Gorgas J., Cardiel N., Sanchez-Blazquez P., Guzmán R., 2002, MNRAS, 332, L59
Simien F., Prugniel Ph., 2002, A&A, 384, 371
van Zee L., Skillman E.D., Haynes M.P., 2004, AJ, 128, 121

FUELLING STARBURSTS AND NUCLEAR RINGS

Johan H. Knapen
Centre for Astrophysics Research, University of Hertfordshire, Hatfield, Herts AL10 9AB, UK
j.knapen@star.herts.ac.uk

Abstract The question of how a starburst is fuelled is closely related to some of the main outstanding problems in our understanding of starbursts, namely which combination of physical conditions leads to the triggering of the burst, and how long the burst can last. There is observational evidence that galactic bars and galaxy-galaxy interactions can lead to starburst activity, but this may be limited to specific and/or extreme cases, and not all bars or interactions lead to starbursts. Nuclear rings are mild circumnuclear starbursts, which are usually (but not always) triggered by bars, and which allow the study of both the dynamical relation of the host galaxy to the starburst, and the physical conditions of the latter. As an example, the case of NGC 278 is briefly reviewed.

1. Introduction

There are many open questions relating to starburst activity in galaxies, and some of the most important of those are what specific combination of physical conditions can lead to the triggering of the burst, and what is the time span over which the starburst can be sustained. These questions are, in turn, closely linked to the question of how a starburst is fuelled. One obvious prerequisite for the occurrence of a starburst is the availability of sufficient gaseous fuel, of the right physical conditions, e.g., temperature, density, and dynamics, and at the right location at the right time.

In this short paper, we will first briefly describe the current observational evidence for fuelling starbursts by means of non-axisymmetries in the galactic potential, specifically bars and interactions. We will then discuss some of the most salient aspects of nuclear rings, a specific morphological class of low-luminosity starbursts, and specifically discuss recent results on the ring in NGC 278.

2. Starburst fuelling

The so-called "fuelling problem" in starbursts does not refer so much to the amount of gas present, but to the issues involved in transporting this gas from the main body of the host galaxy to the central region where it is needed to feed the star formation. Assuming typical starburst gas consumption rates and lifetimes (of, say, a few solar masses per year and of order $10^7 - 10^8$ years), it is easily seen that the outer regions of the host galaxies of starbursts will, in general, contain more than enough gas to fuel the burst. The problem, then, is how to deliver this gas to the starburst region in the centre of the galaxy. To transport the gas radially inward, it must lose most of its angular momentum, for which a number of mechanisms can be invoked. The most important of these are gravitational, driven by a non-axisymmetry in the galactic potential set up by a bar, or a galaxy interaction or merger, and effective on spatial scales of tens of parsecs to kiloparsecs, and possibly even smaller than that. For more complete discussions, the reader is referred to reviews by, e.g., Shlosman, Begelman & Frank (1990) or Knapen (2004, 2005a).

Bars lead to gas concentration toward the central regions of galaxies because gas in the bar can lose angular momentum due to torques and shocks. There is some direct observational evidence that bars instigate central concentration of gas (e.g., Sakamoto et al. 1999, Jogee et al. 2005), and the question is then whether this concentration also leads to central starburst activity. Indeed, nuclear starbursts show a clear preference for barred hosts, although this result is possibly restricted to strong bars (SB) and to early-type galaxies only (Huang et al. 1996, Roussel et al. 2001).

Interactions between galaxies can also lead to angular momentum loss of inflowing material and thus to the fuelling of starbursts. This is most obviously the case for the ultraluminous infrared galaxies (ULIRGs), powered mainly by very powerful starbursts (Genzel et al. 1998), and firmly associated with galaxies involved in mergers or other strong interaction processes (e.g., Sanders & Mirabel 1996). However, although the most extreme of the starbursts are definitely related to galaxy mergers, it is much less clear whether interactions lead to enhancements of the star-formation activity *in general*. For instance, Bergvall, Aalto & Laurikainen (2003) find no evidence for significantly enhanced star-forming activity among interacting/merging galaxies compared to non-interacting galaxies, although they do report a moderate increase in star formation in the very centres of their interacting sample galaxies.

3. Nuclear rings

Nuclear rings and pseudo-rings are common, occurring in around one fifth of all disk galaxies (Knapen 2005b). They can form in the vicinity of resonances in a gaseous disk, usually set up by a bar or other form of non-axisymmetry in the gravitational potential. Gas concentrates near such resonances, can become unstable, and collapse to lead to the massive star formation associated with rings. Nuclear rings, with typical radii of $0.5 - 1$ kpc, form

Fuelling starbursts and nuclear rings

significant numbers of stars, which can help shape a pseudo-bulge, and so drive secular evolution (Kormendy & Kennicutt 2004), but which also contribute a few percent to the overall star-formation rate of the local Universe.

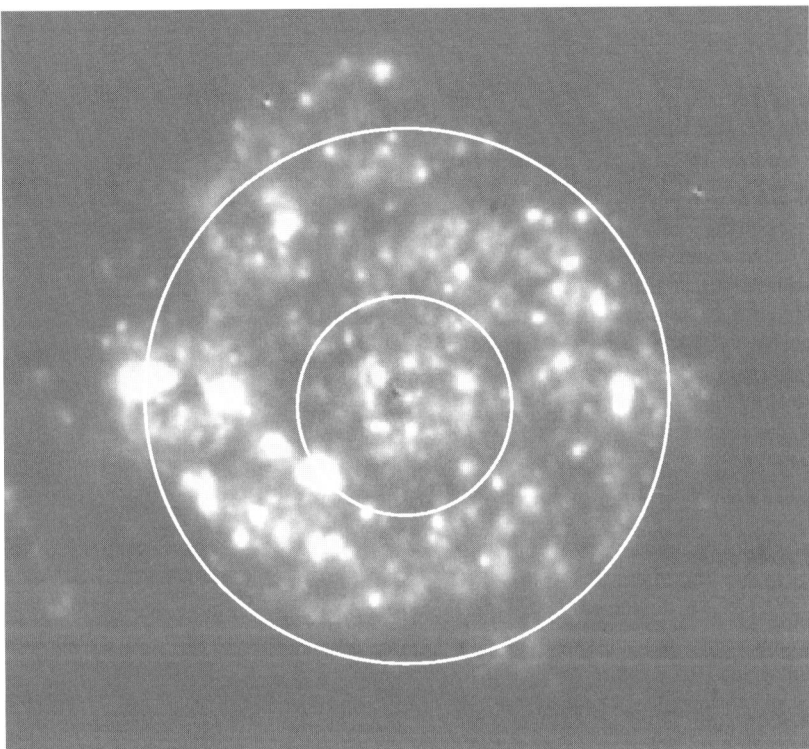

Figure 1. Hα view of the inner region of the galaxy NGC 278. The area shown is about 100 arcsec on a side, or about 5 kpc. The area between the two concentric white circles indicates the nuclear pseudo-ring, with a radius of 1.1 kpc. No massive star formation occurs outside the region shown here. Image data from Knapen et al. (2004).

Although almost all nuclear rings occur in barred galaxies, some do not. A nice example of the latter category is the ring-like zone of enhanced star formation in NGC 278, a small, nearby and isolated spiral galaxy classified as SAB(rs)b in the RC3, but without any evidence for the presence of a bar from either *HST*/WFPC2 or ground-based NIR imaging (Knapen et al. 2004). The optical disk of NGC 278 shows two distinct radial regions, an inner one with much enhanced star formation, shown in Fig. 1, and an outer one which is almost completely featureless, of low surface brightness, and rather red. NGC 278 has an H I disk which is large compared to the optical extent of the galaxy. The H I disk is morphologically and kinematically disturbed (Knapen et al. 2004). These disturbances suggest a recent minor merger with a small gas-rich galaxy, perhaps similar to a Magellanic cloud, and it is this past interaction which we believe has set up a non-axisymmetry in the gravitational potential, which in turn, in a way very similar to the action of a classical bar, leads to the formation of the enhanced star formation in the ring.

4. Conclusions

In this short paper, we have briefly reviewed relevant observational evidence for the fuelling of starbursts by gas from their host galaxies, specifically by exploring the connections between galactic bars and galaxy-galaxy interactions, and the occurrence of starburst activity. Such connections exist, but may be limited to extreme and/or specific cases. We also present an example of a nuclear pseudo-ring, namely the one in the unbarred galaxy NGC 278. By far most nuclear rings occur in barred hosts, but this example shows how a past minor merger event can lead to the formation of a nuclear ring by the same dynamical mechanism.

Acknowledgments

I thank my collaborators, especially those on the paper on NGC 278: Laura Whyte, Erwin de Blok and Thijs van der Hulst.

References

Bergvall N., Laurikainen E., Aalto S., 2003, A&A, 405, 31
Genzel R., et al., 1998, ApJ, 498, 579
Huang J.H., Gu Q.S., Su H.J., Hawarden T.G., Liao X.H., Wu G.X., 1996, A&A, 313, 13
Jogee S., Scoville N.Z., Kenney J., 2005, ApJ, in press (astro-ph/0402341)
Knapen J.H., 2004, in: Penetrating bars through masks of cosmic dust: The Hubble tuning fork strikes a new note, Block D.L., Puerari I., Freeman K.C., Groess R., Block E.K., eds., (Dordrecht: Springer), ASSL, 319, 189 (astro-ph/0407068)
Knapen J.H., 2005a, in: The Evolution of Starbursts, Hüttemeister S., Aalto S., Bomans D.J., Manthey E., eds., (Melville: AIP), in press (astro-ph/0411135)
Knapen J.H., 2005b, A&A, 429, 141
Knapen J.H., Whyte L.F., de Blok W.J.G., van der Hulst J.M., 2004, A&A, 423, 481
Kormendy J., Kennicutt R.C. Jr., 2004, ARA&A, 42, 603
Roussel H., et al., 2001, A&A, 372, 406
Sakamoto K., Okumura S.K., Ishizuki S., Scoville N.Z., 1999, ApJ, 525, 691
Sanders D.B., Mirabel I.F., 1996, ARA&A, 34, 749
Shlosman I., Begelman M.C., Frank J., 1990, Nature, 345, 679

Session V

Star-Formation Rates in Relation to the Host Galaxy Properties

DEMOGRAPHICS AND HOST GALAXIES OF STARBURSTS

Robert C. Kennicutt, Jr.[1], Janice C. Lee[1], Jose G. Funes, S.J.[2], Shoko Sakai[3], and Sanae Akiyama[1]
[1]*Steward Observatory, University of Arizona, Tucson, AZ 85721, USA**
[2]*Vatican Observatory, Steward Observatory, University of Arizona, Tucson, AZ 85721, USA*
[3]*Department of Astronomy, UCLA, Los Angeles, CA 90095, USA*

Abstract   Thanks to an array of major new ultraviolet, infrared, and visible-wavelength imaging and spectroscopic surveys, it has become possible to map the star formation in thousands of nearby galaxies, and derive integrated star-formation rates for hundreds of thousands of objects. This, in turn, will make it possible to quantify robustly the characteristic frequencies, strengths, and durations of starbursts, as well as the dependence of these quantities on the properties and environments of the host galaxies. Here we use results from several recent Hα surveys to characterize the local population of star-forming galaxies, and illustrate how such data can be applied to constrain the properties and incidence of starbursts.

1. Introduction

We are in the midst of a revolution in the study of star-formation rates (SFRs) in galaxies, as data pour in from an array of new surveys in the visible, ultraviolet, and infrared. These promise to revolutionize our ability to measure reliable SFRs for individual galaxies, and provide for the first time the kinds of large and statistically robust samples that are needed to derive the frequency distributions of SFRs (and other SFR properties), and measure the incidence and properties of starbursts in a rigorous manner.

The first wave of these new data has come from the Sloan (SDSS) and 2DF digital sky surveys, which have provided semi-integrated spectra of hundreds of thousands of galaxies out to redshifts of a few tenths, and is continuing with GALEX, which will provide UV-based SFRs for similar-sized samples of galaxies. At the same time, the Spitzer Space Telescope is providing in-

---

*This work is supported in part by the National Science Foundation under Grant No. 0307386, and NASA grant NAG5-8426.

frared imaging and spectra with unprecedented spatial resolution for galaxies spanning the entire range of redshifts. And nearly a dozen large Hα spectroscopic (UCM, KISS, NFGS, K/M) and imaging surveys (e.g., GoldMine, HαGS, SINGG, SINGS, STARFORM, AMIGA, MOSAIC Cluster, 11HUGS, SMUDGES) will provide emission-line based SFRs for thousands of galaxies, and publicly-available image libraries for most. Among these many surveys are a handful that are specifically designed to obtain data across the electromagnetic spectrum, to deal with dust attenuation and wavelength-dependent selection effects in a robust manner. These include three Spitzer Legacy surveys, SINGS, SWIRE, and GOODS. SINGS is also mated to large ground-based HI, CO, and Hα Fabry-Perot surveys, which will enable us to correlate the observed SFRs with the distributions and kinematics of interstellar gas, and thereby probe the complex interplay between star formation and the surrounding ISM (Kennicutt 2003).

The new data are especially powerful for extending our understanding of the nature and populations of starbursts and their host galaxies. The term "starburst" has expanded to encompass an enormous range of scales and phenomena (the current lexicon includes GEHRs, SSCs, HIIGs, ELGs, CNELGs, W-R galaxies, UVCGS, BCGs, BCDs, LCBGs, LIGs and LIRGs, ULIGs and ULIRGs, HLIGs and HLIRGs, HUGs, LEGOs, E+A galaxies, K+A galaxies, nuclear and circumnuclear starbursts, clumpy irregular galaxies, Lyα galaxies, LBGs, DRGs, EROs, submillimeter and SCUBA galaxies!), and with the advent of complete data sets we can now adopt consistent physical definitions for starbursts and study their host environments and evolution quantitatively. Beyond that, one should be able to derive robustly the present-day frequencies of starbursts and their contribution to the total SFR of the local Universe, and trace the evolution in both over cosmic time. Finally, we can begin to reliably constrain the duty cycle of starbursts: the distributions of strengths, durations, and the multiplicity of bursts within individual galaxies, and the fraction of stars in various types of galaxies that are formed in bursts. The data may also allow us to address the more challenging problem of understanding the ubiquity of starbursts – to what extent starbursts are a general characteristic of galaxies as opposed to a phenomenon unique to particular types of galaxies and galaxy environments – and to understand which underlying properties of galaxies make them especially prone to bursts. And by testing and extending the scaling laws that appear to characterize starbursts (e.g., the Schmidt star-formation law, extinction vs. bolometric luminosity relation), we will gain deeper insights into the connections between the various starburst classes and the physical processes responsible for the bursts.

The agenda outlined above is an ambitious one, and a key science driver for several of the ongoing surveys cited above. In this review, we illustrate the kinds of insights that can be gained from these new data sets, by using a few recent Hα surveys to characterize the demographics of local star-forming galaxies, define starburst galaxies, and roughly estimate the fractional contributions of starbursts to the local cosmic star-formation budget. These results are all quite preliminary, and are intended not as definitive products on their own but rather as examples of what will be learned in the near future.

2. Parameterizing Global SFR Properties of Galaxies

Traditionally, authors have defined starbursts in many different ways, and often in a qualitative and subjective manner. If we consider quantitative definitions of a starburst, three possibilities come to mind, and all have been applied at one time or another in the literature (and by other speakers at this conference). One can define a starburst in terms of its absolute SFR. This definition is commonly applied to infrared luminous galaxies, which are defined in terms of an absolute IR luminosity, $\geq 10^{11}$ $L_\odot$ for Luminous Infrared Galaxies (LIGs or LIRGs), and $\geq 10^{12}$ $L_\odot$ for Ultraluminous Infrared Galaxies (ULIGs or ULIRGs). For galaxies whose luminosity is powered by dust-obscured star formation these limits correspond to approximate SFRs of 18 and 180 $M_\odot$ yr^{-1}, respectively (Kennicutt 1998a). One can also define a threshold for a starburst in terms of its SFR surface density or intensity, i.e., the SFR per unit area (e.g., Meurer et al. 1997, Lanzetta et al. 2002, James et al. 2004). Finally, one can define a starburst galaxy as one whose SFR exceeds the average past value by a fixed amount. This definition extends back to the original papers on star-formation bursts by Tinsley and others, and is probably the most physically meaningful definition for galaxies at the present epoch. It usually is characterized in terms of the birth-rate parameter $b \equiv$ SFR / $\langle$SFR$\rangle_{\rm past}$.

Figures 1 and 2 show the distribution of star-formation properties of nearby galaxies in terms of these three parameters, absolute SFR vs. SFR density (Fig. 1) and SFR vs. b (Fig. 2). The data are taken from three recent Hα surveys, a volume-limited survey of all known spiral and irregular galaxies in the local 11 Mpc volume (11HUGS; Kennicutt et al., in prep.), the HαGS survey (James et al. 2004), a survey of early-type spirals by Hameed & Devereux (1999), and a database of SFRs of galaxies in the Virgo cluster from the GoldMine database (Gavazzi et al. 2003). The 11HUGS sample is the only one to fairly represent the relative numbers of galaxies in various parts of the diagram, but we have added the others to fill in the coverage for more massive galaxies and early-type galaxies, which are rare in the local 11 Mpc volume. The data have been corrected for [NII] emission and dust extinction using statistical recipes (updated versions of Kennicutt 1983). The scaling radius used is the diameter of the actively *star-forming* region, rather than the photometric radius of the galaxy, as this seems to be more tightly correlated with the physical properties of the system (Kennicutt 1998b, Martin & Kennicutt 2001).

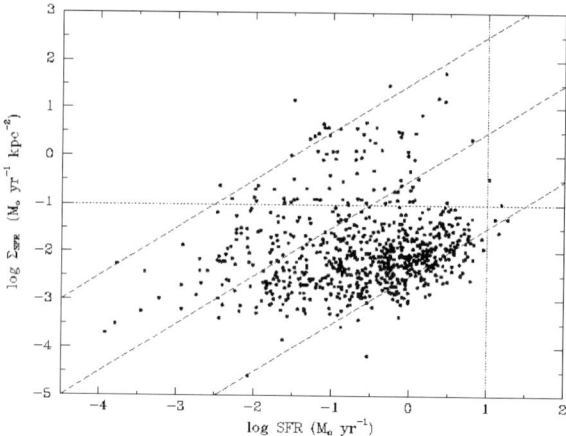

Figure 1. Distribution of SFRs and SFR densities in local star-forming galaxies. Data are taken from our own 11HUGS survey, James et al. (2004), Hameed & Devereux (1999), and the GoldMine database, as described in the text. The diagonal lines denote constant radius of the star-forming regions, from 0.1 to 10 kpc (top to bottom). The horizontal and vertical dashed lines denote the typical values delineating starburst galaxies, as discussed in the text.

Figure 1 illustrates the enormous diversity of the galaxy population, with SFRs and SFR densities each spanning millionfold ranges. Note also that the two parameters are only very loosely correlated with each other; at a fixed SFR there is a variation of more than 10^5 in SFR density and vice versa. Despite the scatter across the entire sample, however, there is a clear bimodality in the distribution of SFR properties. Most galaxies lie on a sequence extending across the lower half of the diagram, with SFR densities of 0.001–0.1 $M_\odot$ yr^{-1} kpc^{-2}. These trace out the normal Hubble sequence of spiral and irregular galaxies. The smattering of galaxies at higher SFR densities represent starbursts, as discussed below.

A similar diversity is seen in Fig. 2, which plots the disk-averaged birth-rate parameter b vs. the SFR. For these data, b was estimated from the equivalent width of the Hα emission line, following the methods given by Kennicutt, Tamblyn & Congdon (1994). Once again, we see a strong concentration of galaxies in a main sequence with an average $b \sim 0.5$, when averaged across all of the types in the sample (but beware, the composite sample plotted is not complete!). We should emphasize that the absolute values of b are somewhat dependent on the recipe that is used to correct the Hα equivalent widths for dust attenuation. Here we have not applied any correction at all, assuming that Hα and the underlying continuum suffer comparable amounts of extinction. A more realistic recipe would be to adopt a higher extinction for the line emission, as done in Kennicutt et al. (1994), and adopting that would increase the values of b in Fig. 2 by about 30 to 50%. However it would not qualitatively change the distributions or the results presented here. The smattering of points in the upper part of the diagram again represent starbursts.

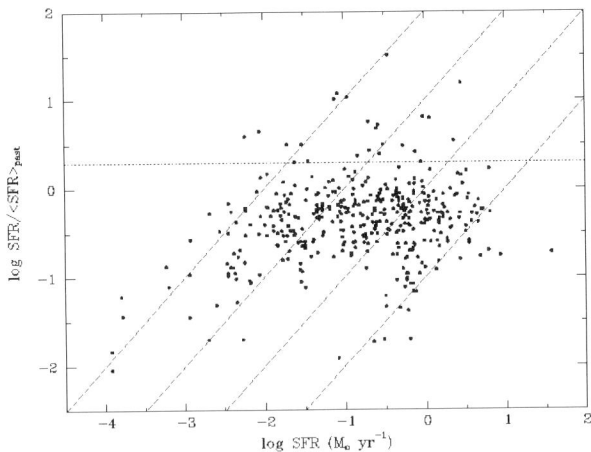

Figure 2. Similar to Fig. 1, except that the SFRs are plotted against the ratio of the current to average past SFR (b). Data from the 11HUGS survey and James et al. (2004). The diagonal lines now denote loci of constant stellar disk mass, increasing from top to bottom from 10^8 to 10^{11} $M_\odot$. The horizontal dashed line is an approximate value delineating starbursts, with $b \geq 2$, as discussed in the text.

If one plots the third dimension of this parameter space (SFR density vs. b), one finds a significant degree of correlation between the two quantities, especially when considering the extremes in the distributions. This correlation has its physical root in the Schmidt law (Kennicutt 1998). In the context of the present discussion, it means that most galaxies that are defined as starbursts in terms of having a high SFR density also generally have a high b value, hence the two definitions of starburst are roughly self-consistent. There are prominent exceptions, however, for example circumnuclear starbursts in barred galaxies, which can have a very high SFR density but represent relatively low SFRs when compared to the global average in the galaxy (Kormendy & Kennicutt 2004).

When one plots these relations for different host galaxy Hubble types, several interesting trends emerge (Kennicutt et al., in prep.). Intermediate-type spiral galaxies (types Sbc–Sd) are remarkably homogeneous in terms of their mean SFR densities and b values, over many orders of magnitude in absolute SFR. We suspect the reason has to do with the influence that star formation has on galaxy classification. By contrast, there is a much higher diversity of properties among the earliest-type disk galaxies (S0/a–Sab) and the Magellanic spiral and irregular galaxies (Sm–Im). In the former case, this is due to a bimodality between low-level disk star formation and circumnuclear star formation (Kennicutt 1998a), whereas in the irregular galaxies the dispersion in SFR properties appears to be due to a stronger role of starbursts in these systems.

3. Application to Starburst Galaxies

We now can apply the three physical criteria described earlier to define the regions in Figs. 1 and 2 that represent starburst galaxies. If one collapses the data in Fig. 1 vertically, the resulting histogram of points is the SFR distribution function (i.e., the Hα luminosity function). The data plotted here are inappropriate for defining this function, because the data are blended from 3 different surveys with varying completeness properties. However, such luminosity functions have been constructed for the UCM and KISS prism surveys (Gallego et al. 1995 and Gronwall 1999, respectively). They are well fitted by Schechter functions with exponential scale SFR* $\simeq$ 5 M$_\odot$ yr^{-1}, when converted to $H_0 = 75$ km s^{-1} Mpc^{-1} and consistent extinction corrections. Therefore, I have defined an arbitrary dividing line for starbursts of 2 SFR*, or 10 M$_\odot$ yr^{-1}, as denoted by the dashed vertical line in Fig. 1. The absence of many points above this limit confirms the rarity of such starburst galaxies in the local Universe. Most of them are heavily dust-obscured galaxies that are very rare locally, as discussed later.

One can use a similar procedure to define a dividing line for starbursts in terms of the SFR intensity $\Sigma_{\rm SFR}$. Collapsing the data in Fig. 2 horizontally yields a distribution function of SFR intensities. This distribution shows a roughly Gaussian distribution, with a tail extending to high intensities. The tail extends approximately above an intensity of 0.1 M$_\odot$ yr^{-1} kpc^{-2}, and we adopt this as our dividing line for starbursts, as denoted by the horizontal dotted line in Fig. 1. This also happens to coincide with the value often adopted by other authors.

Finally, we follow the same procedure to define starbursts in terms of the ratio of present to past SFR (b) in Fig. 2. The distribution function for b also follows a roughly Gaussian distribution, in this case with tails in both directions. The upper tail extends above values of $b \sim 2$, and we have adopted that as our definition for a starburst, as denoted by the horizontal line in the figure. Note that since the typical value of b for quiescent disks is roughly 0.5, this threshold corresponds to a SFR that is typically ~ 4 times higher than the mean SFR among the non-bursting objects. It is a sensible definition from a physical point of view, independent of the distributions in Fig. 2.

It is instructive to plot previously published samples of starbursts in the same diagram, to see how these subjectively defined samples conform to our new physically-based definitions of starbursts. This is illustrated in Fig. 3,

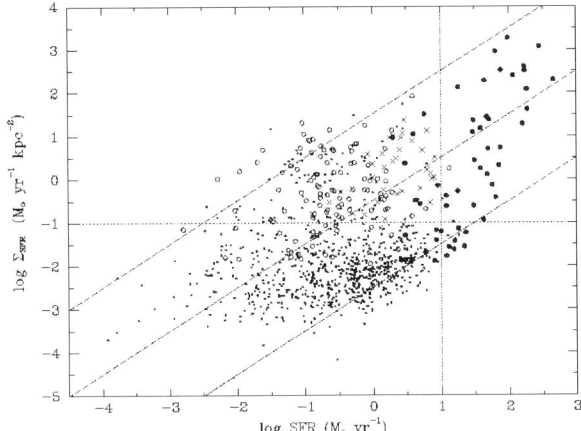

Figure 3. Similar to Fig. 1, except that we have overplotted 3 different types of starburst samples from the literature, luminous and ultraluminous star-forming infrared galaxies (large solid points), blue compact starburst galaxies (open circles), and circumnuclear starbursts identified in the optical (crosses). The reference sample from Fig. 1 is plotted as small solid points. See text for relevant references.

which shows the same reference sample from Fig. 1 (plotted as small solid points), along with 3 sets of starbursts. The large solid points denote infrared luminous and ultraluminous star-forming galaxies from the samples of Soifer et al. (2000, 2001) and Dopita et al. (2002). As anticipated earlier, these objects extend the range of absolute SFRs to values of hundreds of solar masses per year. The open circles are blue compact galaxies (BCGs) from the study of Gil de Paz et al. (2003), and the crosses denote a sample of circumnuclear starbursts compiled from a variety of sources and discussed in Kormendy & Kennicutt (2004). Interestingly, there is reasonable consistency between our physical definitions and these subjective classifications.

As a simple illustration of these applications, we have used the surveys described above in various subsets to estimate the total contribution of starbursts to the total SFR in the local Universe. The 11HUGS data provide a direct estimate of the SFR fraction in systems with $b \geq 2$ or $\Sigma_{SFR} \geq 0.1 \, M_\odot \, yr^{-1} \, kpc^{-2}$; across the entire sample this fraction is $\sim 4–6\%$. Circumnuclear starbursts contribute up to another 3–5% (with the uncertainty arising from the extinction corrections applied), though part of this is double counted with the first contribution. The burst fraction of low-luminosity galaxies is much higher, but the total SFR budget is dominated by intermediate-mass galaxies (Brinchmann et al. 2004). The fraction of star formation from infrared-luminous star formation must be estimated independently, using the infrared luminosity function of galaxies. Soifer et al. (1987) show that LIGs and ULIGs contribute 6% of the total far-infrared luminosity of the local Universe. Depending on how much of the infrared luminosity is associated with star formation at various luminosities this implies that such objects contribute about 4–8% of all star formation locally.

Combining these results and taking into account double counting gives an estimated contribution by starbursts of $\sim 15\%$ to the present-day total star formation. I caution that these numbers are crudely estimated (as indicated by the uncertainties quoted above), and the total could easily change by up to 50% after we perform more careful analyses. This will include a comprehensive analysis of dwarf galaxy burst duty cycles by Janice Lee, as part of her ongoing Ph.D. thesis. However the rough results above are in good agreement with independent estimates using other methods (e.g., Brinchmann et al. 2004). We hope in any case that we have illustrated the kinds of analyses that are being enabled by the new data explosion.

References

Brinchmann J., Charlot S., White S.D.M., Tremonti C., Kauffmann G., Heckman T., Brinkmann J., 2004, MNRAS, 351, 1151
Dopita M.A., Pereira M., Kewley L.J., Capaccioli M., 2002, ApJS, 143, 47
Gallego J., Zamorano J., Aragon-Salamanca A., Rego M., 1995, ApJ, 455, L1
Gavazzi G., Boselli A., Donati A., Franzetti P., Scodeggio M., 2003, A&A, 400, 451
Gil de Paz A., Madore B.F., Pevunova O., 2003, ApJS, 147, 29
Gronwall C., 1999, in: After the Dark Ages: When the Universe Was Young (the Universe at $2 < z < 5$), Holt S., Smith E., eds., AIP Conf. Ser., (AIP Press), vol. 470, p. 335
Hameed S., Devereux N., 1999, AJ, 118, 730
James P.A., 2004, A&A, 414, 23
Kennicutt R.C., 1983, ApJ, 272, 54
Kennicutt R.C., Tamblyn P., Congdon C.W., 1994, ApJ, 435, 22
Kennicutt R.C., 1998a, ARA&A, 36, 189
Kennicutt R.C., 1998b, ApJ, 498, 541
Kennicutt R.C., et al., 2003, PASP, 115, 928
Kormendy J., Kennicutt R.C., 2004, ARA&A, 42, 603
Lanzetta K.M., Yahata N., Pascarelle S., Chen H.-W., Fernández-Soto A., 2002, ApJ, 570, 492
Martin C.L., Kennicutt R.C., 2001, ApJ, 555, 301
Meurer G.R., Heckman T.M., Lehnert M.D., Leitherer C., Lowenthal J., 1997, AJ, 114, 54
Soifer B.T., et al., 2000, AJ, 119, 509
Soifer B.T., et al., 2001, AJ, 122, 1213

STAR-FORMING, RECENTLY STAR-FORMING, AND "RED AND DEAD" GALAXIES AT $1 < Z < 2$

Highlights from the Gemini Deep Deep Survey

Roberto G. Abraham[1], Karl Glazebrook[2], Patrick J. McCarthy[3], David Crampton[4], Sandra Savaglio[2], Stephanie Juneau[4], Damien Le Borgne[1], Hsiao-Wen Chen[6], Raymond G. Carlberg[1], Richard Murowinski[4], Inger Jørgensen[6], Kathy Roth[6], and Ron Marzke[7]

[1]*Department of Astronomy & Astrophysics, University of Toronto, Canada;* [2]*Department of Physics & Astronomy, Johns Hopkins University, USA;* [3]*Observatories of the Carnegie Institution of Washington, USA;* [4]*Herzberg Institute of Astrophysics, National Research Council of Canada, Canada;* [5]*Center for Space Research, Massachusetts Institute of Technology, USA;* [6]*Gemini Observatory;* [7]*Department of Physics & Astronomy, San Francisco State University, USA*

Abstract We summarize some of the key results on red galaxies presented in the first five papers from the Gemini Deep Deep Survey (GDDS). The GDDS is an ultra-deep ($K < 20.6$ mag, $I < 24.5$ mag) spectroscopic redshift survey, which preferentially targeted galaxies in the redshift range $0.8 < z < 2$. The survey was completed in 2004, and the data is publicly available. The primary goal of the GDDS was to make the first direct measurement of the evolving stellar mass function over $0.8 < z < 2$. We find that $\sim 15\%$ of the local stellar mass density is already in place by $z = 1.8$, rising to 40–50% by $z = 1$. Nearly half the stellar mass density at $z = 1.8$ is in massive "red-and-dead" galaxies. Almost all of these red-and-dead galaxies exhibit early-type morphologies, although around 20% of these show tidal distortions consistent with recent (dry?) merger activity. A decomposition of the star-formation history of the Universe into the sum of individual histories for mass-segregated galaxy populations reveals strong evidence for the "down-sizing" paradigm espoused by Cowie et al. (1997). We find that the most massive galaxies form early in the history of the Universe and galaxy formation proceeds from larger to smaller mass scales.

1. The Gemini Deep Deep Survey

The Gemini Deep Deep Survey (GDDS) is an ultra-deep ($K < 20.6$ mag, $I < 24.5$ mag) redshift survey targeting galaxies in the "redshift desert" between $0.8 < z < 2$. The primary goal of the survey is to constrain the space density at high redshift of evolved high-mass galaxies. We obtained 309 spectra in four widely-separated 30 arcmin^2 fields using the Gemini North telescope and the Gemini Multi-Object Spectrograph (GMOS). Integration times of 20–30 hours per field allowed us to obtain high-quality spectra even for faint galaxies near our completeness limit. The spectra define a one-in-two sparse sample of the reddest and most luminous galaxies near the $(I - K)$ vs. I color-magnitude track mapped out by passively evolving galaxies in the redshift interval $0.8 < z < 1.8$. This sample is augmented by a one-in-seven sparse sample of the remaining high-redshift galaxy population.

2. What We've Learned

It is impossible to summarize the first five GDDS papers in the space allotted in these proceedings, so I will focus on some selected highlights. I omit much of the discussion of the chemical evolution in the blue star-forming galaxy population in our survey, since this has recently been reviewed in Savaglio et al. (2004). Based on the principle that a picture is worth a thousand words, I will focus on some key plots generated by the survey team.

The main result from the GDDS is that there is a very significant population of red and massive galaxies at $1.3 < z < 2$, whose integrated light is dominated by evolved stars (Abraham et al. 2004, Glazebrook et al. 2004, McCarthy et al. 2004). The evidence for this is summarized in the left-hand panel of Fig. 1. Note that these galaxies are red because they are old, and not because of dust extinction. Preliminary estimates for the ages of these $z \sim 1.5$ galaxies give a median age of 1.2 Gyr and a median redshift of formation around $z = 2.4$, with 1/4 of the sample having a redshift of formation of $z > 4$ (McCarthy et al. 2004). These galaxies are a major contributor to the stellar mass density at $1 < z < 2$ (Fig. 1, right panel). At $z = 1$ the mass densities for the $M > 10^{10.8}$ M$_\odot$ sample are 38 (± 18)% of their local value (Cole et al. 2001); at $z = 1.8$ this becomes 16 (± 6)% (Glazebrook et al. 2004). This appears consistent with some recent results based on photometric redshifts (Fontana et al. 2003, Bell et al. 2003, Rudnick et al. 2003, Dickinson et al. 2003, Chen et al. 2002). On the basis of these estimates, it is clear that an extinction correction is essential for UV SFR estimates to be consistent with local stellar mass measurements (Cole et al. 2001).

Figure 2 shows the star-formation rate density and its dependence on galaxy stellar mass over the redshift range $0 < z < 2$, using our data Juneau et al. (2005) and that from Brinchmann et al. (2004). The star-formation rate density in the most massive galaxies ($M > 10^{10.8}$ M$_\odot$) was six times higher at $z = 2$ than it is today. It drops steeply from $z = 2$, reaching the present-day value at $z \sim 1$. In contrast, the star-formation rate density of intermediate-mass galaxies ($10^{10.2}$ M$_\odot < M < 10^{10.8}$ M$_\odot$) declines more slowly and may peak or plateau at $z \sim 1.5$. We use the characteristic growth time, defined as the star-

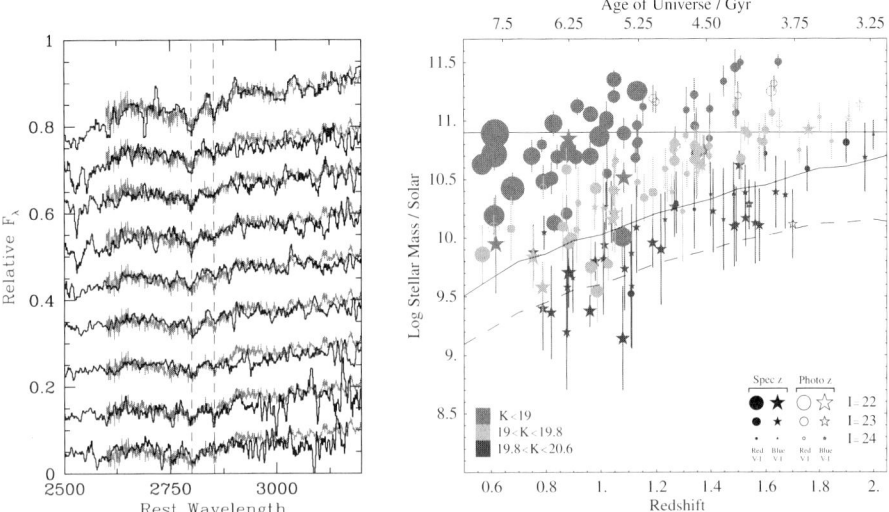

Figure 1. *Left:* Spectra of evolved GDDS galaxies with $z > 1.3$, taken from McCarthy et al. (2004). The SDSS Luminous Red Galaxy composite has been overlaid on each spectrum and an offset has been applied to each, in steps of 10^{-18} erg s^{-1} cm^{-2} Å^{-1}. The locations of the stellar MgIIλ2800 and MgIλ2852 lines are indicated by the dashed lines. *Right:* Mass-redshift distribution for GDDS galaxies, taken from Glazebrook et al. (2004). Points are shaded according to observed K magnitude. Solid and open symbols denote spectroscopic and photometric redshifts, respectively. Circles denote objects redder in $(V - I)$ than a standard model Sbc galaxy template, while objects shown using star symbols are bluer than this standard template. The symbol size is keyed to the I band magnitude. The horizontal line denotes M^* in the local Universe (Cole et al. 2001). The solid curve shows how the K flux limit translates into a mass completeness limit for a maximally old SSP. The dashed curve shows an example mass limit for bluer objects (SFR = constant model).

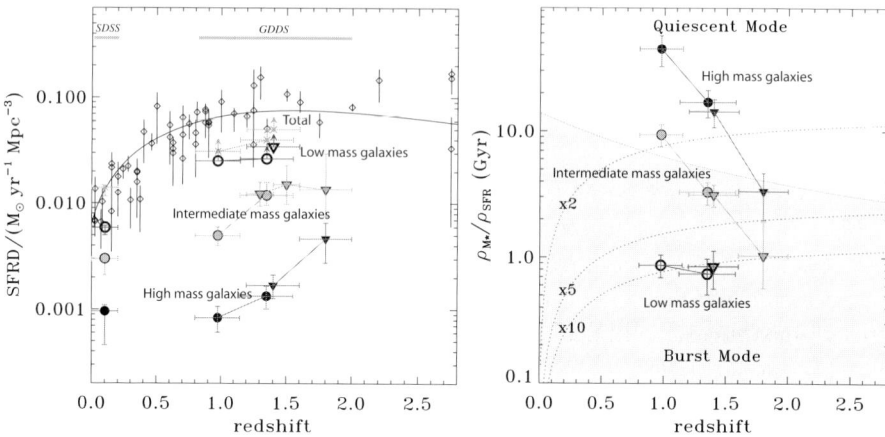

Figure 2. Two figures (based on those in Juneau et al. 2005) that summarize the evidence for galaxy down-sizing in the GDDS. *Left:* Star-formation rate density in different stellar mass intervals. Symbol shading encodes the mass bin in which the object belongs: Open circles correspond to $10^9 \, M_\odot < M_* < 10^{10.2} \, M_\odot$; gray circles correspond to $10^{10.2} \, M_\odot < M_* < 10^{10.8} \, M_\odot$ and solid circles correspond to $10^{10.8} \, M_\odot < M_* < 10^{11.5} \, M_\odot$. Horizontal error bars show the width of the redshift bins used. Vertical error bars combine shot noise and mass-completeness corrections. Symbol shape corresponds to the method used to obtain the star-formation rate (see Juneau et al. 2005 for details). Both the sampling and the spectroscopic completeness corrections given in Abraham et al. (2004) have been applied. Values found locally by Brinchmann et al. (2004) are shown at far left. The compilation made by Hopkins (2004; for all masses), where all the values are converted to a ($\Omega_m = 0.3, \Omega_\Lambda = 0.7, h = 0.7$) cosmology, are overplotted with open diamonds. The line is the fit derived by Cole et al. (2001), assuming $A_V = 0.6$ mag (solid line). *Right:* Characteristic timescale of galaxy growth obtained by dividing the stellar mass density by the star-formation rate density. Symbols are as shown at left. The vertical error bars are statistical. The dashed line shows the age of the Universe in our adopted cosmology. It indicates the transition from quiescent SF mode to burst SF mode (gray area). Massive galaxies transition from burst mode to quiescent mode at around $z = 2$, intermediate-mass galaxies make the transition at around $z = 1$, and low-mass galaxies remain in burst mode at the lowest redshift probed by the GDDS. Along the dotted lines, lookback time to the present allows galaxy stellar mass to increase by a factor of 2, 5 or 10, if the SFR remains constant, as labeled.

formation rate density divided by the stellar mass density, $t_{\rm SFR} \equiv \rho_{M_*}/\rho_{\rm SFR}$, to provide evidence of an associated transition in massive galaxies from a burst to a quiescent star-formation mode at $z \sim 2$. Intermediate-mass systems transit from burst to quiescent mode at $z \sim 1$, while the lowest-mass objects undergo bursts throughout our redshift range. The most massive galaxies formed most of their stars in the first ~ 3 Gyr of cosmic history. Intermediate-mass objects continued to form their dominant stellar mass for an additional ~ 2 Gyr, while the lowest-mass systems have been forming over the whole of cosmic time. This view of galaxy formation clearly supports "down-sizing" in the star-formation rate, where the most massive galaxies form first and galaxy formation proceeds from larger to smaller mass scales.

The morphological evidence supports the view that most of the massive red population in our data are early-type galaxies (Abraham et al., in prep.). Therefore, our results point toward early formation for a significant fraction of present-day *massive* elliptical galaxies (but see the next section for a caveat). The derived ages and most probable star-formation histories suggest a high star-formation rate ($\sim 300 - 500$ $M_\odot$ yr^{-1}) phase in the progenitor population.

3. Predictions, Caveats and Conclusions

Galaxies associated with luminous sub-mm sources have the requisite star-formation rates and space densities to be the progenitor population of the massive red galaxies probed by the GDDS, but current estimates of their redshift distribution suggest such sub-mm sources form at lower redshifts than many of the galaxies in our sample. Our observations lead to a natural prediction that many higher-redshift sub-mm sources are waiting to be discovered, and that their clustering properties, which are on the border of being measurable with surveys such as SHADES, should echo those of early-type galaxies. (See see http://www.roe.ac.uk/ifa/shades/ for information on the SHADES sub-mm program.)

Despite this prediction, it is dangerous to conclude that all early-type galaxies formed at high redshift. The strong trends we see with stellar mass mean that blanket statements about the formation epoch of various aspects of the Hubble sequence are meaningless unless one carefully specifies the mass range of the galaxies being described. The observation that most *very massive* early-type systems formed at high redshifts still leaves open the possibility that most intermediate and low-mass early-type galaxies formed more recently. After all, only 15% of the local stellar mass density is in place at $z \sim 2$, and while this might (or might not, depending on which theorist you ask) pose challenges for hierarchical models, it certainly leaves a lot of room for activity at lower redshifts (Glazebrook et al. 2004). It seems to me that the basic statistical arguments presented by Toomre back in 1977 still remain valid today. Toomre showed that the 11 strong ongoing mergers seen in the New General Catalog correspond to > 250 galaxies in the NGC having participated in major mergers over the lifetime of the Universe. This is pretty similar to the ~ 300 ellipticals in the whole catalog. Despite some obvious weaknesses in this line of reasoning (which Toomre clearly spelled out in his original paper), I find that (on balance) this is still a pretty persuasive argument for forming many early-type galaxies via recent mergers. A key testable prediction that follows from this is that most of these young ellipticals should be low-mass systems.

Finally, I would like to conclude by noting explicitly what should be obvious from the references given above, namely that many of the same conclusions reached by the GDDS team were also reached at more-or-less the same time by other groups, e.g., by the K20 team working with shallower spectroscopic data but over a wider area, and by some groups working with photometric redshifts. For example, compare the GDDS and K20 results for masses (Glazebrook et al. 2004, Fontana et al. 2004), and ages (McCarthy et al. 2004, Cimatti et al. 2004). I think that much of the credibility underlying the mini paradigm shift that has occurred over the last couple of years – in which the view that the most massive galaxies formed early seems to have transformed from heresy to orthodoxy – is due to the way in which observations based on very different experimental designs seem to have converged onto the same basic picture.

References

Abraham R.G., et al., 2004, AJ, 127, 2455
Bell E.F., McIntosh D.H., Katz N., Weinberg M.D., 2003, ApJ, 585, L117
Brinchmann J., Charlot S., White S.D.M., Tremonti C., Kauffmann G., Heckman T., Brinkmann J., 2004, MNRAS, 351, 1151
Chen H.-W., et al., 2002, ApJ, 570, 54
Cimatti A., et al., 2002, A&A, 381, L68
Cimatti A., et al., 2004, Nature, 430, 184
Cole S., et al., 2001, MNRAS, 326, 255
Cowie L.L., Hu E.M., Songaila A., Egami E., 1997, ApJ, 481, L9
Dickinson M., Papovich C., Ferguson H.C., Budavári T., 2003, ApJ, 587, 25
Fontana A., et al., 2003, ApJ, 594, L9
Fontana A., et al., 2004, A&A, 424, 23
Glazebrook K., et al., 2004, Nature, 430, 181
Hopkins A.M., 2004, ApJ, 615, 209
Juneau S., et al., 2005, ApJ, 619, L135
McCarthy P.J., et al., 2004, ApJ, 614, L9
Rudnick G., et al., 2003, ApJ, 599, 847
Toomre A., 1977, in: The Evolution of Galaxies and Stellar Populations, Tinsley B.M., Larson R.B., eds. (New Haven: Yale Univ. Press), p. 401

GLOBAL STAR-FORMATION RATES

Joseph Silk
Department of Physics, University of Oxford, Oxford OX1 3RH, UK
silk@astro.ox.ac.uk

Abstract   I review the basic modes of quiescent star formation and starbursts. Star-formation efficiency is the key to reconciling hierarchical galaxy formation with observed galaxy colours and counts. A unified viewpoint is that all star formation is, at some level, bursty. This is motivated both by local observations and by theory. I describe how self-regulation of star formation provides prescriptions for the star-formation rate and efficiency, and for starburst-driven mass outflows.

1.   Introduction

It has long been argued on the basis of galaxy colours that star formation is bimodal. Disks are blue and have extended star formation over a Hubble time. Ellipticals, S0s and bulges are red, and are required to have formed in a burst of star formation that lasted $\lesssim 1$ Gyr. Indeed, modelling of spectral energy distributions requires the star-formation rate in disks, conveniently parameterised by $b = \mathrm{SFR}/\langle\mathrm{SFR}\rangle$, to vary from ~ 0.1 in early-type disks to ~ 10 in late-type disks. Moreover, the starburst duration is inferred, from studies of ongoing and recent starbursts, to last from $\sim 0.01 - 0.1$ Gyr. Many nearby early-type galaxies, ellipticals and S0s, show traces of ongoing or recent star formation when observed in the FUV. The rates are low, and consistent with $b = 0.01 - 0.001$. Clearly, the concept of two modes of global star formation is likely to be an oversimplification of the physical situation.

2.   Star-formation modes

Cold disks are gravitationally unstable, both to local and global modes, the latter requiring non-axisymmetric perturbations. A hybrid disk containing cold gas and stars is unstable if a suitably generalised Toomre criterion is satisfied, $Q^\mathrm{T} \lesssim 1$. One can very approximately write (Rafikov 2001) $1/Q^\mathrm{T} \approx 1/Q_\mathrm{g}^\mathrm{T} + 1/Q_*^\mathrm{T}$, where $Q_\mathrm{g}^\mathrm{T} \approx \kappa\sigma_\mathrm{g}/G\Sigma_\mathrm{g}$, and $Q_*^\mathrm{T} \approx \kappa\sigma_*/G\Sigma_*$, so that cold gas controls the instability if $\Sigma_\mathrm{g}/\sigma_\mathrm{g} > \Sigma_*/\sigma_*$. Here, κ is the epicyclic frequency, σ_g is the gas velocity dispersion, and σ_* is the stellar radial velocity dispersion. Also, Σ_g is the surface density of cold gas and Σ_* is the surface density of stars.

This condition is satisfied at least initially, since the gas cools and the stellar distribution heats up once the instability begins to operate. For instability, Σ_g must exceed a critical value, but a supply of cold gas is also required to maintain a low value of σ_g. The disk instability is controlled by the cold gas supply. If continuous, as with accretion, the resulting star formation is quasi-steady. If discrete, as would be the case for infalling gas-rich satellites, a more sporadic history of star formation is likely.

There is empirical evidence for a reservoir of cold gas in the high-velocity halo clouds, although the disk accretion rate cannot amount to more than a tenth of the star-formation rate, otherwise the clouds would be too massive and visible around nearby galaxies. However, the circumgalactic ratio of hot ($\sim 300,000$ K) to neutral ($\lesssim 1000$ K) gas may be substantial, as evidenced by OVI absorption studies (Danforth & Shull 2005, Tripp et al. 2005). Infall is also inferred indirectly from chemical evolution of the disk, in order to account for the paucity of metal-poor G dwarfs.

Quiescent mode. Consider first how quiescent star formation might have proceeded. The maximum growth rate for the disk instability is $\sim \Omega$, the rotation rate. A plausible *ansatz* for the star-formation rate is that, at any galactocentric distance where the surface density is above the Toomre threshold,

$$\frac{d\Sigma_*}{dt} = \epsilon \Omega \Sigma_g. \qquad (1)$$

The star-formation efficiency parameter, $2\pi\epsilon$, is defined to be the fraction of gas converted into stars per galactic rotation period. The linear instability criterion cannot yield ϵ. Rather, it must come from feedback and self-regulation, which motivate the proportionality to rotation rate and cold gas surface density (Silk 1997). The observed Kennicutt law, with a universal value of $\epsilon \approx 0.017$, fits this simple expression over a wide range of star-formation rates and galactic disk radii (Kennicutt 1989). The gas e-folding depletion time is enhanced by around 50 per cent when allowance is made for gas return. It requires about 3 e-foldings to arrive at the observed gas fraction, and this takes about 10 Gyr. Cold gas accretion probably plays a minor role in replenishing the disk gas reservoir, at least in the later stages of disk star formation.

The disk instability explanation of the star-formation law is inadequate, not only because the star-formation efficiency is unspecified, but also because the wavelength corresponding to the most rapidly growing mode corresponds to larger masses than are observed, even for giant molecular clouds. Because of this mismatch and the lack of any star-formation physics, the Kennicutt law cannot be considered to provide an *explanation* of global star formation in terms of disk instability. Indeed, star formation is found to occur beyond the threshold value, in the outer disk. A non-linear description involving cloud-cloud collisions, triggered by spiral density waves, is one attempted explanation (Wyse 1986). This can also provide a Kennicutt-like law, and allows star formation below the Toomre threshold. However no prescription for star formation is provided in this approach, that is, again, ϵ is unspecified.

It is clear that other factors must play a role. The most promising of these is interstellar gas turbulence. This is motivated both by observations of edge-on disks and by merger simulations, and will be described below.

Starburst mode. It has long been conjectured that mergers of gas-rich galaxies trigger violent star formation. There is persuasive empirical evidence that the ultraluminous infrared galaxies are undergoing mergers or strong tidal interactions. Near-infrared surface brightness profiles of recent mergers indicate that a de Vaucouleurs profile has been generated, demonstrating that spheroid formation can occur within a merger time-scale. The incidence of close galaxy pairs, potential merger candidates, increases with increasing redshift, consistent with the predictions of hierarchical galaxy formation theory. Detailed merger simulations, while invariably one step or more behind the observations of starbursts, confirm that the gas concentration is strongly augmented by the action of tidal torques and the ensuing enhanced gas dissipation during a merger.

This interpretation of starbursts is necessarily incomplete. There is no detailed modelling of the conversion of gas into stars and hence of the star-formation efficiency. It is remarkable how well the Schmidt-Kennicutt star-formation law, in which the star-formation rate per unit area is proportional to the product of the gas surface density and the disk rotation rate, fits nearby disks and starbursts. Nevertheless, the application of this empirical law to modelling of the star-formation rate in semi-analytical galaxy formation and hydrodynamical merger simulations is inadequate on at least two fronts.

Incorporation of a Schmidt-Kennicutt star-formation law relating the star-formation rate to the local cold gas density fails to explain the extended nature of the star formation observed in merging galaxies (Schweizer, these proceedings). Moreover, adoption of a universal star-formation efficiency results in challenges to the comparison of semi-analytical galaxy formation models with observational data. The predicted colour and age distributions, α element-to-iron abundance ($[\alpha/\text{Fe}]$) ratios w.r.t. solar, and infrared/sub-mm galaxy counts are all in conflict with results from recent surveys (e.g., Thomas et al. 2005).

Nevertheless, merging is well-motivated as a description for starbursts. The challenge is to derive an empirical star-formation law. Theory suggests that both mergers and accretion play important roles in providing the gas supply that regulates star formation. In fact, gas accretion is just the limiting case of minor mergers. In a unified picture, we might just consider the physics of mergers. Indeed, the Milky Way is an interesting case study. There is increasing evidence for past minor mergers from combining chemical and dynamical tracers of high-velocity stars. Chemical signatures associated with the disruption of the Arcturus stream and with ωCen, and kinematical modelling of the Sagitarius dwarf tidal stream, all point towards a merging history which leaves behind relic stellar tracers (Navarro, Helmi & Freeman 2004, Helmi 2004). Such merging events would have injected a sporadic supply of gas that temporarily reinvigorated star formation. Evidence for such "mini starbursts" is seen in the local history of star formation as traced by chromospheric dating of nearby stars (Rocha-Pinto et al. 2000).

Perhaps one of the strongest arguments for a mini-starburst history in galaxies of stellar mass up to $\sim 10^{10}$ $M_\odot$ comes from recent studies of luminous infrared galaxies at $z > 0.4$ (Hammer et al. 2005) and of the mass-metallicity relation in the local Universe (Tremonti et al. 2004). The high frequency of so-called LIRGs argues for an episodic or bursty star-formation history in intermediate-mass galaxies. Moreover, evidence for metal loss via winds is inferred not just for dwarfs, but for moderately massive galaxies. Quiescent disk star formation cannot drive outflows from such galaxies. A series of blow-out events associated with mini starbursts may suffice. Indeed, accumulating evidence from studies of the local Universe suggests that nearby starbursts typically occur in sub-L^* galaxies and have outflows of order the star-formation rate, at least for the handful of well-studied examples (e.g., Martin 2003).

3. A Unified Model for Starbursts and Disk Star Formation

Global simulations of the multiphase interstellar medium have inadequate dynamical range and resolution to be able to trace out the complex interplay of cold gas accretion into clouds, supernova heating, cloud cooling, fragmentation, and star formation. I describe a simple analytic model that appears to incorporate at least some of the crucial physics and that has been tested against simulations of a kiloparsec cube of the interstellar medium.

The basic model is that supernovae explosively blow hot bubbles into the interstellar medium. The bubbles decelerate by sweeping up shells of cold gas and eventually break up when the expansion is halted by the ambient pressure. If the rate of bubble formation is sufficiently high, the bubbles overlap and a multi-phase medium develops of hot shell-shocked gas in which dense cold shell fragments are embedded. If the hot gas permeates through the cold gas scale height, galactic fountains and disk outflows will result.

To develop a simple model, I set the disk outflow rate equal to the product of the supernova rate, the hot gas filling factor, and the mass loading factor. The latter depends on such effects as Kelvin-Helmholtz instabilities that entrain cold gas into the hot phase. The filling factor f_h can be expressed in terms of the hot-gas porosity (which I define as Q) by $1 - e^{-Q}$. If the porosity is low, then $f_h \approx Q$. As the porosity becomes large, the hot-gas filling factor approaches unity. The porosity may be defined as the product of the supernova rate and the maximum 4-volume attained by a supernova-driven bubble,

$$Q \sim (\text{SN bubble rate per unit volume}) \times (\text{maximum bubble 4-volume}). \tag{2}$$

Now the radius of a bubble is computed from the approximately momentum-conserving phase, when the shell shock velocity has dropped well below the critical velocity at which radiative cooling sets in, namely $v_c \approx 400 \text{ km s}^{-1}$. One finds that $R_a^3 t_a$, where R_a is the maximum bubble size achieved at time t_a, is proportional to $p_{\text{turb}}^{-1.4}$, where p_{turb} is the ambient, mostly turbulent, interstellar gas pressure that limits the bubble expansion. I conclude that the star-formation rate is proportional to $Qp_{\text{turb}}^{1.4}$. Including the correct normalisation, I can now write

$$\text{SFR} = Q\epsilon\Omega\rho_{\text{gas}}, \tag{3}$$

where $\epsilon = (\sigma_{\text{gas}}/\sigma_f)^{2.7}$, $\sigma_f \approx 20 \text{ km s}^{-1} (E_{\text{SN}}/10^{51}\text{erg})^{0.6} (200 M_\odot/m_{\text{SN}})^{0.4}$, and Ω is the disk rotation rate. Here, E_{SN} is the kinetic energy output of the supernova and m_{SN} is the mass in stars formed per Type II supernova (approximately 200 $M_\odot$ for a Kroupa IMF).

The analogy with the empirical Schmidt-Kennicutt law is suggestive. Firstly, we may anticipate that the porosity self-regulates as long as the gas supply is maintained, in such a way that $Q\epsilon \approx$ constant. The argument is simple: if $Q \gg 1$, the cold gas supply is exhausted and massive star formation is quenched, and if $Q \ll 1$, there is no impedance to runaway star formation via supernova-induced turbulent pressure. This latter argument does assume that supernovae initially exert positive feedback, that is when the porosity is low. Of course, once the porosity is high, the disk will vent hot gas and the feedback will be negative. Moreover, the star-formation rate cannot exceed the gas supply, so $Q\epsilon \lesssim 1$. Secondly, we recognise a new addition to the phenomenological star-formation rate: a factor ϵQ that we can identify with star-formation efficiency and that is proportional to the product of the porosity and the turbulent velocity dispersion of the ambient gas raised to the 2.7th power.

It is plausible that the interstellar turbulence driven by the supernova ejecta approximately conserves momentum. I write the Kennicutt coefficient as $\alpha_* = \dot{\Sigma}_*/\Sigma_{\text{gas}}\Omega$. Momentum balance suggests that $\alpha_* \approx \sigma_{\text{gas}} v_c m_{\text{SN}} E_{\text{SN}}^{-1}$, and

$$\alpha_* = 0.02(\sigma_{\text{gas}}/10 \text{ km s}^{-1})(v_c/400 \text{ km s}^{-1})(m_{\text{SN}}/200 M_\odot)(10^{51}\text{erg}/E_{\text{SN}}). \tag{4}$$

The observed value of α_* is 0.017. Our derived star-formation rate suggests $Q = (H_{\text{gas}}/H_*)\alpha_* \epsilon^{-1}$, where H_{gas} and H_* are the cold gas and stellar scale heights, respectively. Thus, $Q \sim 0.5$ in a disk but $Q \ll 1$ in a starburst.

We infer that in extreme turbulence, the porosity is likely to be small, and that in typical disks where self-regulation yields $Q \sim 1$, turbulence enhances star formation. On the scale of individual star-forming clouds and young stellar associations, there is empirical evidence that star formation is accelerating (Palla & Stahler 2000), and this has been interpreted as evidence for mini starbursts (Silk 2005). In this regime, the feedback is positive. Globally, however, disk venting results in negative feedback. Hence, enhanced turbulence in disks effectively controls star formation by reducing the cold gas reservoir.

Turbulence-driven star formation has further implications. There is positive feedback if turbulence is injected, for example, in a galaxy merger. The resulting star formation is distributed more broadly than if the star-formation rate depends only on the local density. The characteristic star-formation time $t_{\rm sf}$ is defined by $\rho_*/$SFR and is $f_*/(Q\epsilon)$ dynamical times, where $f_* \equiv \rho_*/\rho_{\rm gas}$. We may say that a starburst occurs (but this is not a unique definition!) when $t_{\rm sf}$ is less than a dynamical time, or in a gas-rich environment where $f_* < Q\epsilon$. Star formation ceases, largely due to gas consumption in star formation and to the onset of outflows.

The estimated value of $\sigma_{\rm f}$ assumes spherically symmetric bubbles in a homogeneous medium. This will inevitably be an underestimate, since both geometrical effects and the inclusion of Rayleigh-Taylor instabilities will increase the porosity and hence the effective value of $\sigma_{\rm f}$. Disk galaxies typically have $\sigma_{\rm gas} \approx 10$ km s^{-1}, and the gas turbulence is limited by the escape velocity from the kinematically cold disk. Hence we may draw two preliminary conclusions. Star formation is quenched in dwarf galaxies. Also in merging galaxies, where $\sigma_{\rm gas} \gg \sigma_{\rm f}$, and a starburst occurs, the porosity is low, which means that (at least nuclear) outflows are suppressed.

Simulations generally confirm these analytical results. Hitherto, a cubic kiloparsec of the interstellar medium has been modelled at 10 parsec resolution, including a multiphase treatment of star formation and supernova feedback (Slyz et al. 2005). The computed star-formation rate follows the naive porosity-based prediction. The only adjustable parameter is the star-formation efficiency, to which the results seem relatively insensitive. The gas is thermally unstable as supernova feedback maintains approximate pressure balance. The adopted periodic boundary conditions mean that it is not straightforward to generalise this result to an entire galaxy. The following results, however, emerge from the simulations.

The analytic porosity formulation gives a good match to the simulated star-formation rate. Insight into why there initially is positive feedback indicates that this arises because the computational resolution allows one to track the stellar kinematics and to explode stars in regions less dense than their natal material. Over the lifetime of an OB star, the star drifts out of the dense core where it formed and explodes in a region of lower density. The resulting explosions further compress the gas and induce further star formation. This process is limited by the availability of cold gas. The star-formation rate peaks in a burst that declines as the gas supply is exhausted.

The preceding model applies equally well to disks and to nuclear starbursts. Disk star formation is envisaged as a series of mini starbursts. In a disk, $\sigma_{\rm gas} \approx 10$ km s^{-1}, so that $Q\epsilon \approx 0.03$. The observed efficiency, according to Kennicutt, is inferred from $\dot{\Sigma}_* = 0.017\Sigma_{\rm gas}\Omega$, and is globally about 2 per cent. This means that $Q \approx 0.5$, as is observed for the Milky Way.

In a starburst, however, the star formation efficiency is necessarily higher because $\epsilon \propto \sigma_{\rm gas}^{2.7}$. The remarkable phenomenological result is that starbursts also lie on the same Kennicutt fit, $\dot{\Sigma}_* = Q\epsilon\Sigma_{\rm gas}\Omega$, with $Q\epsilon = 0.017$. However there are two differences with quiescent disk star formation. The porosity is low, since ϵ is necessarily large. And the overall efficiency of star formation is high because of the enhanced stellar scale height associated with the starburst. Spatially extended star formation is seen in mergers. This has been successfully modelled for a pair of merging galaxies (the Mice) by including a turbulence-motivated prescription for the star formation law (Barnes 2004).

To better understand the Kennicutt law implications, we may write $\dot{M}_* = Q\epsilon(H_*/H_{\rm gas})^2 M_{\rm gas}\Omega$, so that in a starburst one might plausibly expect that $\dot{M}_* \lesssim M_{\rm gas}\Omega$. Because $H_{\rm gas} \ll H_*$, this would allow extremely efficient star formation, with no feedback since the porosity is low; gas depletion occurs on a dynamical time-scale. This might be the case for ULIRGs, or more generally for nuclear starbursts. This is in contrast to the inefficient star formation and long-lived gas supply at high porosity in disks where $H_* \approx H_{\rm gas}$.

The difference in star-formation characteristics between massive and low-mass disks has received attention in a recent study of edge-on disks (Dalcanton, Yoachim & Bernstein 2004). The case is made that the scale height increases sharply (by a factor of about 2) as the rotational velocity drops below 120 km s^{-1}, simultaneously with an increase in disk gravitational stability to axisymmetric perturbations. This could be due to an increase in gas turbulence or to a decrease in disk surface density. One would expect star-formation efficiency to globally increase with lower turbulence, due to the reduced effects of disk venting via chimneys and fountains. This trend may be seen in the edge-on disk sample, although detailed simulations are needed in order to explore the theoretical implications that involve the interaction of disk gravitational instabilities, molecular cloud evolution, disk outflows and star formation.

Lower-mass disks are generally gravitationally stable and not self-regulating. The star-formation rate is low presumably because the gas supply provided by gravitational instability is reduced. Systematically lower star-formation efficiency in low-rotation, low-mass disks seems to be indicated by the data.

4. Conclusions

Starbursts are ubiquitous at high redshift. In fact, starbursts are limited by the local gas supply. If the gas supply is stochastic, this favours a series of (mini)bursts. Infalling satellites yield a stochastic gas supply. A simple self-regulation hypothesis yields the star-formation rate in a range of physical situations, thereby accounting for the Schmidt-Kennicutt law. One can regard quiescent disk star formation as a series of mini starbursts, which seamlessly progresses into the major starburst regime as the gas-supply rate is enhanced. The fact that the phenomenological star-formation law accounts both for quiescent disks and starbursts can therefore be accommodated.

Acknowledgments

I am grateful to the Kapteyn Astronomical Institute at Groningen and to the Director, Prof. Piet van der Kruit, for hospitality when I was preparing this review, and to KITP and the organisers of the Galaxy-IGM Interactions Program, where this review was completed.

References

Barnes J., 2004, MNRAS, 350, 798
Dalcanton J., Yoachim P., Bernstein R., 2004, ApJ, 608, 189
Danforth C., Shull J.M., 2005, in: Astrophysics in the Far Ultraviolet: Five Years of Discovery with FUSE, Sonneborn G., Moos W., Andersson B.-G., eds., ASP Conf. Ser., (ASP: San Francisco), in press (astro-ph/0408262)
Hammer F., Flores H., Elbaz D., Zheng X.Z., Liang Y.C., Cesarsky C., 2005, A&A, 430, 115
Helmi A., 2004, ApJ, 610, L97
Kennicutt R.C., 1989, ARA&A, 36, 189
Martin C., 2002, ApJ, 574, 663
Navarro J., Helmi A., Freeman K.C., 2004, ApJ, 601, 43
Palla F., Stahler S., 2000, ApJ, 540, 255
Rafikov R., 2001, MNRAS, 323, 445
Rocha-Pinto H., Scalo J., Maciel W., Flynn C., 2000, ApJ, 531, L115
Silk J., 1997, ApJ, 481, 703
Silk J., 2005, in: IMF@50: The Stellar Initial Mass Function Fifty Years Later, Corbelli E., Palla F., Zinnecker H., eds., Astrophysics and Space Science Library, (Kluwer: Dordrecht), in press (astro-ph/0410080)
Slyz A., Devriendt J., Bryan G., Silk J., 2005, MNRAS, 356, 737
Thomas D., Maraston C., Bender R., Mendes de Oliveira C., 2005, ApJ, in press (astro-ph/0410209)
Tremonti C.A., et al., 2004, ApJ, 613, 898
Tripp T., et al., 2005, in: Astrophysics in the Far Ultraviolet: Five Years of Discovery with FUSE, Sonneborn G., Moos W., Andersson B.-G., eds., ASP Conf. Ser., (ASP: San Francisco), in press (astro-ph/0411151)
Wyse R.F.G., 1986, ApJ, 311, L41

STAR FORMATION EFFICIENCIES AND STAR CLUSTER FORMATION

Uta Fritze – v. Alvensleben
Universitätssternwarte Göttingen, Geismarlandstr. 11, 37083 Göttingen, Germany
ufritze@uni-sw.gwdg.de

Abstract Starbursts produce large numbers of Young Star Clusters (YSCs). Multi-color photometry, in combination with a dedicated SED analysis tool, allows to derive ages, metallicities, $E(B-V)$, and masses, including 1σ uncertainties, for individual clusters and, hence, mass functions for YSC systems. The mass function, known to be Gaussian for old Globular Cluster (GC) systems, is still controversial for YSC systems. GC formation is expected in massive gas-rich spiral–spiral mergers because of their high global star-formation efficiencies, and observed in ≥ 1 Gyr-old merger remnants. Yet, it has not been possible to identify young GCs among YSC populations. We suggest a compactness parameter involving masses and half-light radii of YSCs to investigate if young GCs are formed in starbursts and if the ratio of young GCs to more loosely bound star clusters depends on galaxy type, mass, burst strength, etc.

1. Star Formation Efficiencies & Star Cluster Formation

Both burst strengths, b, defined by the relative increase of the stellar mass in the course of starbursts, $b \equiv \frac{\Delta S_{\rm burst}}{S_{\rm total}}$, and Star Formation Efficiencies (SFEs), defined as the total stellar mass formed from the available mass of gas, SFE $\equiv \frac{\Delta S_{\rm burst}}{G}$, are difficult to determine. Reasonable estimates are only possible in young post-starbursts. As long as a burst is active, only lower limits can be given. Once a burst is over, or if a burst lasts longer than the lifetimes of the most massive stars, the amount of stars already died needs to be accounted for. The stellar and gaseous masses before the burst can only be estimated on the basis of Hubble types, H I observations, etc. The strongest bursts are reported in mergers of massive gas-rich galaxies, with total burst durations on the order of a few 100 Myr. Bursts in massive interacting galaxies are much stronger and last much longer than those in isolated dwarf galaxies. Blue Compact Dwarf galaxies (BCDs), for instance, feature bursts with durations of the order of a few Myr, $b \ll 0.1$, SFE ≤ 0.01, and a trend of decreasing burst strengths for increasing total galaxy masses (including H I) (Krüger et al. 1995). Massive interacting galaxies feature bursts that are stronger and more efficient by one to two orders of magnitude, similar to the progenitors of E+A galaxies in clusters, ULIRGs, SCUBA galaxies, and optically identified starburst galaxies in the early Universe. The post-burst spiral–spiral merger remnant NGC 7252, with two long, gas-rich tidal tails pointing at an age of $\lesssim 1$ Gyr after the onset of the strong interaction, and its blue and radially constant colors and very strong Balmer absorption line spectrum, must have experienced a very strong and global starburst, increasing its stellar mass by as much as ~ 40 % between 600 and 1,000 Myr ago. Conservative estimates still lead to a very high SFE $\geq 30\%$ (Fritze–v. Alvensleben & Gerhard 1994). A large number of Star Clusters (SCs) formed throughout the main body, many of them apparently so strongly bound that they managed to survive for $500 - 900$ Myr the violent relaxation phase that restructured the remnant into a de Vaucouleurs profile (cf. Fritze–v. Alvensleben & Burkert 1995; see Schweizer [2002] for a recent review). Most of these star clusters are young GCs, based on their ages, luminosities, and radii. How many clusters were already destroyed since the onset of the burst? An analogous system at a younger age is NGC 4038/39 where the two spiral disks are just starting to overlap. Its burst around the two nuclei, along the tidal structures, and – strongest – in the optically obscured disk–disk overlap region is in its initial stage. Thousands of bright YSCs are seen, with luminosities ranging from those of individual red supergiant stars to $M_V \geq -15$. How many of these will survive for $\gg 1$ Gyr and become GCs?

Hydrodynamic modelling shows that the formation of long-lived strongly bound GCs requires SFEs $\gg 10\%$, originally thought to only occur in the early Universe. In normal SF in spirals, irregulars, and starbursting isolated dwarfs, like BCDs, GC formation should not be possible. In the high-pressure ISM with its strong shocks in spiral–spiral mergers, however, GC formation is observed in reality, as well as in high-resolution hydrodynamical models (Yuexing et al. 2004). Hence, young and intermediate-age GCs are tracers of high SFE periods in their parent galaxies. A number of very fundamental questions are still open at present: Does the amount of SF that goes into massive, compact, long-lived SCs scale with burst strength and/or (local/global) SFE? Or is there a threshold in SFE, below which only field stars and weakly bound, less massive SCs or OB-associations can be formed that dissolve on timescales of 10^8 yr, and above which GCs can be formed or even become the dominant component? Does the same star and SC formation mechanism work in vastly different environments and scale over a huge dynamical range, or are there two different modes of SF, like "normal" and "violent"? SCs are seen to form in many environments, from normal Irrs and spirals through dwarf starbursts, spiral mergers, and ULIRGs, constrained to nuclear regions (e.g., in ULIRGs), over their main body (e.g., NGC 4038/39), and all along tremendous tidal tails (e.g., the Tadpole; cf. de Grijs et al. 2003). The spatial extent of a starburst probably depends on the orbit and relative orientations of the interacting galaxies, on whether or not they had massive bulges and/or dark matter halos. Are all these YSC systems similar, or systematically different in terms of masses, mass functions, sizes, compactness or degree of binding and, hence, survival times?

SCs are Simple Stellar Populations (SSPs), with all stars having the same age and chemical composition. Evolutionary synthesis models like GALEV describe the evolution of SCs over a Hubble time, from the youngest stages of 4 Myr all the way to the oldest GC ages ≥ 14 Gyr for 5 different metallicities $-1.7 \leq [\text{Fe}/\text{H}] \leq +0.4$. The TP-AGB phase is very important for age-dating SCs between 100 Myr and a few Gyr on the basis of their $(V - I)$ colors (cf. Schulz et al. 2002). Gaseous emission in terms of lines and continuum for the respective metallicities makes important contributions to broad-band fluxes and colors at young ages (Anders & Fritze–v. Alvensleben 2003). Lick absorption indices significantly help disentangle ages and metallicities of older SCs (Lilly & Fritze–v. Alvensleben, in prep.). GALEV models yield the detailed spectral evolution of SCs from 90 Å through 160 μm, luminosities, M/L ratios, and colors in many filter systems (Johnson, *HST*, Washington, Strömgren, ...) and can be retrieved from http://www.uni-sw.gwdg.de/~galev/.

2. Analysing Star Cluster Systems

The time evolution of luminosities, colors, and M/L ratios significantly depends on metallicity in a way that is different in different wavelength regimes. For young SC systems, such as in NGC 4038/39, extinction is an important issue. Older starbursts, like in NGC 7252, are significantly less extinced. An ESO/ASTROVIRTEL project provides us with *HST* and VLT multi-λ photometry for SC systems from young to old that have been observed in 4 or more passbands. A dedicated Spectral Energy Distribution (SED) analysis tool, ANALYSED, compares observed SC SEDs with an extensive grid of 117,000 SSP model SEDs for 5 different metallicities, 1,170 ages from 4 Myr to 14 Gyr, and 20 extinction values $0 \leq E(B - V) \leq 1$. We use Calzetti et al.'s (2000) starburst extinction law, since internal extinction is only an issue in ongoing starbursts. A probability $p(n) \sim \exp(-\chi^2)$ is assigned to each model SED by a maximum likelihood estimator $\chi^2 = \sum_\lambda (m_\lambda^{\text{obs}} - m_\lambda^{\text{model}})^2 / \sigma_{\text{obs}}^2$. The best-fit model is the one with the highest probability. Probabilities are normalised to $\sum_n p(n) = 1$. Summing models with decreasing probabilities until $\sum_n p(n) = 0.68$ provides $\pm 1\sigma$ uncertainties for ages, metallicities, extinction values, and masses of individual SCs (Anders et al. 2004a). Testing ANALYSED with artificial SCs, we found that there are good and bad passband combinations, slightly depending on the ages and metallicities of the clusters, and we identified a combination of 4 passbands U, B, V or I, and H or K with observational photometric accuracy ≤ 0.2 mag as optimal for YSC systems.

We agree with the independent investigation by Cardiel et al. (2003), that for typical photometric accuracies broad-band photometry with useful passband combinations is as powerful in disentangling ages and metallicities as is spectroscopy with typical S/N. The ANALYSED tool is currently extended to also include Lick indices for analyses of intermediate-age and old GC systems (Lilly et al., in prep.; see also Lilly et al., these proceedings). In the dwarf starburst galaxy NGC 1569 we identify 169 YSCs on the ASTROVIRTEL images, the bulk of them with ages ≤ 25 Myr, low extinction and metallicities. Their masses are typically in the range from 10^3 to 10^4 M$_\odot$, only the two Super SCs have masses in the mass range of GCs (see Fig.1 and Anders et al. 2004b).

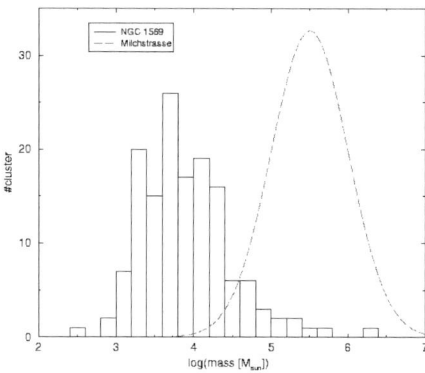

Figure 1. Distribution of YSC masses in NGC 1569 (histogram), compared to the Milky Way GC mass function, normalised to the same number of clusters (Gaussian curve).

We conclude that the starburst in the dwarf galaxy NGC 1569 did not form many GCs, whereas the starburst in the spiral–spiral merger NGC 7252 did. Does SC formation produce a continuum in masses and binding energies, or are there different modes of SC formation that produce open and globular clusters, respectively? With increasing burst strength, an increasing number of SCs is formed. Does the statistical effect of having a higher chance to have a more massive cluster within a larger sample explain the difference between cluster masses in NGC 1569 and NGC 7252? Probably not. Mass Functions (MFs) are power laws for open cluster systems and Gaussians for old GC systems. Initial MFs derived from models for survival and destruction of GCs in galactic potentials are controversial. Vesperini (2001) favors an initially Gaussian shape for GC MFs that is essentially conserved by the competing destruction effects of low-mass GCs through tidal disruption and of high-mass GCs by dynamical friction. Gnedin & Ostriker (1997) favor an initial power-law GC MF that is secularly transformed into a Gaussian by higher destruction rates for lower-mass GCs. The MF of the young SC system in NGC 4038/39, that very probably comprises a mixture of OB associations, open, and GCs, as derived from *HST* photometry, is also controversial. Whereas Zhang & Fall (1999) derive a power-law MF using reddening-free Q_1–Q_2 indices, we find a Gaussian MF (Fritze–v. Alvensleben 1998, 1999). Both approaches have their drawbacks. Zhang & Fall (1999) excluded a significant number of clusters from the ambiguous age range in the Q_1–Q_2 plot, while we assumed a uniform average reddening in the WF/PC *UVI* data. If we exclude the same SCs as Zhang & Fall (1999), we also find a power-law. The dust distribution in NGC 4038/39 is clearly patchy and a re-analysis of ASTROVIRTEL *UBVIK* data from *HST*/WFPC2 and VLT/ISAAC is underway (Anders et al., in prep.). The shape of the MF need not correspond to the shape of the Luminosity Function (LF) for a young SC system, as age spread effects can distort the shape of the LF with respect to that of the underlying MF – to the point of transforming a Gaussian MF into a power-law LF up to the observational completeness limit. The key to survival or destruction is the strength of a SC's internal gravitational binding, as measured for Galactic GCs by their concentration parameter, c. By definition, $c \equiv \log \frac{r_t}{r_c}$ involves the tidal and core radii. Very young clusters need not yet be tidally truncated, and tidal radii could not be measured for the bulk of the YSCs on top of the bright galaxy background in NGC 4038/39 anyway. Therefore, we define the compactness of a young SC by the ratio between its mass and half-light radius (Anders et al., in prep.), a robust quantity that can be measured reliably, and is predicted by dynamical SC evolution models to not significantly change over a Hubble time. To this aim, we first have to improve upon the determination of SC radii by using appropriate aperture corrections. Improved SC radii, in turn, lead to improved SC photometry, and, hence, to improved photometric masses (see Anders et al., these proceedings).

3. Conclusions and Perspective

From the ages, masses, and radii of their SCs we know that major gas-rich mergers can form significant secondary populations of GCs in their strong and global starbursts. SFEs in mergers are higher by 1–2 orders of magnitude than in normal SF galaxies and (non-interacting) dwarf galaxy starbursts. Comparing good precision photometry in at least 4 reasonably chosen passbands (e.g., *UBVK*) to GALEV evolutionary synthesis models for SCs with a given age, metallicity, extinction, and mass by means of a dedicated SED analysis tool (ANALYSED) allows us to reliably determine individual SC ages, metallicities, extinction values, and masses, including their respective 1σ uncertainties. The first dwarf galaxy starburst analysed in detail in this way shows only very few clusters with masses in the range of GCs among its $\sim$ 170 YSCs. Clearly, both more major merger and dwarf galaxy starbursts need to be analysed in detail. Pixel-by-pixel analyses (de Grijs et al. 2003) or integral-field spectroscopy can provide burst strengths and SFEs. From a comparison with *HST* multi-λ

imaging of their YSC systems the relative ratios of SF in field stars, short-lived open clusters, and long-lived GCs, respectively, can be determined. A key question is whether or not these quantities, as well as the intrinsic properties of the YSCs, like masses and half-mass radii, depend on environment, in a smooth way or with some threshold. A comparison of starbursts in dwarf and massive, interacting and non-interacting starbursts should tell if SF and SC formation are universal processes or depend on environment. GC age and metallicity distributions will allow to trace back a galaxy's violent (star) formation history and constrain galaxy formation scenarios (Fritze–v. Alvensleben 2004). This requires B through NIR photometry and medium-resolution spectra to measure Lick indices. With only one observed color, we cannot disentangle the age–metallicity degeneracy of intermediate-age and old stellar populations, and see if more than one GC population is hidden in the red peak of many elliptical galaxies' bimodal color distributions.

Acknowledgments

I gratefully acknowledge travel support, in part from the DFG (FR 916/10-2), and in part from the organisers.

References

Anders P., Fritze-v. Alvensleben U., 2003, A&A, 401, 1063
Anders P., Bissantz N., Fritze-v. Alvensleben U., de Grijs R., 2004a, MNRAS, 347, 196
Anders P., de Grijs R., Fritze-v. Alvensleben U., Bissantz N., 2004b, MNRAS, 347, 17
Calzetti D., Armus L., Bohlin R.C., Kinney A.L., Koornneef J., Storchi-Bergmann T., 2000, ApJ, 533, 682
Cardiel N., Gorgas J., Sánchez-Blázquez P., Cenarro A.J., Pedraz S., Bruzual G., Klement J., 2003, A&A, 409, 511
Fritze-v. Alvensleben U., 1998, A&A, 336, 83
Fritze-v. Alvensleben U., 1999, A&A, 342, L25
Fritze-v. Alvensleben U., 2004, A&A, 414, 515
Fritze-v. Alvensleben U., Burkert A., 1995, A&A, 300, 58
Fritze-v. Alvensleben U., Gerhard O.E., 1994, A&A, 285, 775
Gnedin O.Y., Ostriker J.P., 1997, ApJ, 474, 223
de Grijs R., Lee J., Mora Herrera M.C., Fritze-v. Alvensleben U., Anders P., 2003, New Astron., 8, 155
Krüger H., Fritze-v. Alvensleben U., Loose H.-H., 1995, A&A, 303, 41
Schulz J., Fritze-v. Alvensleben U., Möller C.S., Fricke K.J., 2002, A&A, 392, 1
Schweizer F., 2002, in: Extragalactic Star Clusters, Geisler D., Grebel E.K., Minniti D., eds., IAU Symp. 207, (ASP: San Francisco), p. 630
Vesperini E., 2001, MNRAS, 322, 247
Yuexing L., MacLow M.-M., Klessen R.S., 2004, ApJ, 614, L29
Zhang Q., Fall S.M., 1999, ApJ, 527, L81

STAR CLUSTER POPULATIONS IN NEARBY STARBURST GALAXIES

Jason Harris[1], Daniela Calzetti[2], Denise A. Smith[2,3], John S. Gallagher, III[4], and Christopher J. Conselice[5]

[1] *Steward Observatory, USA*
[2] *Space Telescope Science Institute, USA*
[3] *Computer Sciences Corporation, USA*
[4] *University of Wisconsin, USA*
[5] *California Institute of Technology, USA*

jharris@as.arizona.edu, calzetti@stsci.edu, dsmith@stsci.edu, jsg@astro.wisc.edu, cc@astro.caltech.edu

Abstract The cluster populations of nearby starburst galaxies provide a valuable tool for unlocking the detailed star-formation histories of these systems, and for shedding light on the processes governing star formation in starburst events. We have obtained broad and narrow-band *HST*/WFPC2 images of the active regions of three nearby starburst galaxies (M83, NGC 3077, and NGC 5253). Our analysis of their cluster populations reveals sharp, short-duration starburst events in each galaxy. In NGC 5253 we find no clusters older than 20 Myr, perhaps evidence for accelerated cluster destruction processes in that galaxy.

1. Introduction

It is now clear that throughout the history of the Universe, a significant fraction of the stellar mass has formed in rapid starburst events. These events play important roles in the ionization, chemical enrichment, and overall evolution of their host galaxy. Because starburst events are often observed in interacting galaxies and merger remnants, it is reasonable to infer that gravitational interactions trigger starburst activity. In the context of the hierarchical merger scenario of galaxy formation, it is probable that a large fraction of a typical galaxy's stellar content was formed during a starburst event.

Despite their ubiquity and significance for understanding the origins of stellar populations, we currently have only a rudimentary understanding of the cause, nature and evolution of starburst events. Nearby starburst galaxies provide an important opportunity to study galaxy-scale star formation in great detail. We have obtained broad (F300W, F547M, and F814W) and narrow-band (F487N, F507N, and F656N) images of the active regions of three nearby starburst galaxies: M83, NGC 3077, and NGC 5253, in order to study their star cluster populations and interstellar media.

2. Measuring the Cluster Populations

Cluster Identification

To construct catalogs of cluster candidates in each galaxy, we use the DAOPHOT FIND routine to identify point-like sources in each image. However, not all of the pointlike sources are star clusters; some may be foreground stars or very luminous stars in the target galaxy. The target galaxies are 3.5–4 Mpc away; at these distances, typical star clusters are slightly resolved in the WFPC2 images. We therefore measure the profile size of each candidate object and reject objects whose profiles are not larger than the angular size of known stars in the images. We also require the objects to be detected in each of the three broad-band filters.

Photometry and Extinction

These data present challenges to both of the standard methods of measuring photometry of point-like sources. Aperture photometry suffers from the crowded environments of the images, and from the variable background galaxy light. Profile-fitting photometry suffers from the fact that the clusters have variable profile shapes: some are larger than others, making it difficult to fit the sample with a single profile model. To mitigate these problems, we have adopted a hybrid approach. We use the profile-fitting technique to clean crowding neighbors from the images, and then perform aperture photometry on the neighbor-cleaned images (see Harris et al. 2004 for details).

To correct the photometry for the highly-variable interstellar extinction in these galaxies, we employ our F487N and F656N images to construct a map of the $H\alpha/H\beta$ ratio, which gives a reliable estimate of the extinction at the position of each cluster.

3. Cluster Age and Mass Estimates

Once we have the extinction-corrected photometry for each cluster, we determine the mass and age of each cluster by comparing its photometry to a STARBURST99 population synthesis model (Leitherer et al. 1999) whose metallicity matches that of the galaxy. We plot the cluster photometry in a two-color diagram, and identify the point along the model curve which most closely matches the colors of each cluster (Fig. 1). The age of the matching model point is adopted as the age of the cluster. Mass estimates then follow by simply assuming the data/model flux ratio is equal to their mass ratio (the models adopt a mass of 10^6 $M_\odot$). By acccounting for the photometric and extinction uncertainties, we actually determine plausible ranges for the age and mass of each cluster.

The mass and age estimates for the cluster populations in each galaxy are shown in Fig. 2. Each galaxy shows evidence for a recent, short-duration burst of cluster formation activity. However, only in NGC 5253 do we see a complete absence of clusters older than 10–20 Myr.

4. Summary

We have measured the photometry of star clusters in three nearby starburst galaxies (M83, NGC 3077, and NGC 5253), and used this information to determine the age and mass of each cluster. We find short burst durations in each of the galaxies. Furthermore, we find no clusters in NGC 5253 older than 10–20 Myr. While it is possible that the starburst event in NGC 5253 did not begin until $\sim$ 20 Myr ago, we suggest that cluster destruction rates may be accelerated in NGC 5253, due to a stronger tidal field at its center.

Acknowledgments

This work has been supported by NASA LTSA grant NAG5-9173 and by NASA *HST* grant GO-9144.01-A. This work was based on observations obtained with the NASA/ESA *Hubble Space Telescope* at the Space Telescope Science Institute, which is operated by the Association of Universities for Research in Astronomy, Inc., under NASA contract NAS5-26555.

References

Harris J., Calzetti D., Gallagher J.S., Smith D.A., Conselice C.J., 2004, ApJ, 603, 503
Leitherer C., et al., 1999, ApJS, 123, 3

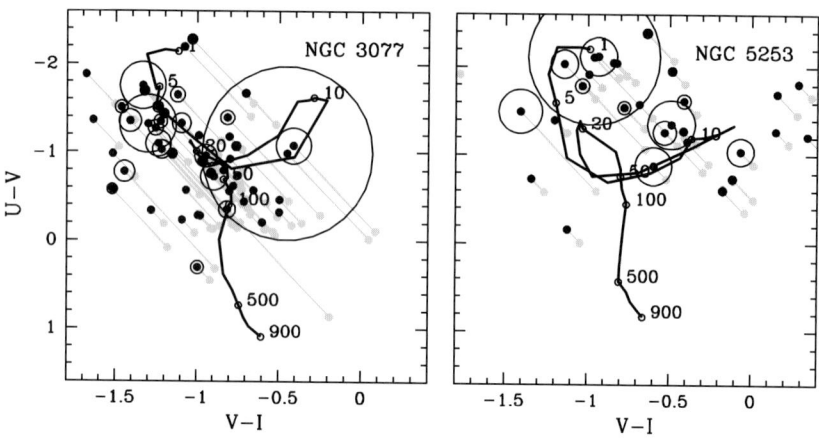

Figure 1. Two-color diagrams for the cluster populations in the two dwarf starburst galaxies, NGC 3077 (left) and NGC 5253 (right). For each cluster, the observed colors are shown in grey, and the extinction-corrected colors are shown in black. A circle whose size is proportional to the F547M flux of each cluster is centered on each extinction-corrected point. The thick curve is the STARBURST99 model track for the metallicity of the galaxy, and this curve is labeled with the ages of several points along the model, in Myr.

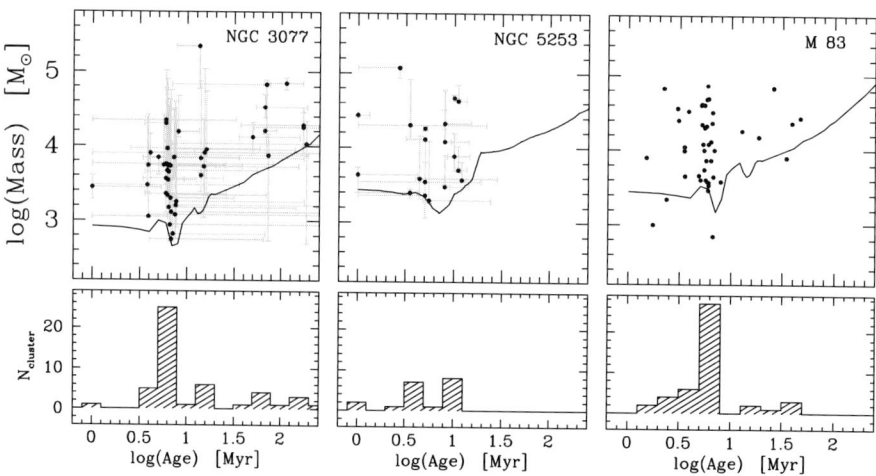

Figure 2. Mass and age estimates for the cluster populations of the three starburst galaxies. The top panels show mass vs. age, and the bottom panels show the age histograms for each cluster population. The curves in the top panels represent the 50% completeness limits for each galaxy.

YOUNG MASSIVE CLUSTERS IN NON-INTERACTING GALAXIES

Søren S. Larsen[1], Jean P. Brodie[2], Deidre A. Hunter[3], and Tom Richtler[4]
[1]*ESO / ST-ECF, Germany;* [2]*UCO / Lick Observatory, USA;* [3]*Lowell Observatory, USA;*
[4]*Universidad de Concepción, Chile*

Abstract   Young star clusters with masses well in excess of 10^5 $M_\odot$ have been observed not only in merging galaxies and large-scale starbursts, but also in fairly normal, undisturbed spiral and irregular galaxies. Here, we present virial mass estimates for a sample of 7 such clusters and show that the derived mass-to-light ratios are consistent with "normal" Kroupa-type stellar mass distributions.

1. Introduction

Young star clusters with masses in the range $10^4 - 10^6$ $M_\odot$ ("young massive clusters"; YMCs) are frequently observed in a wide variety of external galaxies (see review in Larsen 2004). Determining the mass-to-light (M/L) ratios of YMCs is a matter of considerable interest, because this may lead to constraints on the stellar mass function (MF). The MF, in turn, is of interest not only for the general question as to whether or not there is a universal MF, but also for the long-term survival of YMCs. Here, we discuss new results for a sample of 7 YMCs, for which we have attempted to constrain the MF using dynamical M/L ratios.

2. Data

Three of our target clusters are located in spiral galaxies (one in NGC 6946 and two in NGC 5236), and were selected from the sample of Larsen & Richtler (1999). The other four are in dwarf irregulars (2 each in both NGC 4214 and NGC 4449) and were selected from Billett et al. (2002) and Gelatt et al. (2001). Archival *HST*/WFPC2 imaging is available for all of our targets, and the clusters are all free of crowding, and appear superimposed on a reasonably smooth background. The host galaxies have estimated distances in the range $2-6$ Mpc, making the clusters appear well resolved on *HST* images, although a correction for instrumental resolution is still necessary when measuring the sizes. Based on photometry, ages were estimated to be in the range $11 - 800$ Myr and all

Table 1. Data for the YMCs. R_{hl} is the half-light radius, M_{vir} is the virial mass, v_x is the line-of-sight velocity dispersion and ρ_0 is the estimated central density.

ID	R_{hl} [pc]	v_x [km s^{-1}]	log(Age) [yr]	M_{vir} [10^5 M$_\odot$]	ρ_0 [M$_\odot$ pc^{-3}]
N4214-10	4.33 ± 0.14	5.1 ± 1.0	8.3 ± 0.1	2.6 ± 1.0	$(2.5 \pm 1.0) \times 10^3$
N4214-13	3.01 ± 0.26	14.8 ± 1.0	8.3 ± 0.1	14.8 ± 2.4	$(1.9 \pm 0.6) \times 10^5$
N4449-27	3.72 ± 0.32	5.0 ± 1.0	8.9 ± 0.3	2.1 ± 0.9	$(1.9 \pm 0.8) \times 10^3$
N4449-47	5.24 ± 0.76	6.2 ± 1.0	8.5 ± 0.1	4.6 ± 1.6	$(6.8 \pm 2.4) \times 10^3$
N5236-502	7.6 ± 1.1	5.5 ± 1.0	8.0 ± 0.1	5.2 ± 0.8	$(2.8 \pm 1.0) \times 10^3$
N5236-805	2.8 ± 0.4	8.1 ± 1.0	7.1 ± 0.2	4.2 ± 0.7	$(1.6 \pm 1.1) \times 10^4$
N6946-1447	10.2 ± 1.6	8.8 ± 1.0	7.05 ± 0.1	17.6 ± 5	$(2.3 \pm 0.8) \times 10^4$

clusters have photometric mass estimates greater than 10^5 M$_\odot$. The expected velocity dispersions are of order 5 – 10 km s^{-1}.

The two clusters in NGC 5236 were observed with the UVES echelle spectrograph on the ESO/Very Large Telescope, while the remaining clusters were observed with HIRES and NIRSPEC on the Keck I telescope. The spectral resolution was between $\lambda/\Delta\lambda = 25000$ and $\lambda/\Delta\lambda = 60000$ and the S/N ratio was typically about 20–30, or better, per resolution element. Velocity dispersions were measured using the cross-correlation technique of Tonry & Davis (1979). In brief, the cluster spectra were first cross-correlated with the spectrum of a suitable (red supergiant) template star. The template star spectrum was then cross-correlated with the spectrum of another template star. The velocity dispersion of the cluster spectrum was then essentially given as the difference in quadrature between the Gaussian dispersions of the peaks of the two cross-correlation functions. The cluster sizes were derived from the *HST*/WFPC2 images by convolving EFF models (Elson, Fall & Freeman 1987) with the WFPC2 point-spread function and solving for the best fit to the observed images (Larsen 1999). Cluster ages and reddenings were estimated by comparing multi-colour photometry with Bruzual & Charlot (2003) simple stellar population (SSP) models. For further details regarding the data reduction and analysis we refer to Larsen et al. (2004; for NGC 4214, NGC 4449 and NGC 6946) and Larsen & Richtler (2004; for NGC 5236).

3. Virial mass-to-light ratios and the MF

Our results for the 7 YMCs are summarised in Table 1. As expected from the photometry, all clusters have virial masses in excess of 10^5 M$_\odot$, and two have masses $> 10^6$ M$_\odot$. Interestingly, there is little, if any, correlation between cluster mass and half-light radius. Central densities, estimated from the EFF

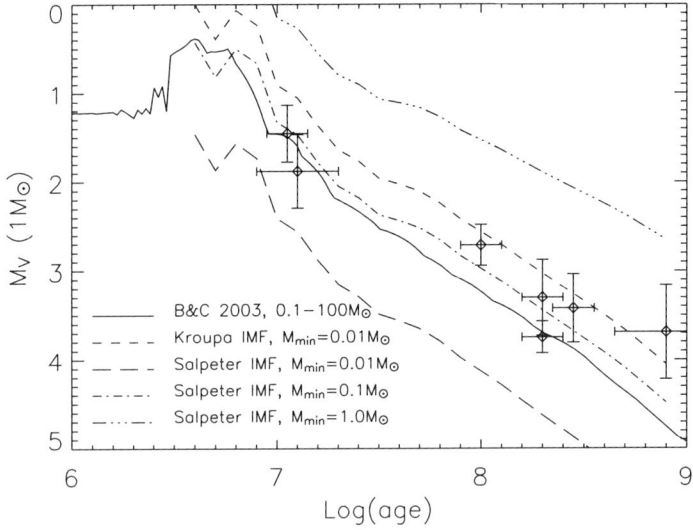

Figure 1. Comparison of observed M/L ratios with SSP models

fits, are listed in the last column and range from $\sim 1000\ M_\odot\ pc^{-3}$ to greater than $10^5\ M_\odot\ pc^{-3}$.

In Fig. 1 we compare the observed M/L ratios with predictions by SSP models for various MFs. The solid line shows the M_V magnitude per solar mass for solar metallicity and a Salpeter (1955) MF with a lower mass limit of 0.1 $M_\odot$, from Bruzual & Charlot (2003). The other curves show our calculations for Salpeter MFs with lower mass limits of 0.01, 0.1 and 1.0 $M_\odot$ (long dashed, dot-dashed and triple dot-dashed lines, respectively) and a Kroupa (2002) MF (short-dashed line), obtained by populating solar-metallicity isochrones from Girardi (2000) according to the various MFs. It is not strictly correct to put all YMCs on the same plot, since the clusters in NGC 4214 and NGC 4449 may have metallicities of only 1/4–1/3 solar (Larsen et al. 2004). However, the V-band M/L ratios are predicted to change by less than 0.2 for models of one-fifth solar metallicity (shifting the curves in Fig. 1 upwards).

The comparison in Fig. 1 suggests that our data are mostly consistent with a Kroupa-type or a Salpeter MF extending down to 0.1 $M_\odot$. At the present level of accuracy, we cannot distinguish between these possibilities. It should also be noted that there is a degeneracy between the MF slope and the lower mass limit. However, there is no suggestion that any of the YMCs studied here may have significantly top-heavy MFs. It should be kept in mind that the virial mass estimates are subject to a number of uncertainties which are not easily quantified. In order to assess the role of macroturbulence in the template stars, we used several stars in the cross-correlation analysis, and found no strong dependence on the choice of template star. However, it is difficult to find local red supergiants that are as luminous as those expected to be present in the youngest clusters, and if the macroturbulence varies significantly with luminosity, then this may lead to systematic errors. Mass segregation (primordial or dynamical) can also lead to erroneous results if the virial theorem is applied blindly. Recent calculations by Lançon & Boily (these proceedings) suggest that mass segregation will typically lead to the cluster masses being underestimated, i.e. the data points would shift downwards in Fig. 1, and the case for top-heavy MFs would then seem even weaker.

4. Summary

Using a combination of ground-based high-dispersion spectroscopy and *HST* imaging, we have derived virial M/L ratios for a sample of 7 YMCs with masses in the range $10^5 - 10^6$ $M_\odot$ in nearby spiral and irregular galaxies. By comparing the M/L ratios with predictions from SSP models, we conclude that our data are consistent with "normal" (e.g., Kroupa [2002]-type) stellar MFs, suggesting that the clusters may eventually evolve into objects which will be very similar to the old globular clusters commonly observed around galaxies.

Acknowledgments

JPB and SSL acknowledge support from National Science Foundation grant AST-0206139 and *HST* archival grant AR-09523-01-A. DAH acknowledges National Science Foundation grant AST-0204922. TR gratefully acknowledges support from the Chilean Center for Astrophysics under FONDAP No. 15010003.

References

Billett O.H., Hunter D.A., Elmegreen B.G., 2002, AJ, 123, 1454
Bruzual G., Charlot S., 2003, MNRAS, 344, 1000
Elson R.A.W., Fall S.M., Freeman K.C., 1987, 323, 54
Gelatt A.E., Hunter D.A., Gallagher J.S., 2001, PASP, 113, 142
Girardi L., Bressan A., Bertelli G., Chiosi C., 2000, A&AS, 141, 371
Kroupa P., 2002, Science, 295, 82
Larsen S.S., 1999, A&AS, 139, 393
Larsen S.S., 2004, in: Planets to Cosmology: Essential Science in *Hubble*'s Final Years, Livio M., ed., (STScI: Baltimore), in press (astro-ph/0408201)
Larsen S.S., Brodie J.P., Hunter, D.A., 2004, AJ, 128, 2295
Larsen S.S., Richtler T., 1999, A&A, 345, 59
Larsen S.S., Richtler T., 2004, A&A, 427, 495
Salpeter E.E., 1955, ApJ, 121, 161
Tonry J., Davis M., 1979, AJ, 84, 1511

NASCENT STARBURSTS IN SYNCHROTRON-DEFICIENT GALAXIES

Hélène Roussel[1], George Helou[2,1], James Condon[3], and Rainer Beck[4]
[1]*Caltech, USA;* [2]*IPAC, USA;* [3]*NRAO, USA;* [4]*MPIfR, Germany*
hroussel@irastro.caltech.edu

Abstract We have identified a rare category of galaxies deviating from the universal infrared-radio correlation of star-forming galaxies in being significantly deficient in synchrotron radiation at 20 cm. The selected objects also have high dust temperatures, indicating intense radiation fields. From a detailed study of the prototype of this class, NGC 1377, the most likely scenario accounting for their properties is a starburst just breaking out in a host previously quiescent for at least 100 Myr. We have selected a statistical sample of candidate nascent starbursts from the cross-correlation of the IRAS Faint Source Catalog with the NVSS and FIRST VLA radio surveys, and discuss the first results obtained from a recent multi-wavelength VLA campaign.

1. Introduction

The infrared-radio correlation of star-forming galaxies (Helou et al. 1985) suffers very few exceptions. At centimeter wavelengths, the radio continuum is dominated by synchrotron emission, from cosmic rays previously accelerated in supernova remnants, propagating in the interstellar magnetic field, and decaying in less than 10^8 yr (Condon 1992). The far-infrared continuum measures the peak of the dust emission, and is a fair tracer of the instantaneous star formation for starburst galaxies. The infrared-radio correlation can hold only if strong coupling mechanisms operate. The production rate of cosmic rays has to be roughly proportional to that of dust-heating photons (which is achieved for a fixed initial mass function), and in addition, starbursts seem to constantly adjust their magnetic field, so that the ratio of magnetic energy density to radiation energy density remains constant (Lisenfeld et al. 1996). Yet, cosmic rays are released only about 4 Myr after the birth of their progenitors. This time delay implies that the relative amounts of infrared and radio emission are expected to vary significantly, and can in principle be exploited to constrain the age of individual star-forming regions. Galaxies, however, host young stellar populations that span an extended age range, and star formation, even in the

form of bursts, goes on for significantly longer periods than 4 Myr; radio continuum deficiency is thus exceedingly hard to achieve on large spatial scales. We term candidate nascent starbursts a class of galaxies selected in a specific way: they deviate upward from the average infrared-to-radio flux ratio by more than 3 times the standard deviation observed in star-forming galaxies; and they have high far-infrared color temperatures, to ensure that dust is heated by intense radiation and that any cirrus component is negligible, so that the radio weakness cannot be explained by decayed star formation. We use
$\bar{q} = \log\left[1.26 \times (2.58\, F_{60} + F_{100}) / 3.75\, (\text{THz}) / S(20\, \text{cm})\right] = 2.34 \pm 0.19$
(Roussel et al. 2003). In our interpretation, these galaxies host an intense star-formation episode younger than a few Myr, and they have been previously quiescent for more than ~ 100 Myr (otherwise, the cosmic rays generated by previous star-formation episodes would not have decayed completely). Such objects seem ideal to constrain the initial conditions, triggering mechanisms and early development of starbursts in galaxies, since they offer the setting to observe the onset of a burst unconfused by previous star-formation episodes, a very rare occurrence.

2. The most extreme case: NGC 1377

NGC 1377 is the only member of the IRAS Bright Galaxy Sample undetected in the 20 cm continuum (Condon et al. 1990), and it deviates from the infrared-radio correlation by more than 8σ. Its dust content is very hot, with $F_{60}/F_{100} = 1.2$ indicating a black-body temperature of 80 K, or a dust temperature of 54 K, using a realistic emissivity law. NGC 1377 is a low-mass lenticular galaxy ($M_* \sim 8 \times 10^9\, M_\odot$), with $L_{\text{FIR}} = 4 \times 10^9\, L_{\odot,\text{bol}}$. We review here briefly the essential aspects discussed by Roussel et al. (2003); readers are referred to this paper for more details and a discussion of alternative scenarios. We observed NGC 1377 at 3.6 and 6.3 cm, and obtained upper limits from which we conclude that the galaxy is not only devoid of synchrotron radiation, but also deficient in free-free emission. Using the star-formation rate derived from the far-infrared, at least 70% of the expected ionizing photons are missing, which suggests that most of the energetic radiation is absorbed by dust. The reservoir of molecular gas is at least ten times more massive than what is required by the starburst hypothesis. It is very compact, possibly overcritical, and both the high infrared to CO(1-0) flux ratio (7 times the average for normal galaxies) and the low CO(2-1)/CO(1-0) brightness temperature (0.53 ± 0.14) are similar to values encountered in starbursts. The only emission line detected in the near-infrared is that of H_2 (1-0) S(1) at 2.12μm, and the limits obtained on the ratios of three H_2 transitions indicate slow shock excitation with $T \leq 1500$ K. From preliminary results of the SINGS survey (Kennicutt et al. 2003), the only bright emission line in the $10-35\mu$m range arises from $T \sim 200$ K H_2 gas, and no high-excitation lines are present. Finally, the ISOPHOT mid-infrared spectrum (Laureijs et al. 2000) is very unusual, containing a broad emission feature between 6 and $8.5\mu m$ in place of aromatic bands universally found in metal-rich star-forming galaxies; modelling of the infrared SED by a pure continuum with deep silicate absorption presents critical difficulties for the energy balance accounting for the moderate far-infrared power. Although it is often claimed that non-stellar activity is required to explain unusually hot and compact infrared sources, we have shown from energetics arguments that this is very unlikely for NGC 1377 (Roussel et al. 2003). The analogy between NGC 1377 and a scaled-up version of Becklin-Neugebauer objects, which represent the transition stage between protostars and ultracompact HII regions, may offer an interesting route to understand its many peculiarities.

3. A statistical sample of nascent starbursts

We have defined a new sample of synchrotron-deficient galaxies from the IRAS Faint Galaxy Sample and the NVSS and FIRST 20 cm VLA surveys, above a declination of -35 degrees. They were selected in infrared flux ($F_{60} > 0.7$ Jy), in infrared color ($F_{60}/F_{100} > 0.7$) and in infrared-to-radio flux ratio ($q \geq \bar{q} + 3\sigma_q$). They are very rare objects, constituting only on the order of 1% of an infrared flux-limited sample, but make up a non-negligible fraction of systems brighter at $60\mu m$ than at $100\mu m$ ($\sim 16\%$). This sample contains an overwhelming majority of compact systems, many of which show morphological disturbances such as plumes or shells, or double nuclei. The galaxies which are classified are generally S(B)0-a, and of the HII or LINER types. These galaxies are not akin to dwarfs, which sometimes may be assimilated to a single giant HII region: they have near-infrared colors similar to those of massive galaxies, have high infrared-to-optical flux ratios, and compact reservoirs of CO gas, suggesting a normal metal content. All galaxies were observed by us with the VLA at 20 cm in the C configuration, in order to obtain deeper maps. In addition, the galaxies with already robust limits at 20 cm were mapped at 3.5 and 6 cm, in order to constrain their radio emission mechanisms. Among this subsample, we found that 23 galaxies (59%) are intrinsically deficient in synchrotron emission. Their weakness at 20 cm is confirmed at shorter wavelengths. Sixteen other galaxies have spectral indices between 3.5 and 6 cm significantly steeper than between 6 and 20 cm, likely caused by high thermal opacity at 20 cm. If we correct for that effect by extrapolating the 3.5–6 cm spectral index, their deviation from the infrared-to-radio correlation is reduced, ranging between 0.6σ and 2.8σ. They may represent a later evolutionary stage than the galaxies deviating from the correlation by more than 3σ at 6 cm. Six of the 23 galaxies genuinely poor in cosmic rays are still undetected at the

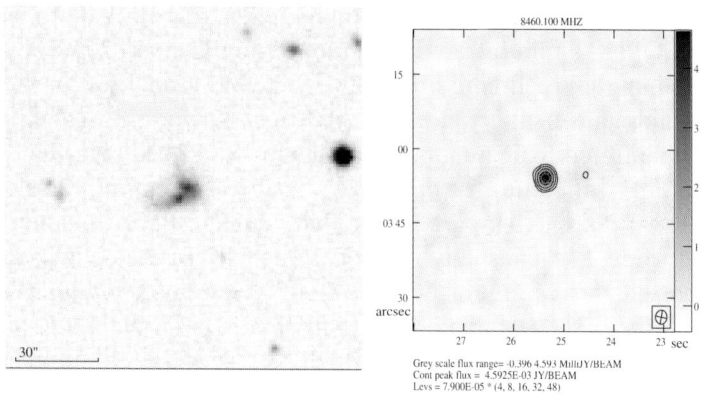

Figure 1. The more luminous of the two systems whose spectral index between 3.5 and 6 cm is close to -0.1. To the left is an optical image from the Digitized Sky Survey; to the right is the 3 cm VLA map.

0.2 – 0.3 mJy level at 6 cm, and some may be considered faint analogs of NGC 1377. The radio spectrum of 11 other galaxies, steeper than $F_\nu \propto \nu^{-0.4}$, implies that the synchrotron component constitutes a significant fraction of their radio power, either the residue of an older star-formation episode, or the product of a developing burst. Finally, 6 of the 23 truly synchrotron-deficient galaxies have 3.5 – 6 cm spectral indices flatter than -0.4. At least two seem completely dominated by thermal emission; they are fairly luminous in the far-infrared (2×10^{10} $L_{\odot,\mathrm{bol}}$ and 5×10^{11} $L_{\odot,\mathrm{bol}}$, respectively). They are both classified as LINERs, one is surrounded by faint shells and the other is a merging system (Fig. 1). This preliminary analysis of the radio data underscores the diversity of the high infrared-to-radio ratio galaxies. The most useful information about their energy source, geometry and degree of evolution is expected to be brought by mid-infrared spectroscopy. A detailed description of the sample, data and analysis will be provided in a forthcoming paper.

References

Condon J.J., Helou G., Sanders D.B., Soifer B.T., 1990, ApJS, 73, 359
Condon J.J., 1992, ARA&A, 30, 575
Helou G., Soifer B.T., Rowan-Robinson M., 1985, ApJ, 298, L7
Kennicutt R.C. Jr., et al., 2003, PASP, 115, 928
Laureijs R.J., et al., 2000, A&A, 359, 900
Lisenfeld U., Völk H.J., Xu C., 1996, A&A, 314, 745
Roussel H., et al., 2003, ApJ, 593, 733

HST/STIS SPECTROSCOPY OF THE STARBURST CORE OF M82

L.J. Smith[1], M.S. Westmoquette[1], J.S. Gallagher[2], R.W. O'Connell[3] and R. de Grijs[4]
[1]*Department of Physics & Astronomy, University College London, London, UK*
[2]*Department of Astronomy, University of Wisconsin, Madison, WI, USA*
[3]*Department of Astronomy, University of Virginia, Charlottesville, VA, USA*
[4]*Department of Physics & Astronomy, University of Sheffield, Sheffield, UK*

Abstract We present *HST*/STIS spectroscopy for one slit position crossing the starburst core of M82, sampling both clusters and ionized gas at optical wavelengths. The spectra have a spatial resolution of 2 pc at the distance of M82, allowing us to probe the conditions within a starburst environment in unprecedented detail. We find that the ionized gas has high turbulent and thermal pressures ($P/k \sim 10^7$), reflecting the intense energy input from supernovae concentrated in super star clusters. The data also provide new insights into young clusters and their immediate environments. We identify a 6–7 Myr old cluster, M82-A1, which is still embedded in a compact H II region.

1. Introduction and Observations

M82 is the nearest example of a giant starburst galaxy, at a distance of 3.6 Mpc (Freedman et al. 1994). An intense burst of nuclear star formation was triggered by a close encounter with M81 a few hundred million years ago (Yun, Ho & Lo 1993, 1994). The active starburst region has a diameter of 500 pc and is defined optically by the high surface brightness clumps, denoted A, C and E (O'Connell & Mangano 1978), as shown in Fig. 1. Region A is of special interest because it contains a remarkable complex of young super star clusters (SSCs; O'Connell et al. 1995, Melo et al. 2005) which appear to be feeding the well-known galactic-scale wind (Shopbell & Bland-Hawthorn 1998, Westmoquette et al. [these proceedings], Gallagher et al. [in prep.]). The closeness of M82 (0.1 arcsec = 1.8 pc) means that individual clusters are easily resolved by *HST* and are bright enough to obtain spectroscopy. In this paper, we present spectroscopy obtained with the Space Telescope Imaging Spectrograph (STIS) for one slit position crossing the starburst core of M82 in regions A and C (Fig. 1), sampling both clusters and ionized gas at optical wavelengths.

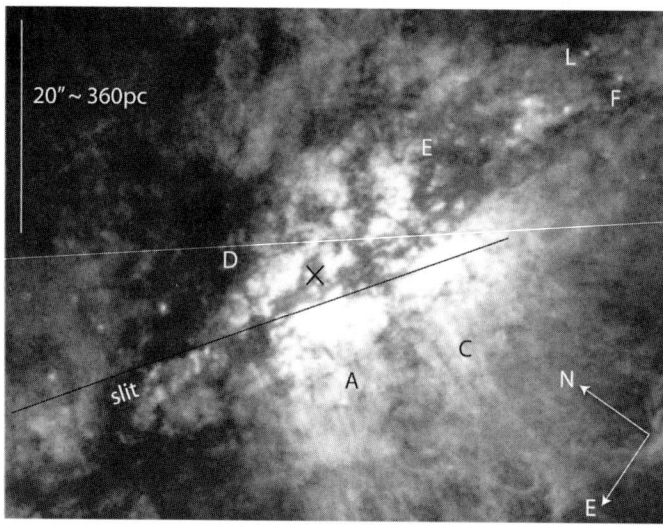

Figure 1. Mosaic of *HST*/WFPC2 F656N images showing the starburst core of M82, and the position of the 52 arcsec long STIS slit. The optically bright regions A, C, D, and E, and the SSCs F and L (O'Connell & Mangano 1978) are labelled. The position of the 2.2μm nucleus is marked with a cross.

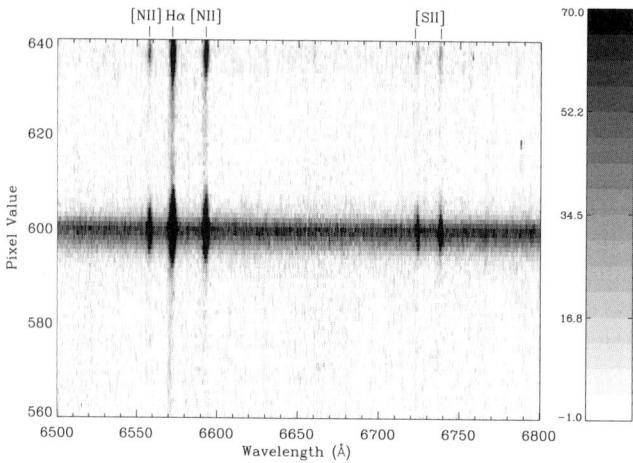

Figure 2. *HST*/STIS G750M rectified two-dimensional spectral image, showing the cluster M82-A1 and its compact H II region over the wavelength region 6500–6800 Å. The nebular emission lines are marked; the surface brightness scale is in units of 10^{-16} erg cm^{-2} s^{-1} Å^{-1} arcsec^{-2}.

The STIS observations were obtained as part of a larger programme devoted to spectroscopy of SSCs in M82 (GO 9117; P.I. O'Connell). We used the G430L ($\lambda\lambda 2900$–5700) and G750M ($\lambda\lambda 6295$–6865) gratings to record two-dimensional spectra on the STIS CCD, with a plate scale of 0.05 arcsec pixel^{-1} over the 52 arcsec long slit. In Fig. 2, a small portion of the reduced two-dimensional image for the G750M grating is shown. The slit was centred on the brightest isolated cluster in region A, which we designate M82-A1. Figure 2 shows that M82-A1 is clearly resolved and is surrounded by a compact HII region. One-dimensional spectra were extracted over the extent of the cluster, sampling the ionized gas extending over regions A and C.

2. Results

Concerning the properties of the ionized gas in regions A and C, we see spatially extended emission along most of the STIS slit. The ([OII] + [OIII])/Hβ nebular line ratio indicates that the gas has an approximately solar abundance (see also Origlia et al. 2004). The velocity field is smooth, suggesting that the ionized gas is not disturbed by the emerging superwind. The emission lines are, however, very broad (FWHM $\sim$ 120 km s^{-1}) compared to diffuse gas found in normal galaxies (Hunter & Gallagher 1997). For [SII], we find that $\lambda 6731$ is stronger than $\lambda 6717$ throughout the STIS data set, implying electron densities of ~ 1000 cm^{-3}, and therefore a high pressure in the starburst core of $P/k \sim 10^7$. This suggests that either the ionized gas is being compressed and confined by the winds of the SSCs, or that it is in pressure equilibrium with a pervasive hot phase of the ISM, as observed in the central regions by Chandra (Griffiths et al. 2000, Strickland et al. 2004).

For the super star cluster M82-A1, we find that the G430L spectrum contains Balmer absorption lines (partially filled in with Balmer emission) with a weak Balmer jump, indicative of a young age. We derive a reddening $E(B-V) = 1.5 \pm 0.1$ from HST B, V, I photometry, and an absolute magnitude $M_V = -15.3$ mag. To determine the half-light radius of M82-A1, we fitted Gaussian profiles to the WFPC2 images and derive a mean value of 115 ± 5 mas (corrected for the PSF) or 2.0 ± 0.1 pc. This value agrees well with that determined by McCrady et al. (2003), of 119 ± 14 mas from a NICMOS/NIC2 F160W image.

To estimate the age for M82-A1, we have compared the observed equivalent width of the Hα emission line with the predicted values for an instantaneous burst with a Salpeter IMF from the spectral synthesis code STARBURST99 (Leitherer et al. 1999). The observed equivalent width of 30Å gives an age of 6.7 ± 0.3 Myr. At this age, the ionizing flux from O stars in the cluster is rapidly decreasing, and the equivalent width is thus very sensitive to age. The derived age is therefore quite accurate if all the hydrogen-ionizing photons are absorbed. We can check this by measuring the properties of the resolved compact HII region surrounding M82-A1. We find that it has a radius of 3.9 pc and an electron density $N_e = 1500$ cm^{-3}. We now require the mass of the cluster to determine the number of Lyman ionizing photons Q_0. Combining the age range we have measured with the absolute magnitude of M82-A1, we find it has a photometric mass of $(1.5 \pm 0.6) \times 10^6$ M$_\odot$ for a standard Salpeter IMF. This then gives a Strömgren radius of 2.5–3.7 pc, in good agreement with the measured radius of the HII region. It is curious that this compact HII region has survived despite the supernova-driven winds from its central cluster. M82-A1 appears to be spatially separated from the main concentration of SSCs in clump A, and thus from the strongest wind zone.

Acknowledgments

We appreciate support for this programme from STScI, including funding for JSG and RWO through grants GO-9117 and GO-9455. JSG also thanks the Wisconsin Alumni Research Foundation for partial support.

References

Freedman W., et al., 1994, ApJ, 427, 628
Griffiths R.E., Ptak A., Feigelson E.D., Garmire G., Townsley L., Brandt W.N., Sambruna R., Bregman J.N., 2000, Science, 290, 1325
Hunter D.A., Gallagher J.S., 1997, ApJ, 475, 85
Leitherer C., et al., 1999, ApJS, 123, 3
McCrady N., Gilbert A.M., Graham J.R., 2003, ApJ, 596, 240
Melo V.P., Muñoz-Tuñón C., Maíz-Apellániz J., Tenorio-Tagle G., 2005, ApJ, 619, 270
O'Connell R.W., Mangano J.J., 1978, ApJ, 221, 62
O'Connell R.W., Gallagher J.S. III, Hunter D.A., Colley W.N., 1995, ApJ, 446, L1
Origlia L., Ranalli P., Comastri A., Maiolino R., 2004, ApJ, 606, 862
Shopbell P., Bland-Hawthorn J., 1998, ApJ, 493, 129
Strickland D.K., Heckman T.M., Colbert E.J.M., Hoopes C.G., Weaver K.A., 2004, ApJ, 606, 829
Yun M.S., Ho P.T.P., Lo K.Y., 1993, ApJ, 411, L17
Yun M.S., Ho P.T.P., Lo K.Y., 1994, Nature, 372, 530

Session VI

Starburst Tracers: Gas, Dust and Star Formation

STARBURST GALAXIES: AN INFRARED PERSPECTIVE

Natascha M. Förster Schreiber
Max-Planck-Institut für extraterrestrische Physik, Garching, Germany
forster@mpe.mpg.de

Abstract The infrared regime is a key range for studies of starburst galaxies. It contains a rich variety of tracers of the gas, dust, and star-formation activity and enables us to probe deep into dust-enshrouded star-forming regions. Selected results based on diagnostics at near and mid-infrared wavelengths are highlighted.

1. Introduction

In the past decade, our understanding of the starburst phenomenon has dramatically improved thanks to progress in instrumentation and observational techniques at all accessible wavelengths. In parallel, substantial efforts have been devoted in the development of empirical and theoretical tools for the interpretation and modeling of the observations. Together with multi-wavelength approaches, this provides unprecedented constraints to assess the relationships between the various stellar and interstellar components in starbursts, characterize quantitatively the star-formation process in starburst environments, and establish consistent scenarios for the triggering, evolution, and feedback of starburst activity.

The infrared (IR) regime is particularly important in studies of starburst galaxies. IR observations allow one to probe deep into the highly-obscured starburst regions inaccessible at optical and UV wavelengths, and the IR range contains a rich variety of diagnostics of the stellar populations and the ISM. The near-IR ($\lambda = 1 - 2.5\,\mu$m) and mid-IR ($\lambda = 2.5 - 45\,\mu$m) sections have been best explored. Figure 1 shows the near and mid-IR spectrum of the archetypal starburst galaxy, M82, which exhibits the typical signatures of starburst activity. H and He recombination lines, as well as fine-structure lines of Ne, Ar, and S for instance, trace the H II regions. [Fe II] lines originate primarily in iron-enriched, partially-ionized shocked gas. Near-IR ro-vibrational and mid-IR rotational H_2 emission lines trace warm molecular material. The near-IR continuum contains a wealth of molecular and atomic absorption fea-

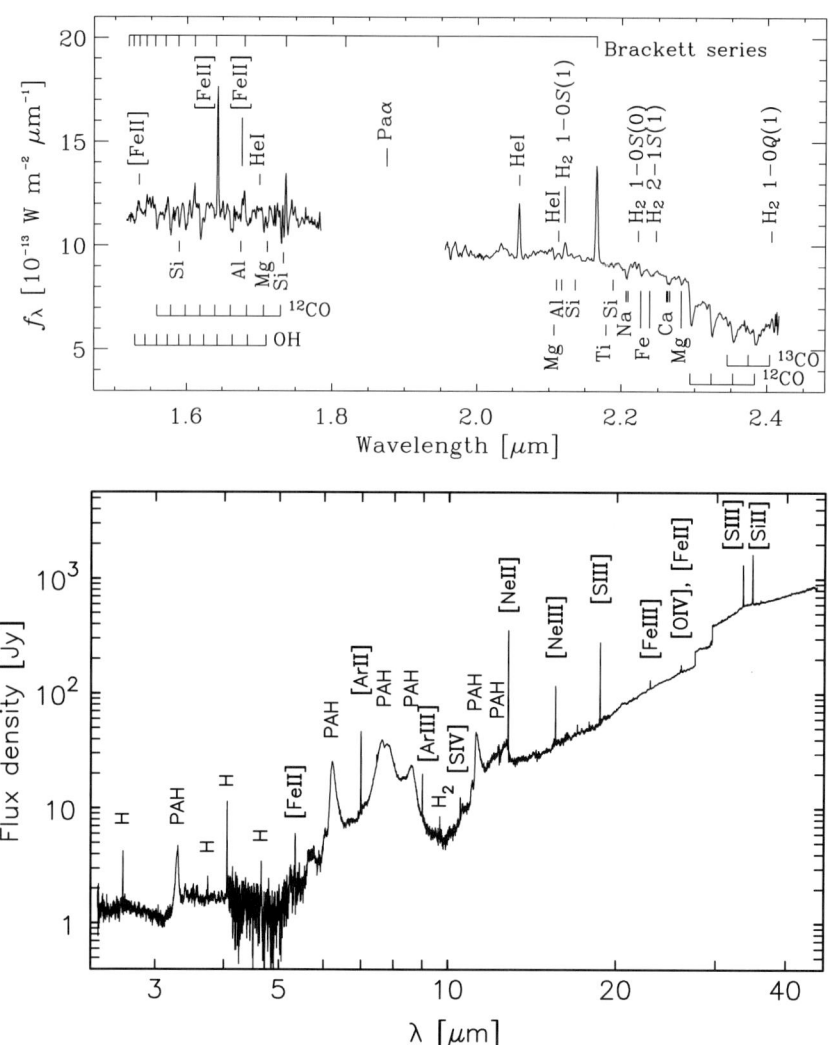

Figure 1. Near to mid-IR spectrum of the starburst core of M82. The top panel shows the H and K-band ranges and the bottom panel the full mid-IR range (from the MPE 3D and ISO SWS instruments, respectively; Förster Schreiber et al. 2001, Sturm et al. 2000).

tures, produced in the atmosphere of cool stars, which are generally stronger in red giants and supergiants. The mid-IR spectral energy distribution (SED) of starbursts exhibits broad emission features commonly attributed to polycyclic aromatic hydrocarbons (PAHs) and arising primarily in photodissociation regions (PDRs), and a continuum rising steeply towards long wavelengths at $\lambda > 10$ μm due to very small dust grains (VSGs) in the vicinity of young massive stars. It also includes absorption features by dust and even ices residing in cold molecular clouds. These spectral features provide valuable diagnostics to investigate the properties of starbursts. In the following, selected recent results from their application to studies of nearby starbursts are highlighted (for more comprehensive reviews see, e.g., Genzel & Cesarsky 2000, Alonso-Herrero et al. 2001, Förster Schreiber et al. 2001, 2003a, and references therein)

2. Evolution of starburst activity

The most widely used near and mid-IR diagnostics for age-dating starbursts include the following. The absolute and relative intensities of H and He lines (e.g., Brγ, Paβ, Brα, Paα from space, HeI at 2.058 and 1.083 μm) and of fine-structure lines (e.g., [NeII] 12.81 μm, [NeIII] 15.56 μm) are sensitive to the intensity and shape of the ionizing radiation from the hot massive stars. For a given IMF and star-formation history, the measurements can be translated into the age of the OB stellar population and the instantaneous star-formation rate (SFR), using predictions of starburst models incorporating evolutionary synthesis and photo-ionization modeling. The relative strength of H lines to the near-IR stellar continuum (e.g., the Brγ equivalent width) provides a measure of the relative importance of massive main-sequence stars with respect to red giants/supergiants, and thus of the evolutionary state of the starburst. The strength of the stellar absorption features (e.g., the CO(2,0) and CO(6,3) band heads at 2.29 and 1.62 μm, combined with the SiI feature at 1.59 μm) characterizes the cool evolved stars. Mid-IR emission can also be used to estimate the bolometric luminosity of the stars, assuming that it traces well the bulk of the total IR luminosity and that the bulk of the star light is absorbed and re-radiated at IR wavelengths by dust.

Imaging and imaging spectroscopy of an increasingly large number of starbursts in the past few years have resolved complex substructure in the spatial distribution of the HII regions and of the red giants/supergiants, with generally important differences or even anticorrelations between the two, down to the smallest angular scales achieved by current instruments at ground-based telescopes or from space ($\sim 1 - 10$ pc in the nearest starbursts; e.g., Satyapal et al. 1995, Puxley et al. 1997, Engelbracht et al. 1998, Alonso-Herrero et al. 2001, among numerous others). Qualitatively, this indicates spatial progression of starburst activity and can be taken as evidence that the burst at a given location has not lasted for much longer than the lifetime of the most massive stars ($\sim 1 - 10$ Myr), presumably because of the strong feedback from massive stellar winds and supernova explosions rapidly disrupting the ISM locally after the onset of star formation (see, e.g., Förster Schreiber et al. 2003a for a discussion of M82 and further references). An immediate implication of this complexity is that simple models for the interpretation of the properties of starbursts (in particular the assumption of a SFR varying monotonically with time) do not provide a good description, making it difficult to reconcile the various diagnostics simultaneously. More complex models involving multiple burst events have been applied, but the problem becomes rapidly underconstrained and solutions are not unique, but rather there are families of models that can equally match the set of available observational constraints (e.g., Rieke et al. 1993, Engelbracht et al. 1998, Alonso-Herrero et al. 2001).

Another important issue is the surface brightness bias, which introduces a bias towards preferential ages or locations depending on the specific indicator or the dust-star geometry. It is important to bear in mind that any diagnostic based on the emission properties is a luminosity-weighted quantity tracing the population that dominates the luminosity at the corresponding wavelength range and/or suffers least extinction along the line of sight. Again, this complicates the interpretation and modeling of the data; simple models may not be adequate, and one may miss important stellar populations that are undetected at the observed wavelengths or are outshined by more luminous populations.

To reconstruct more accurately the global picture, one would ideally need to probe the evolution of starburst activity continuously in space and time, by combining a large collection of indicators probing various stellar evolutionary stages, and on the scales of $\sim 1 - 10$ pc typical of individual stellar clusters and H II regions, or complexes of such systems. Integral-field spectroscopy is ideally suited for this purpose and the populations of compact luminous stellar clusters found ubiquitously in starburst systems provide an interesting means of probing the evolution of starburst activity (see Förster Schreiber et al. 2003a, de Grijs et al. 2001, Mengel et al. [in prep.] for examples involving near-IR observations). This level of detailed investigation is obviously challenging, but state-of-the-art near-IR multi-object and integral-field spectrographs (such as SINFONI on the VLT, OSIRIS and KIRMOS on Keck) promise to enable progress in this direction in the near future.

Obviously, there are many uncertainties involved in constraining the evolution of starburst activity, and it is impossible to be exhaustive here (further discussions and references can be found in the literature cited in this contribution). One aspect that deserves emphasis concerns the CO band heads: Origlia and co-workers (Origlia et al. 1999, Origlia & Oliva 2000) cautioned that model predictions of the CO band heads can be significantly underestimated at ages greater than ~ 10 Myr if the thermally-pulsing AGB phase is not accounted for, and because of the possible inadequacy of current stellar tracks for the evolution of massive stars in the red-supergiant phase, most severely at sub-solar metallicities. Another aspect concerns whether stellar clusters provide a reliable probe of the overall starburst activity. The outcome of current work on the dissolution timescales and gravitational boundedness of clusters (e.g., de Grijs et al. 2003, Mengel et al. [in prep.], and references therein), which also bear on the issues of globular cluster formation and the low-mass IMF in starburst environments, will be very important in this context. Further studies of the more diffuse background population should also provide essential constraints on the evolution of the clusters, as well as of starburst activity in general (e.g., Tremonti et al. 2001, Förster Schreiber et al. 2003a).

3. The high-mass IMF in dusty starbursts

The HeI $2.058\,\mu$m/Brγ ratio has often been applied to constrain the upper IMF in starbursts (e.g., Doherty et al. 1995), but the HeI $2.058\,\mu$m transition is sensitive to resonance and collisional effects, and to the local physical conditions, which affect its reliability (e.g., Shields 1993). The physically simpler HeI $1.701\,\mu$m transition suffers little from these effects and, combined with the nearby Br10 line, provides an alternative constraint (e.g., Vanzi et al. 1996), although the lines are fainter and lie in a spectral region where strong stellar absorption features may be present in the underlying continuum. The mid-IR range provides very sensitive diagnostics based on fine-structure line ratios, in particular [NeIII] $15.56\,\mu$m/[NeII] $12.81\,\mu$m (e.g., Thornley et al. 2000).

It has long been known that dust-rich starburst galaxies frequently exhibit a low-excitation nebular-line spectrum at IR wavelengths. A possible interpretation is that the formation of very massive stars ($\sim 50 - 100$ M$_\odot$) may be suppressed in starburst environments. Studies of larger samples in recent years have actually shown quite a large range in diagnostic ratios among starbursts, consistent with a range of ages and showing a strong correlation with metallicity (e.g., Thornley et al. 2000, Verma et al. 2003, Rigby & Rieke 2004). The strong degeneracy between burst age, burst timescale, and upper mass end of the IMF make it difficult to uniquely constrain these parameters. In view of the evidence for the spatially complex evolution of starburst activity on timescales that can be comparable to the main-sequence lifetimes of massive stars, for the ubiquitous formation of very massive stars in Galactic and nearby extragalactic templates of high-mass star-forming regions, and other direct or indirect indications of the presence of such stars in extragalactic starbursts, the consensus now seems to favor the scenario in which very massive stars do form in starbursts and rapid ageing of the starbursts accounts most plausibly for the low nebular excitation observed (e.g., Thornley et al. 2000, Rigby & Rieke 2004).

It is worth emphasizing that comparisons between different optical and IR diagnostics often yield significantly different results regarding the OB star populations. Along with the uncertainties in the model ingredients and assumptions, this outlines other factors influencing the observed properties but which have been little explored so far (such as the nebular geometry, and the contribution from ultra-compact H II regions and/or of the diffuse ionized medium; e.g., Rigby & Rieke 2004).

4. Dust and aromatic bands as star-formation tracers

The low-resolution mid-IR SED of pure star-forming galaxies is remarkably similar in terms of features present. It is characterized by strong PAH features arising predominantly in PDRs and showing little variations in their relative intensities, and VSG continuum redwards of ~ 10 μm from regions closer to the ionizing OB stars, with varying intensity and spectral slope (e.g., Laurent et al. 2000, Roussel et al. 2001a, Förster Schreiber et al. 2003b, Lu et al. 2003). A variety of results on Galactic and extragalactic star-forming regions and starbursts suggest a close connection between bright PAH and VSG emission and star-formation activity (as summarized, e.g., in the above references). However, few attempts have been made to constrain this connection quantitatively, in part because of the large uncertainties about the nature of the emitting particles and the emission processes involved. Recently, the combination of sizeable samples of normal spiral galaxies and starburst systems observed with ISO together with available optical-to-IR H line data has been used to establish empirically-calibrated relationships between the 7 and 15 μm emission and the SFR, as quantified by the inferred Lyman continuum luminosity ($L_{\rm Lyc}$) from the H lines (Roussel et al. 2001b, Förster Schreiber et al. 2004). Per unit projected area, the 7 μm emission scales essentially linearly with $L_{\rm Lyc}$ over six orders of magnitude; a similar relation is observed for the 15 μm, emission with the distinction that there is a break distinguishing the quiescent spiral disks regime where the long-wavelength continuum is presumably produced by particles akin to PAHs, and the more active regime in circumnuclear regions of spirals and starburst galaxies where the VSG emission becomes dominant. These empirical relationships provide useful estimators of the SFR in deeply embedded star-forming regions as well as important constraints for theoretical models.

5. [Fe II] emission and supernova activity

Near-IR [FeII] emission is conspicuous in starbursts, with the brightest lines usually detected at 1.644 and 1.257 μm. [FeII] emission is enhanced in extended partially-ionized transition zones, making thermal excitation in shocks more efficient than UV fluorescence. Various lines of evidence have suggested a link between supernova activity and [FeII] line emission in starbursts, and several attempts have been made to calibrate the [FeII] line luminosity against the rate of supernova explosions as derived from the non-thermal radio emission (see, e.g., Vanzi & Rieke 1997, Alonso-Herrero et al. 2003, and references therein). There are, however, complications outlined notably by the offset in the correlation between the integrated radio 6cm and [FeII] emission of starbursts, compared to individual supernova remnants in the Milky Way and Large Magellanic Cloud, and by the generally poor spatial coincidence between compact [FeII] sources and radio supernova remnants (e.g., Forbes & Ward 1993, Lumsden & Puxley 1995, Greenhouse et al. 1997; see also Morel et al. 2002). These have been attributed to different lifetimes of SNRs in the [FeII] and radio-emitting phases, and to large-scale shocks associated with a starburst wind producing a significant fraction of the total [FeII].

The work of Alonso-Herrero et al. (2003) based on *HST*/NICMOS imaging of M82 and NGC 253 is particularly relevant for these issues, as the data resolve the starburst regions on scales of ≈ 5 pc. They found a spatial correspondence between bright compact [FeII] sources and radio SNRs for only $\approx 30 - 50\%$ of the radio SNRs. Moreover, the individual compact [FeII] sources account for $\sim 30\%$ of the integrated [FeII] emission. The more diffuse dominant component in the disks could be unresolved emission from very closely packed SNRs or expanding SNRs that have merged while the emission at higher galactic latitudes may be associated with the starburst wind, ultimately also powered by supernova explosions. As such, one may argue that the [FeII] emission can provide a reasonable estimate of the supernova activity in starbursts, albeit perhaps not a robust quantitative indicator. It may also be more meaningful to derive calibrations of the [FeII] luminosity versus the radio supernova rate based on integrated quantities rather than individual SNRs. It remains, however, to be determined whether a "universal" calibration exists among starbursts. An interesting avenue to explore in the future will be the nature of the diffuse [FeII] component and whether it holds clues about the physics of starburst feedback and supernova-driven galactic outflows.

6. Summary

IR observations provide us with essential constraints and valuable insights into the nature and evolution of starburst activity. Much progress has been made in the past years, but much remains to be done before we have a comprehensive and quantitative description of the starburst phenomenon. Despite the current uncertainties and limitations from the observational and interpretation/modeling points of view, the picture has emerged that starburst activity is a spatially and temporally complex phenomenon, the evolution of which follows remarkably diverse courses among starburst systems. This hints at the importance of internal and external factors in determining the evolution of starburst activity, including the exact triggering mechanism, the properties of the host galaxy (nature, dynamics, stellar and gas content and distribution) and its environment, and the interplay between the pre-existing stars and gas, the starburst population, and the starburst feedback. Starburst galaxies remain, as before, fascinating objects.

Acknowledgments

I would like to thank E. Sturm, A. Verma, M. Lehnert, S. Mengel, A. Gilbert, R. Davies, and D. Lutz for many interesting discussions, as well as for their help and for providing material for the purpose of this contribution.

References

Alonso-Herrero A., Engelbracht C.W., Rieke M.J., Rieke G.H., Quillen A.C., 2001, ApJ, 546, 952
Alonso-Herrero A., Rieke G.H., Rieke M.J., Kelly D.M., 2003, AJ, 125, 1210
de Grijs R., Bastian N., Lamers, H.J.G.L.M., 2003, MNRAS, 340, 197
de Grijs R., O'Connell R.W., Gallagher J.S. III, 2001, AJ, 121, 768
Doherty R.M., Puxley P.J., Lumsden S.L., Doyon R., 1995, MNRAS, 277, 577
Engelbracht C.W., Rieke M.J., Rieke G.H., Kelly D.M., Achtermann J.M., 1998, ApJ, 505, 639
Forbes D.A., Ward M.J., 1993, ApJ, 416, 150
Förster Schreiber N.M., Genzel R., Lutz D., Kunze D., Sternberg A., 2001, ApJ, 552, 544
Förster Schreiber N.M., Genzel R., Lutz D., Sternberg A., 2003a, ApJ, 599, 193
Förster Schreiber N.M., Roussel H., Sauvage M., Charmandaris V., 2004, A&A, 419, 501
Förster Schreiber N.M., Sauvage M., Charmandaris V., Laurent O., Gallais P., Mirabel I.F., Vigroux L., 2003b, A&A, 399, 833
Genzel R., Cesarsky C.J., 2000, ARA&A, 38, 761
Greenhouse M.A., et al., 1997, ApJ, 476, 105
Laurent O., Mirabel I.F., Charmandaris V., Gallais P., Madden S.C., Sauvage M., Vigroux L., Cesarsky C., 2000, A&A, 359, 887
Lu N., et al., 2003, ApJ, 588, 199
Lumsden S.L., Puxley P.J., 1995, MNRAS, 276, 723
Morel T., Doyon R., St-Louis N., 2002, MNRAS, 329, 398
Origlia L., Goldader J.D., Leitherer C., Schaerer D., Oliva E., 1999, ApJ, 514, 96
Origlia L., Oliva E., 2000, A&A, 357, 61
Puxley P.J., Doyon R., Ward M.J., 1997, ApJ, 476, 120
Rieke G.H., Loken K., Rieke M.J., Tamblyn P., 1993, ApJ, 412, 99
Rigby J.R., Rieke G.H., 2004, ApJ, 606, 237
Roussel H., et al., 2001a, A&A, 369, 473
Roussel H., Sauvage M., Vigroux L., Bosma A., 2001b, A&A, 372, 427
Satyapal S., et al., 1995, ApJ, 448, 611
Shields J.C., 1993, ApJ, 419, 181
Sturm E., Lutz D., Tran D., Feuchtgruber H., Genzel R., Kunze D., Moorwood A.F.M., Thornley M.D., 2000, A&A, 358, 481
Thornley M.D., Förster Schreiber N.M., Lutz D., Genzel R., Spoon H.W.W., Kunze D., Sternberg A., 2000, ApJ, 539, 641
Tremonti C.A., Calzetti D., Leitherer C., Heckman T.M., 2001, ApJ, 555, 322
Vanzi L., Rieke G.H., Martin C., Shields J., 1996, ApJ, 466, 150
Vanzi L., Rieke G.H., 1997, ApJ, 479, 694
Verma A., Lutz D., Sturm E., Sternberg A., Genzel R., Vacca W., 2003, A&A, 403, 829

DUSTY STARBURSTS AS A STANDARD PHASE IN GALAXY EVOLUTION

David Elbaz
CEA Saclay/DAPNIA/SAp F-91191 Gif-sur-Yvette Cedex, France
delbaz@cea.fr

Abstract   We discuss the implications of the ISO, SCUBA and Spitzer extragalactic surveys, combined with the COBE cosmic infrared background, on the star-formation history of galaxies. A class of galaxies is emerging, similar to local luminous infrared galaxies, that appears to play a dominant role in the shaping of present-day galaxies. Instead of a class, these observations suggest that these "dusty starbursts" should be considered a common phase experienced by most if not all galaxies, once or even several times over their lifetimes.

1. Introduction

It is widely accepted that in the local Universe stars form in giant molecular clouds (GMCs), where their optical and – mostly – their UV light is strongly absorbed by the dust surrounding them. Whether extinction did already occur in the more distant Universe where galaxies are less metal rich was less obvious ten years ago. Galaxies forming stars at a rate greater than about 20 $M_\odot$ yr^{-1} were known, thanks to IRAS, to radiate the bulk of their luminosity above 5μm, the so-called luminous (LIRGs, $12 > \log(L_{\rm ir}/L_\odot) \geq 11$) and ultraluminous (ULIRGs, $\log(L_{\rm ir}/L_\odot) \geq 12$) infrared (IR) galaxies. In the following, we will call these galaxies "dusty starbursts". The bolometric luminosity of galaxies experiencing such large star-formation rates is dominated by the radiation of their young and massive stars. In the local Universe, such objects are very rare and, indeed, "dusty starbursts" radiate only 2% of the bolometric luminosity of galaxies at $z \sim 0$. In the past, galaxies were more gaseous and formed the bulk of their present-day stars, hence we may expect to find more of these violent star-formation events. Already IRAS observations indicated a rapid decline of the co-moving number density of ULIRGs since $z \sim 0.3$ (Kim & Sanders 1998, see also Oliver et al. 1996), but this was over a small redshift range and affected by small number statistics. However, the idea that "dusty starbursts" should have been common in the past was not accepted until a combination of observations arose during the past decade. Distant galaxies were expected to be only marginally affected by dust extinction, with respect to local galaxies and because they were less metal rich. Star-formation rates were commonly measured from optical emission lines uncorrected for extinction, such as [OII] or Hα, or the UV continuum. The first version of the cosmic star-formation history of the Universe (Madau et al. 1996) was published without accounting for any extinction effect. However, several independent observations converged towards another scenario, where most star formation taking place in the Universe was obscured by dust, such as:

1 Extragalactic source counts at 15μm (Elbaz et al. 1999, Metcalfe et al. 2003, Gruppioni et al. 2003, Rodighiero et al. 2004, Fadda et al. 2004) exhibit a slope that cannot be reconciled with model expectations, unless strong evolution is advocated, either in luminosity and/or density of the mid-infrared luminosity function, hence of the amount of star formation hidden by dust (e.g., Chary & Elbaz 2001).

2 The nearly simultaneous discovery of the cosmic infrared background (CIRB, Puget et al. 1996, Fixsen et al. 1998, Hauser & Dwek 2001, and references therein), at least as strong as that in the UV–optical–near-IR, whereas local galaxies only radiate about 30% of their bolometric luminosity in the IR above $\lambda \sim 5\mu$m.

3 The 850μm number counts from the SCUBA sub-millimeter bolometer array at the JCMT (Hughes et al. 1998, Barger et al. 1998, Smail et al. 2002, Chapman et al. 2003, and references therein), which also indicate a strong excess of faint objects in this wavelength range, implying that even at high redshifts dust emission must have been very significant in at least the most active galaxies.

4 The most distant galaxies, individually detected thanks to the photometric redshift technique using their Balmer or Lyman break signature showed signatures of strong dust extinction. The so-called "β-slope" technique (Meurer et al. 1999) used to derive the intrinsic luminosity of these galaxies and correct their UV luminosity by factors of a few (typically between 3 and 7, Steidel et al. 1999, Adelberger & Steidel 2000) was later shown even to underestimate the SFR of LIRGs/ULIRGs (Goldader et al. 2002).

5 The slope of the sub-mJy deep radio surveys (Haarsma et al. 2000).

6 More recently, extragalactic source counts at 24μm with MIPS onboard Spitzer confirmed the strong evolution found at 15μm (Chary et al. 2004, Papovich et al. 2004).

It has now become clear that the cosmic history of star formation based on rest-frame UV or emission-line indicators of star formation such as [OII] or Hα

strongly underestimate the true activity of galaxies in the past if not corrected by large factors to account for dust extinction (Flores et al. 2004, Liang et al. 2004, Cardiel et al. 2003). Although distant galaxies were less metal-rich and much younger than their local counterparts, they must have produced dust rapidly in order to efficiently absorb the UV light of their young stars.

2. Towards a coherent picture of the cosmic star-formation history

We have compiled published versions of the co-moving density of star formation for different star-formation indicators, using references that we could find at the time of this conference, in Fig. 1. Figure 1a shows the very large dispersion in all of these measurements, leading the reader to get the impression that they provide no valuable constraint on what really happened. However, once we take out data points providing only lower limits because they are not corrected for dust extinction, we get a much sharper scenario (Fig. 1b).

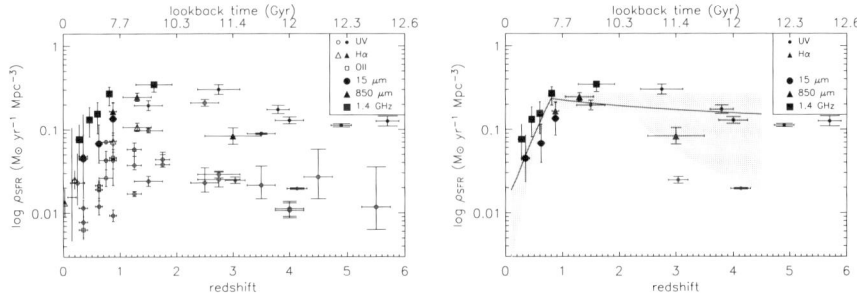

Figure 1. Density of star formation per unit co-moving volume as a function of redshift (or lookback time, upper axis), i.e. cosmic star-formation history. Cosmology: $H_0 = 70$ km s^{-1} Mpc^{-1}, $\Omega_m = 0.3, \Omega_\Lambda = 0.7$. Data are from 1500 Å (Massarotti et al. 2001, Madau et al. 1998, Pascarelle et al. 1998), 1700 Å (Steidel et al. 1999), 2000 Å (Treyer et al. 1998), 2800 Å (Connolly et al. 1997, Lilly et al. 1996, Cowie et al. 1999), 3000 Å (Sawicki et al. 1997), [OII] (Hammer et al. 1997), Hα (Gallego et al. 1995, Tresse & Maddox 1998, Glazebrook et al. 1999, Yan et al. 1999), 15μm (Flores et al. 1999), 850μm (Hughes et al. 1998), 21 cm (Haarsma et al. 2000). *Left (a):* empty symbols are only modestly corrected for dust extinction (except for Hα, uncorrected) following the recipe of Ascasibar et al. (2002): $A(1500-2000$ Å$) = 1.2$ mag and $A(2880$ Å, 3000 Å, [OII]$) = 0.625$ mag, i.e., by factors of 3 and 1.8, respectively. The filled symbols are corrected for extinction, or do not require any correction (as for the 15, 850μm and 1.4 GHz data). We used the corrections quoted by the authors, except for Hα, for which no correction was available. We applied a correction of a factor of 2.3 to this indicator, i.e., half that observed for ISOCAM galaxies (see Hammer et al. 2004, Liang et al. 2004). *Right (b):* only the filled points are represented and compared to the range of possible star-formation histories derived from source counts in the mid-IR, far-IR, sub-mm and the CIRB by Chary & Elbaz (2001).

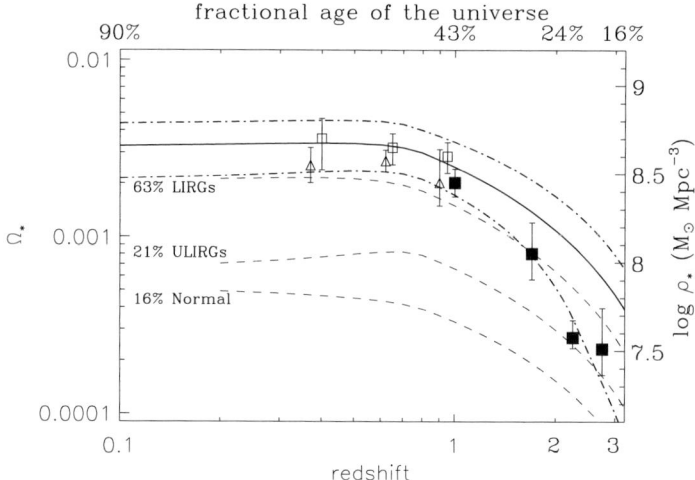

Figure 2. Redshift evolution of the co-moving density of stars, or Ω_* (divided by the critical density of the Universe). Data from Dickinson et al. (2003). The solid line is the best-fitting model of Chary & Elbaz (2001), including both luminosity and density evolution. The dashed lines represent the range of possible models within 1σ of the observational constraints.

The grey area represents the range of possible scenarios from Chary & Elbaz (2001), fitting the combination of mid-IR, far-IR, sub-mm counts, the cosmic infrared background (CIRB), and the redshift distribution of the 15μm ISO-CAM sources. A revised version including Spitzer MIPS 24μm counts will soon be submitted. It does not require a major revision of this scenario. These data suggest that strong evolution took place below $z \sim 2$ that gave rise to a large fraction of the present-day stars. The main actors of this scenario are the "dusty starbursts", which are required to explain the source counts, as well as the CIRB. An interesting test for this scenario consists of integrating the cosmic star-formation history and to compare it to the observed evolution of the cosmic density of stars in the Universe. This integral (for a Gould IMF, Gould et al. 1996) is compared to the compilation of measurements from Dickinson et al. (2003) in Fig. 2. We have assumed a fraction of 20% of the counts and of the infrared background to be due to active galactic nuclei (AGN, see Fadda et al. 2002). Although the best-fitting model from Chary & Elbaz (2001) slightly overpredicts the co-moving stellar mass density above $z \sim 1$, the data are consistent with the broad region permitted by the range of valid models (i.e., the region within the two dot-dashed lines). Note that the LIRGs by themselves provide about 63% of the present-day stars in this framework and fit the observed stellar mass density by themselves.

The excess of stellar mass derived from the integration of the cosmic star-formation history (Fig. 1b) could be produced in several ways:

(i) A larger fraction of AGN above $z \sim 1.5$; (ii) A top-heavy initial mass function in dusty starbursts; (iii) A change of the IR spectral energy distribution of galaxies that would imply that the SFR that we derived is overestimated for distant galaxies.

However, the global agreement of the two cosmic histories – of star formation and of stellar mass density – suggests that we are getting close to a coherent picture in which dusty starbursts play a major role in shaping present-day galaxies. In this scenario, SCUBA galaxies of a few mJy with redshifts measured around $z \sim 2.5$ (Chapman et al. 2003) are the tip of the iceberg (i.e., ULIRGs), while most of the evolution is due to LIRGs.

Many questions still remain unsolved, among others:

(i) What is the triggering mechanism for the distant dusty starbursts? Major mergers are not frequent enough to explain such numbers of dusty starbursts in the past. Other dynamical processes such as close encounters, may represent an important alternative to major mergers that should be carefully considered in hierarchical simulations. A recent study by Moy et al. (in prep.) indeed suggests that clustering plays a major role in triggering these phases.

(ii) What are the present-day counterparts of distant LIRGs? A discussion of the relative contribution of dusty starbursts as a function of morphological type is given by Hammer et al. (2004), who suggest that the disks of spiral galaxies may have been destroyed and thence rebuilt recently.

(iii) Have we underestimated the contribution of Compton-thick AGN to infrared counts? A population of such objects might not have been detected by XMM-Newton or Chandra but could still provide an important contribution to the peak of the cosmic X-ray background at 30 keV (see Worsley et al. 2004).

Acknowledgments

We thank the Centre National d'Etudes Spatiales for their financial support.

References

Adelberger K.L., Steidel C.C., 2000, ApJ, 544, 218
Ascasibar Y., Yepes G., Gottlöber S., Müller V., 2002, A&A, 387, 396
Barger A.J., Cowie L.L., Sanders D.B., Fulton E., Taniguchi Y., Sato Y., Kawara K., Okuda H., 1998, Nature, 394, 248
Cardiel N., Elbaz D., Schiavon R.P., Willmer C.N.A., Koo D.C., Phillips A.C., Gallego J., 2003, ApJ, 584, 76
Chapman S.C., Blain A.W., Ivison R.J., Smail I., 2003, Nature, 422, 695
Chary R.R., Elbaz D., 2001, ApJ, 556, 562
Chary R.R., et al., 2004, ApJS, 154, 80
Connolly A.J., Szalay A.S., Dickinson M., Subbarao M.U., Brunner R.J., 1997, ApJ, 486, L11
Cowie L.L., Songaila A., Barger A.J., 1999, AJ, 118, 603

Dickinson M., Papovich C., Ferguson H.C., Budaári T., 2003, ApJ, 587, 25
Elbaz D., et al., 1999, A&A, 351, L37
Fadda D., Flores H., Hasinger G., Franceschini A., Altieri B., Cesarsky C.J., Elbaz D., Ferrando P., 2002, A&A, 383, 838
Fadda D., Lari C., Rodighiero G., Franceschini A., Elbaz D., Cesarsky C., Pérez-Fournon I., 2004, A&A, 427, 23
Fixsen D.J., Dwek E., Mather J.C., Bennett C.L., Shafer R.A., 1998, ApJ, 508, 123
Flores H., et al., 1999, ApJ, 517, 148
Flores H., Hammer F., Elbaz D., Cesarsky C.J., Liang Y.C., Fadda D., Gruel N., 2004, A&A, 415, 885
Gallego J., Zamorano J., Aragón-Salamanca A., Rego M., 1995, ApJ, 455, L1
Giavalisco M., et al., 2004, ApJ, 600, L103
Glazebrook K., Blake C., Economou F., Lilly S., Colless M., 1999, MNRAS, 306, 843
Goldader J.D., et al., 2002, ApJ, 568, 651
Gould A., Bahcall J. N., Flynn C., 1996, ApJ, 465, 759
Gruppioni C., Pozzi F., Zamorani G., Ciliegi P., Lari P., Calabrese E., La Franca F., Matute I., 2003, MNRAS, 341, L1
Haarsma D.B., Partridge R.B., Windhorst R.A., Richards E.A., 2000, ApJ, 544, 641
Hammer F., et al., 1997, ApJ, 481, 49
Hammer F., Flores H., Elbaz D., Zheng X.Z., Liang Y.C., Cesarsky C., 2004, A&A, 430, 115
Hauser M., Dwek E., 2001, ARA&A, 37, 249
Hughes D.H., et al., 1998, Nature, 394, 241
Kim D.-C., Sanders D.B., 1998, ApJS, 119, 41
Liang Y., Hammer F., Flores H., Elbaz D., Marcillac D., Cesarsky C.J., 2004, A&A, 423, 867
Lilly S.J., Le Fèvre O., Hammer F., Crampton D., 1996, ApJ, 460, L1
Madau P., Ferguson H.C., Dickinson M.E., Giavalisco M., Steidel C.C., Fruchter A., 1996, MNRAS, 283, 1388
Madau P., Pozzetti L., Dickinson M., 1998, ApJ, 498, 106
Massarotti M., Iovino A., Buzzoni A., 2001, ApJ, 559, L105
Metcalfe L., et al., 2003, A&A, 407, 791
Meurer G.R., Heckman T.M., Calzetti D., 1999, ApJ, 521, 64
Oliver S., et al., 1996, MNRAS, 280, 673
Papovich C., et al., 2004, ApJS, 154, 70
Pascarelle S.M., Lanzetta K.M., Fernández-Soto A., 1998, ApJ, 508, L1
Puget J-L., Abergel A., Bernard J.-P., Boulanger F., Burton W.B., Desert F.-X., Hartmann D., 1996, A&A, 308, L5
Rodighiero G., Lari C., Fadda D., Franceschini A., Elbaz D., Cesarsky C., 2004, A&A, 427, 773
Sawicki M.J., Lin H., Yee H.K.C., 1997, AJ, 113, 1
Smail I., Ivison R.J., Blain A.W., Kneib J.-P., 2002, MNRAS, 331, 495
Steidel C.C., Adelberger K.L., Giavalisco M., Dickinson M., Pettini M., 1999, ApJ, 519, 1
Tresse L., Maddox S.J., 1998, ApJ, 495, 691
Treyer M.A., Ellis R.S., Milliard B., Donas J., Bridges T.J., 1998, MNRAS, 300, 303
Yan L., McCarthy P.J., Freudling W., Teplitz H.I., Malumuth E.M., Weymann R.J., Malkan M.A., 1999, ApJ, 519, L47
Worsley M.A., Fabian A.C., Barcons X., Mateos S., Hasinger G., Brunner H., 2004, MNRAS, 352, L28

IS THE INTERSTELLAR GAS OF STARBURST GALAXIES WELL MIXED?

Vianney Lebouteiller and Daniel Kunth
Institut d'Astrophysique de Paris, CNRS, 98 bis Boulevard Arago, F-75014 Paris, France
leboutei@iap.fr, kunth@iap.fr

Abstract   The extent to which the interstellar medium (ISM) in galaxies is well mixed is not yet settled. Metal abundances measured in the diffuse neutral gas of star-forming gas-rich dwarf galaxies are deficient with respect to those of the ionized gas. The reasons, if real, are not clear and need to be based on firm grounds. Far-UV spectroscopy with FUSE of giant HII regions, such as NGC 604 in the spiral galaxy M33, allows us to investigate possible systematic errors in the metallicity derivation. We still find underabundances of nitrogen, oxygen, argon, and iron in the neutral phase by a factor of ~ 6. This could be explained either by the presence of less chemically evolved gas pockets along the lines of sight, or by dense clouds from which HII regions form. Those could be more metal rich than the diffuse medium.

1.   Introduction

The fate of metals released by massive stars in HII regions (where stars are forming) is not yet settled. Kunth & Sargent (1986) suggested that the HII gas can enrich itself with metals expelled by supernovae and stellar winds over the time-scale of a starburst. However, ionized regions in the Large and Small Magellanic Clouds (LMC, SMC) exhibit very little dispersion in their metal content, while X-ray studies show that metals reach the halo of galaxies in a hot phase before they cool down and eventually mix within the ISM in a few $\times 10^9$ yr. While HII region abundances derived from optical emission lines are usually believed to be representative of the metallicity of extragalactic regions, the derived abundances would not necessarily reflect the actual abundances of the ISM, if HII regions are self-polluted.

Blue compact dwarf galaxies (BCDs) are prime targets for the study of their neutral gas using far-UV absorption lines. These galaxies are thought to be chemically unevolved, and the outskirts of their neutral cloud could still be pristine. The fate of the newly-produced metals in these objects is not clear. A possibility is that, once released by massive stars, they remain in a hot phase, thus being unobservable immediately through optical and UV emission lines (Tenorio-Tagle 1996). On the other hand, Kunth & Sargent (1986) suggested that heavy elements released by supernovae lead to the prompt self-enrichment of HII regions on the time-scale of a burst of star formation. This is supported by Recchi et al.'s (2001) model, in which most of the newly-synthetized metals mix within the cold gas phase in a few $\times 10^6$ yr. The Tenorio-Tagle (1996) and Recchi et al. (2001) models differ in the delay between the release and the final mixing, which can take several $\times 10^9$ yr in the former case, but only several $\times 10^6$ yr in the latter.

A real surprise came from a recent FUSE study of four BCDs, I Zw 18 (Lecavelier et al. 2004, Aloisi et al. 2003), Markarian 59 (Thuan et al. 2002), I Zw 36 (Lebouteiller et al. 2004), and SBS 0335−052 (Thuan et al., subm.). In these objects, nitrogen is systematically underabundant in the neutral phase compared to nitrogen abundances derived from the ionized gas (see Fig. 1). Oxygen is either identical in both the ionized and neutral phases or deficient in the HI gas. The overall picture suggests a new scenario for the chemical evolution of the ISM in a galaxy. However, it is possible that these results suffer from many uncertainties, such as ionization corrections, depletion effects or systematic errors due to both multiple lines of sight, and multiple HII regions within the slit.

In this context, nearby giant HII regions provide a much simpler case, since only one region falls into the aperture, thus reducing possible systematic errors. The study of NGC 604 presented here is part of a larger project involving several nearby giant HII regions.

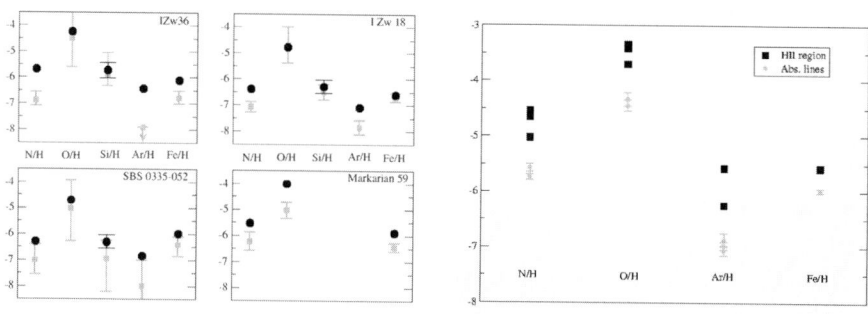

Figure 1. Comparison of log(X/H) in BCDs *(left)* and in NGC 604 *(right)* between the neutral gas abundances, derived from absorption lines with FUSE (in grey), to the HII gas abundances derived from optical emission lines (in black). Error bars indicate the 2σ uncertainties. References for ionized gas values: Izotov et al. (1997) for BCDs, Esteban et al. (2002) for NGC 604. References for neutral gas values: Lecavelier et al. (2004) for I Zw 18, Lebouteiller et al. (2004) for I Zw 36, Thuan et al. (in prep.) for SBS 0335−052, Thuan et al. (2002) for Markarian 59 and Lebouteiller et al. (in prep.) for NGC 604.

2. The case of NGC604

The FUSE observations (LWRS and MDRS apertures) allow us to determine the chemical composition, while the *HST*/STIS spectrum makes it possible to map the neutral gas inhomogeneities and investigate possible multiple line-of-sight effects.

Data analysis

Data analysis has been performed using the profile fitting procedure OWENS developed at the Institut d'Astrophysique de Paris by Martin Lemoine and the French FUSE Team. This program returns the most likely values of many free parameters, such as heliocentric velocities, turbulent velocities, or column densities, via a χ^2 minimization of absorption-line profiles. The associated errors include uncertainties in all of the free parameters, in particular in the position and shape of the continuum.

We checked the shape of the adopted continuum by comparing the observed spectrum with a theoretical model of young stellar populations. No significant difference is found between the model and the continuum we adopted for the profile fitting. Moreover, by comparing the data with the model, we find no significant contamination of neutral interstellar lines by stellar atmospheres.

By analyzing the two FUSE observations, we show that an additional broadening of the lines can account for the spatial distribution of the bright sources within the slit. The column densities we derive account for this extension. For the first time, we could check saturation effects for O I and Fe II lines by analyzing the lines independently. We find no correlation between column densities derived from either line with the oscillator strength, implying that saturated O I and Fe II lines in those spectra do not contribute to significant systematic errors.

Neutral gas inhomogeneities

The high spatial resolution of the *HST*/STIS spectrum of NGC 604 provides the possibility to extract spectra towards individual stars of the ionizing cluster (Bruhweiler et al. 2003). So far, we have analyzed three lines of sight from which we have measured the H I column density using the Lyα line. We find spatial variations (of up to 0.4 dex), suggestive of inhomogeneities in the diffuse neutral gas. This could be a source of systematic errors when determining global column densities from a spectrum of an entire cluster.

We built the global spectrum (i.e., the mean spectrum of all sightlines, weighted by stellar magnitudes) of the *HST*/STIS observation to mimic the spectrum of a cluster, in order to compare the real mean column density we want to determine with the weighted mean we actually measure. Preliminary results show that we tend to underestimate the actual column density when analyzing a global spectrum of several sightlines toward clouds having different physical properties.

3. Results and conclusions

Within the error bars, we derive similar column densities with the two FUSE apertures (see Fig. 1). By modelling the ionization structure of the H II gas with the photoionization code CLOUDY, we find that N I and O I are good tracers of the neutral gas, contrary to Ar I, Fe II, and Si II, which require ionization corrections to obtain the final abundances shown in Fig. 1.

We find that N, O, Ar, and Fe are underabundant by approximately the same factor of ~ 6 in the neutral phase of NGC 604. Whatever the correct interpretation is, the fact that all species are equally deficient compared to that of the H II gas, regardless of their stellar origin (primary or secondary) does not favor the self-pollution explanation. These results could, alternatively, favor the presence of less chemically evolved gas pockets along the lines of sight, which would tend to dilute the metallicity measured toward the H II region, or imply that dense clouds from which H II regions form could be more metal rich than the diffuse ISM.

References

Aloisi A., Savaglio S., Heckman T.M., Hoopes C.G., Leitherer C., Sembach K.R., 2003, ApJ, 595, 760
Bruhweiler F.C., Miskey C.L., Smith Neubig M., 2003, AJ, 125, 3082
Esteban C., Peimbert M., Torres-Peimbert S., Rodríguez M., 2002, ApJ, 581, 241
Kunth D. Sargent W.L.W., 1986, ApJ, 300, 496
Lebouteiller V., Kunth D., Lequeux J., Lecavelier des Etangs A., Désert J.-M., Hébrard G., Vidal-Madjar A., 2004, A&A, 415, 55
Lecavelier des Etangs A., Désert J.-M., Kunth D., Vidal-Madjar A., Callejo G., Ferlet R., Hébrard G., Lebouteiller V., 2004, A&A, 413, 131
Recchi S., Matteucci F., D'Ercole A., 2001, MNRAS, 322, 800
Tenorio-Tagle G., 1996, AJ, 111, 1641
Thuan T.X., Lecavelier des Etangs A., Izotov Y.I., 2002, ApJ, 565, 941

STAR CLUSTERS IN M51: CONNECTION BETWEEN MOLECULAR GAS, STARS AND DUST

Eva Schinnerer[1], Axel Weiß[2], Susanne Aalto[3], Nicolas Z. Scoville[4], Michael P. Rupen[5], Robert C. Kennicutt[6], Rainer Beck[7], and Andrew Fletcher[7]
[1]*MPIA, Heidelberg, Germany;* [2]*IRAM, Granada, Spain;* [3]*OSO, Onsala, Sweden;* [4]*CalTech, Pasadena, USA;* [5]*NRAO, Socorro, USA;* [6]*Steward Observatory, Tucson, USA;* [7]*MPIfR, Bonn, Germany*

Abstract We have mapped key molecular line probes (^{12}CO, ^{13}CO, C^{18}O, HCN, HCO$^+$) in two distinct regions in the spiral arms of the Whirlpool galaxy, M51. Line Velocity Gradient (LVG) analysis performed at a linear resolution of 135–210 pc (2.9–4.5 arcsec) suggests physical conditions in the Giant Molecular Cloud complexes (GMCs) of M51 very similar to those in the Milky Way: cold ($T_{\rm kin} \sim 15$ K) clouds with moderate H$_2$ density ($n({\rm H}_2) < 10^{2.7}$ cm^{-3}). We find indications for a galactocentric trend, with higher kinetic temperature for smaller radii. The data show little evidence for cloud heating by the massive star formation at our resolution of 135 pc. Our new deep ~ 2 arcsec radio continuum images at 3.6 and 6 cm reveal the presence of some highly dust-obscured young star-forming regions within the molecular spiral arms.

1. Introduction

The Whirlpool galaxy, M51, is one of the closest (9.6 Mpc; 1 arcsec ~ 46.5 pc), nearly face-on grand-design spiral galaxies. Most of its molecular gas is found in the spectacular spiral arms (e.g., Scoville & Young 1983, Aalto et al. 1999). Therefore, it offers a unique opportunity to study the physical properties of the molecular gas within the spiral arms. In addition to the large number of Giant Molecular Cloud Associations (GMAs) (up to 16 within one spiral arm, Aalto et al. 1999), numerous OB star clusters reside in the spiral arms, as is obvious in high-resolution *HST* data (Scoville et al. 2001). Nevertheless, there is no simple correlation between the gas reservoir and the star-forming regions which indicates varying star-formation efficiencies, and thus varying physical condition of individual GMAs. This might be partially due to the patchy extinction within the spiral arms which could block the direct view of the youngest star-forming regions.

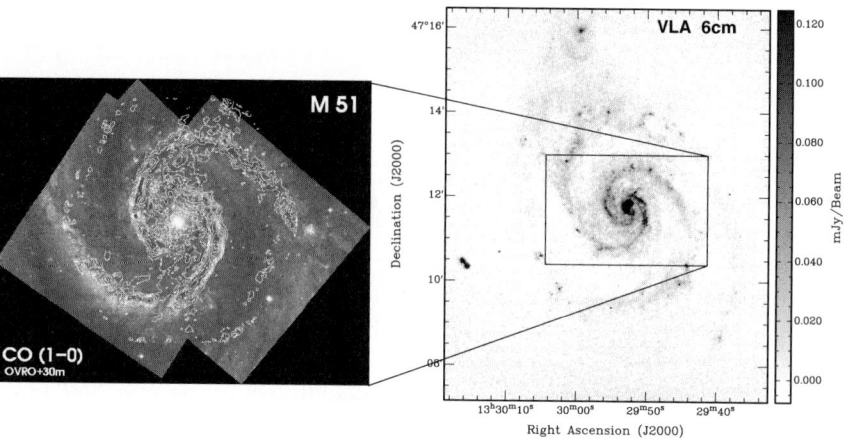

Figure 1. Left: Integrated ^{12}CO(1-0) line emission (contours) overlaid onto a composite optical *HST* color image (from Scoville et al. 2001). The OVRO CO data of Aalto et al. (1999) were short spacing corrected (SSC) using the 30m single dish data from García-Burillo et al. (1993). At a current resolution of ∼ 150 pc, the CO emission is well-correlated with the more extincted regions in the spiral arms. *Right:* The VLA BCD array image at 6 cm reveals numerous compact emission regions. A significant number of these regions shows evidence for thermal emission, suggesting that they are young star-forming regions.

2. Observations

Molecular Gas

The OVRO M51 mosaic in the CO(1-0) line (Aalto et al. 1999) and the IRAM 30m maps of the CO(1-0) and CO(2-1) lines (García-Burillo et al. 1993) were combined with new OVRO and IRAM 30m data for analysis (see Fig. 1). Two distinct regions in the spiral arms of M51 were observed in the CO(2-1), ^{13}CO(1-0), and C^{18}O(1-0) lines with the Owens Valley Radio Observatory (OVRO) mm-interferometer. In addition, we obtained IRAM 30m data to apply short spacing corrections (SSC), to avoid missing flux problems. The final short spacing corrected data cubes were binned to two common resolutions of 2.9 arcsec (135 pc) and 4.5 arcsec (210 pc) for Line Velocity Gradient (LVG) analysis. The line ratios were measured from spectra coming from the same position.

Radio Continuum

We combined VLA archival data at 3.6 and 6 cm by Fletcher et al. (in prep.) with new sensitive and high angular resolution observations in the C and B array, respectively. All data have been short spacing corrected using single dish data from the Effelsberg 100m telescope.

3. Results

LVG analysis: To constrain the physical conditions of the GMCs we use an isothermal, spherical large velocity gradient model (for details see Weiß et al. 2001). $C^{18}O(1-0)$ was not included in the analysis because of its marginal detection even at the strongest positions. The LVG models suggest that the physical conditions in the GMCs of M51 are very similar to those in the Milky Way: The bulk of the CO emission arises from cold ($T_{\rm kin} \sim 15$ K) clouds with moderate H_2 density ($n(H_2) < 10^{2.7}$ cm^{-3}; Fig. 2). The most obvious spatial variation of the gas physical condition is a decrease of the kinematic temperature with increasing galactocentric radius along the western spiral arm. Temperatures decrease from about 20 K at 1.3 kpc (Pos W1) to 5 K at 2.2 kpc (Pos W9). In the southern region (below the zone shown in Fig. 2), with almost constant galactocentric radius, no such gradient is detected. A comparison of the LVG results from star-forming clouds (as traced by Hα) and non-star-forming regions suggests that the gas temperature in the star-forming regions might be on average lower ($T_{\rm kin} \sim 10$ K) than the gas temperature in non-star-forming regions ($T_{\rm kin} \sim 23$ K). The mean density for both environments is about $n(H_2) \approx 10^{2.4}$ cm^{-3}. This suggests that the CO emission from the star-forming regions is dominated by cold gas surrounding the HII regions. *The impact of star formation on the molecular clouds (e.g., heating at the cloud surfaces) is not visible at a linear resolution of 210 pc* (or even for the higher 135 pc resolution data).

High-resolution CO(2-1) data: The molecular gas distribution closely follows the dust lanes seen in the optical. The average H_2 volume densities of $\sim 10^{2.5}$ cm^{-3} inferred from the LVG analysis imply an extinction A_V of ~ 15 mag (assuming a scale of 15 pc as the GMC height and a screen geometry). This number is quite large compared to the average A_V of 3 mag found for the HII regions (Scoville et al. 2001), and suggests that a significant fraction of on-going star formation could be hidden in the dust. We find a weak correlation between the presence of HII regions as traced by their Hα and/or Paα line emission and peaks in the molecular gas distribution: All HII regions within the spiral arms are close to, but usually not coinciding with molecular gas peaks. The internal velocity dispersion changes dramatically across the western spiral arm. The observed large difference might explain why – despite the peak in the intensity – no HII regions are observed/formed on the inner side of the spiral arm.

Radio Continuum (RC): RC emission is widely accepted as an unbiased tracer of recent star formation, both for local galaxies as well as at high z.

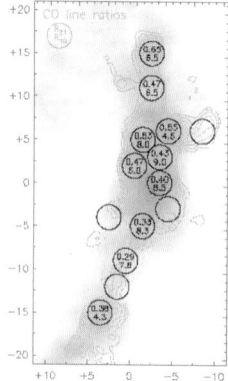 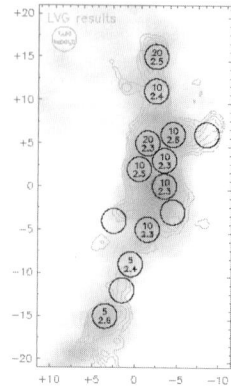

Figure 2. Western region studied in M51: CO(2-1) emission (contours) overlaid onto the CO(1-0) emission (gray-scale). Circles indicate the positions of the spectra used for the LVG analysis at 135 kpc linear resolution. *Left:* Values derived for the CO line ratios: R_{21} = CO(2-1)/CO(1-0) (top) and R_{10} = CO(1-0)/^{13}CO(1-0) (bottom). *Right:* Results of the LVG analysis: T_{kin} [K] (top) and $\log(n(H_2))$ (bottom).

The RC emission consists of (i) the dominant non-thermal (NT) emission attributed to synchrotron emission from cosmic ray electrons, and (ii) the less prominent thermal (T) free-free emission emerging directly from very young HII regions and heated dust (Condon 1992). Our new 3.6/6 cm data reveal numerous compact regions of which about 2/3 appear to have significant thermal emission evidenced by their flatter spectral indices. Several of these regions have no optical counterpart and are located within the dusty spiral arms. This is consistent with the fact that most optically (Hα, Paα) selected HII regions show a low extinction of up to $A_V \sim 3$ mag. New Spitzer mid to far-IR data of M51 also show numerous star-forming regions located in the spiral arms, offset from the optical and UV-traced sites (Kennicutt, Calzetti; priv. comm.).

References

Aalto S., Hüttemeister S., Scoville N.Z., Thaddeus P., 1999, ApJ, 522, 165
Condon J.J., 1992, ARA&A, 30, 575
García-Burillo S., Guélin M., Cernicharo J., 1993, A&A, 274, 123
Scoville N.Z., Polletta M., Ewald S., Stolovy S.R., Thompson R., Rieke M., 2001, AJ, 122, 3017
Scoville N.Z., Young J.S., 1983, ApJ, 265, 148
Weiß A., Neininger N., Hüttemeister S., Klein U., 2001, A&A, 365, 571

Session VII

Starbursts at Intermediate Redshifts and the Starburst versus AGN Paradigm

STARBURSTS IN NEARBY RADIO GALAXIES

Clive Tadhunter
Department of Physics & Astronomy, University of Sheffield, UK
c.tadhunter@sheffield.ac.uk

Abstract Young stellar populations (YSPs) are present in 30–50% of all radio source host galaxies in the local Universe. The analysis of high-quality spectra of the YSPs demonstrates that they comprise post-starburst populations that are relatively old (0.05–2.5Gyr) and massive ($1 \times 10^9 < M_{YSP} < 2 \times 10^{10}$ M$_\odot$), representing a significant proportion (1–50%) of the total stellar mass in the regions sampled by the spectroscopic slits. These results are consistent with the idea that radio sources are triggered in major galaxy mergers, but relatively late in the merger sequence, following a starburst phase in which the host galaxies appear as luminous or ultraluminous infrared galaxies (LIGs, ULIGs). Thus, the nearby radio galaxies with YSPs form a subset of the population of early-type galaxies that is evolving most rapidly in the local Universe.

1. Introduction

There is growing evidence that nuclear activity in galaxies is intimately linked to the evolution of their host galaxies. For example, the redshift evolution of the global rate of star formation in the Universe (e.g., Madau et al. 1996) shows similarities with the evolution of the population of powerful radio galaxies (Dunlop & Peacock 1990). Such links may be a consequence of the triggering of both the quasar and starburst activity as galaxies merge together in a hierarchical galaxy formation scenario (e.g., Kauffmann & Haehnelt 2000). Indeed any process that leads to a substantially increased density of gas in the nuclear regions of a galaxy is likely to enhance both the AGN activity and the rate of star formation. Therefore, we expect AGN activity to be closely linked to star formation in the nuclear regions of galaxies.

In recent years, much observational effort has been put into investigating the relationship between starbursts and AGN, and there is now clear evidence for links between star formation and nuclear activity in low-luminosity active galaxies such as Seyferts and LINERS (Kauffmann et al. 2003, González Delgado et al. 2001, 2004). However, the situation for higher-luminosity active galaxies – quasars and radio galaxies – has, until recently, been much less clear. In this article I summarise new results for powerful radio galaxies that have a direct bearing on our understanding of the link between star formation and nuclear activity in powerful active galaxies.

2. Young stellar populations in radio galaxies

Powerful radio galaxies are invariably associated with early-type galaxies. However, these are no ordinary early-type galaxies: not only do deep optical images show the presence of tidal tails, arcs, fans and double nuclei characteristic of mergers in $\sim 50\%$ of nearby powerful radio galaxies (Heckman et al. 1986, Smith & Heckman 1989), but photometric and spectroscopic observations show the presence of significant UV excesses relative to quiescent elliptical galaxies (Lilly & Longair 1984, Smith & Heckman 1989, Tadhunter et al. 2002). Although the UV excesses may be taken as evidence of recent star formation, various activity-related components, including scattered quasar light, direct AGN light from weak quasar nuclei, and nebular continuum can hamper our interpretation of the optical/UV colours (see Tadhunter et al. [2002] for a full discussion). A major advance over the last decade is that it has become possible to quantify the level of AGN contamination of the optical/UV continua of radio galaxies using careful spectroscopic and polarimetric observations. With the AGN contamination accounted for, it is then possible to investigate the stellar populations. In this way, the occurrence rate of significant young stellar populations (YSPs) has been quantified in three complete samples of powerful, nearby radio galaxies. The results are as follows.

- **Complete sample of powerful, intermediate-redshift radio galaxies** ($0.15 < z < 0.7$; Tadhunter et al. = 2002). This sample of 22 objects comprises mainly powerful FR II, CSS or GPS radio sources; 30–50% of the complete sample show evidence for significant young stellar populations at rest-frame optical/UV wavelengths.

- **Complete sample of low-redshift 3C radio galaxies** ($0.05 < z < 0.2$; combining Aretxaga et al. 2001 with Wills et al. 2002). This sample of 12 objects comprises mainly powerful FR II radio sources; 33% of the sample show evidence for recent star-formation activity.

- **Complete sample of low-power southern 2Jy radio sources** ($z < 0.06$; Wills et al. 2004). This sample of 11 objects comprises mainly low-power FR I radio sources; 30% of the sample show evidence for recent star-formation activity.

Therefore, on the basis of these studies, roughly one third of powerful radio galaxies show evidence for significant star formation in their optical/UV continuum spectra.

The other interesting result to emerge from these studies is that the radio galaxies with the clearest evidence for recent star-formation activity at optical/UV wavelengths tend to be those with the most luminous far-IR emission as measured by the IRAS satellite (Tadhunter et al. 2002, Wills et al. 2002, 2004). This is consistent with the idea that the cool dust detected at far-IR and sub-mm wavelengths is heated by young stars rather than AGN, and supports the use of the far-IR/sub-mm excess as a probe of star formation in active galaxies in the high-redshift Universe (e.g., Archibald et al. 2001). However, relatively few ($\sim 25\%$) powerful radio galaxies in the local Universe were detected by IRAS, and deeper observations of complete samples of radio galaxies with Spitzer are required to put this result on a firmer footing.

3. The detailed properties of the young stellar populations

Having established the existence of YSPs in a significant fraction of powerful radio galaxies, it is then important to investigate the detailed properties of the YSPs. In this way it is possible to gauge the evolutionary status of the host galaxies, and understand the link between the YSPs and the quasar/jet activity. With this in mind, we have undertaken a programme of deep, wide wavelength coverage spectroscopy of the 21 known radio galaxies at $z < 0.7$ with significant YSPs, using the WHT, ESO/NTT and ESO/VLT telescopes. Because many of the main diagnostic absorption lines (e.g., the Balmer lines) are contaminated by emission lines associated with the activity, our main approach to analysing the spectra is to model the continuum SEDs, taking full account of activity-related components, and old as well as young stellar populations. However, where possible, we also use the information provided by the (relatively few) diagnostic absorption features that are not significantly contaminated by line emission (e.g., CaII K, G band).

The key results from our modelling work are as follows (see Tadhunter et al. [2002, 2005] for a discussion of the results for a subset of the sample).

- **The ages of the YSPs.** In general, we find that the optical/UV SEDs are well-modelled with a combination of an old (~ 12.5 Gyr) elliptical galaxy component plus a YSP. The YSPs have post-starburst ages in the range $0.05 < t_{YSP} < 2.5$ Gyr, with a preference for ages $t_{YSP} > 0.1$ Gyr.

- **The masses of the YSPs.** The masses of the YSPs in individual radio galaxies fall in the range $10^9 < M_{YSP} < 2 \times 10^{10}$ M$_\odot$, making up a significant fraction (1–50%) of the total stellar mass in the regions covered by the slits.

- **Spatial extent of the YSPs.** We find that the YSPs are often spatially extended on radial scales of up 20 kpc from the nuclei of the host galaxies, and the ages of the YSPs in the extended regions are often similar to those deduced for the near-nuclear YSPs.

- **Reddened nuclear starbursts.** Despite their significant spatial extent, the YSPs are often concentrated towards the nuclear regions of the host galaxies, and some of the nuclear YSPs show evidence for significant reddening ($0.2 < E(B-V) < 1.0$). Such reddening can cause the continuum colours of the galaxies to appear relatively red, despite the evidence for significant YSPs provided by strong Balmer-line absorption and a relatively small 4000 Å, break.

- **Morphogies of the host galaxies.** In all cases for which we have the requisite deep imaging observations, we find that the radio galaxies in which we detect YSPs show morphological evidence for recent galaxy interactions/mergers.

Overall, the properties of the YSPs are consistent with the triggering of the activity in major galaxy mergers, in which a significant fraction of total stellar mass has been formed via merger-induced star formation.

A particularly interesting aspect of these results is that the post-starburst YSPs tend to be older than the typical ages estimated for the current radio jet activity ($10^5 < t_{\mathrm{radio}} < 10^8$ yr). Given that the age of a merger-induced starburst represents a *lower limit* on the time since the start of the merger, this suggests that the jet activity occurs relatively late in the merger sequence, following the major burst of star formation associated with the merger (i.e., the starburst and jet activity are not coeval). This, in turn, supports the general idea that galaxies with major merger-induced starbursts can *evolve into* luminous active galaxies (e.g., Sanders et al. 1988). Indeed, assuming that the starbursts that formed the YSPs were not too prolonged ($\Delta t_{\mathrm{sb}} < 0.1$ Gyr), we deduce that some of the radio galaxies would have appeared as luminous or ultraluminous infrared galaxies ($L_{\mathrm{IR}} > 10^{11}$ L$_\odot$) in the past (see Wills et al. 2002, Tadhunter et al. 2005).

4. Comparisons with other studies

The only other comparable studies of the YSPs in the host galaxies of luminous AGN are those by Nolan et al. (2001) and Canalizo & Stockton (2001) of the host galaxies of luminous quasars. Both studies found evidence for significant young stellar populations in the extended regions of the quasars, although Nolan et al. (2001) concluded that the quasar host galaxies are dominated by old stellar populations, with the YSPs making up 1% or less of the stellar mass of the host galaxies. However, rather than measuring the YSP ages explicitly Nolan et al. (2001) assumed a uniform age of 0.1 Gyr. Therefore, if the YSPs in the quasar hosts are older than 0.1 Gyr – as found in many radio galaxies – the mass and proportional contribution of the YSPs estimated by Nolan et al. (2001) will be substantially underestimated. On the other hand, Canalizo & Stockton (2001) do find relatively young ages ($t_{\mathrm{YSP}} < 0.25$ Gyr) for the dominant YSPs in the extended regions around their sample of quasars with "transition" far-IR colours. It is interesting that the YSPs in the radio galaxy hosts appear to be significantly older, although it is important to add the caveat that, because they are not affected by the bright nuclear point sources, many of the observations of the radio galaxies sample regions closer to the nuclei of the host galaxies.

Finally I note that, in terms of a comparison with lower-luminosity active galaxies, the intermediate ages of the YSPs in radio galaxies are similar to those deduced for LINER-type AGN (González Delgado et al. 2004), but are older than those deduced for many Seyfert galaxies (e.g., González Delgado et al. 2001).

5. Ongoing star formation?

The dominant YSPs *detected* in the optical/UV spectra of the radio galaxies have intermediate post-starburst ages. It is interesting to consider whether the detected YSPs represent the full extent of the star-formation activity, or whether there are also signs of more recent, ongoing star formation. The evidence for ongoing star formation in radio galaxies includes the following.

- **Luminous far-IR emitters.** Although in some cases heating by the detected YSPs can explain the far-IR luminosities of the sources (e.g., 3C305, 3C293: Tadhunter et al. 2005), in other cases the heating of the cool dust by the detected, intermediate-age YSPs is insufficient to explain the prodigious far-IR emission of the sources (e.g., 3C459, PKS 1549−79: Tadhunter et al. 2002). Although in the latter cases it is difficult to entirely rule out the idea that the cool dust is heated by the hidden quasar nuclei, heating by highly extinguished nuclear starburst components seems more likely.

- **Stellar photo-ionized H II regions**. Recently, sensitive observations have detected stellar photo-ionized H II regions in at least some powerful radio galaxies. Examples include 3C277.3 (Solorzano-Innarea & Tadhunter 2003), PKS1345+12 (Holt 2005) and PKS1932−46 (Villar-Martín et al. 2005). PKS1932−46 is a particularly interesting case because it shows a string of H II regions extending over a total diameter of $\sim$ 160 kpc in a direction almost perpendicular to the radio axis.

These observations demonstrate that in some cases the star-formation histories of radio galaxies are more complex than a single burst, and that the star formation may continue through epochs at which we observe the jet activity.

6. Conclusions

From studies of their YSPs it is clear that the activity in a significant subset of the nearby radio galaxies is triggered in major galaxy mergers. Therefore these objects form part of the population of early-type galaxies that is evolving most rapidly in the local Universe. Clearly, detailed observations of the YSPs have great potential for enhancing our understanding of the nature of the triggering mergers, the order of events, and the relationship between radio galaxies and other classes of merging galaxies.

Acknowledgments

I thank my collaborators Rosa González Delgado, Raffaella Morganti, Karen Wills, Montse Villar-Martín, Jo Holt, Tim Robinson and Bob Dickson for their important contributions to this work.

References

Archibald E.N., Dunlop J.S., Hughes D.H., Rawlings S., Eales S.A., Ivison R.J., 2001, MNRAS, 323, 417
Aretxaga I., Terlevich E., Terlevich R.J., Cotter G., Díaz A.I., 2001, MNRAS, 325, 636
Canalizo G., Stockton A., 2001, ApJ, 555, 719
Dunlop J.S., Peacock J.A., 1990, MNRAS, 247, 19
González Delgado R.M., Leitherer K., Heckman T.M., 2001, ApJ, 546, 845
González Delgado R.M., Cid Fernandes R., Pérez E., Martins L.P., Storchi-Bergmann T., Schmitt H., Heckman T., Leitherer C., 2004, ApJ, 605, 127
Heckman T.M., Smith E.P., Baum S.A., van Breugel W.J.M., Miley G.K., Illingworth G.D., Bothun G.D., Balick G.D., 1986, 311, 526
Holt J., 2005, Ph.D. thesis, University of Sheffield, UK
Kauffmann G., Haehnelt M., 2000, MNRAS, 311, 576
Kauffmann G., et al., 2003, MNRAS, 346, 1055
Lilly S.J., Longair M.S., 1984, MNRAS, 211, 833
Madau P., Ferguson H.C., Dickinson M.E., Giavalisco M., Steidel C.C., Fruchter A., 1996, MNRAS, 283, 1388
Nolan L.A., Dunlop J.S., Kukula M.J., Hughes D.H., Boroson T., Jiménez R., 2001, MNRAS, 323, 308
Sanders D.B., Soifer B.T., Elias J.H., Madore B.F., Matthews K., Neugebauer G., Scoville N.Z., 1988, ApJ, 325, 74
Smith E.P., Heckman T.M., 1989, ApJ, 341, 658
Solorzano-Innarea C., Tadhunter C.N., 2003, MNRAS, 340, 705
Tadhunter C., Dickson R., Morganti R., Robinson T.G., Wills K., Villar-Martín M., Hughes M., 2002, MNRAS, 300 977
Tadhunter C., Robinson T.G., González Delgado R.M., Wills K., Morganti R., 2005, MNRAS, 356, 480
Villar-Martín M., Tadhunter C., Morganti R., Holt J., 2005, MNRAS, in press (astro-ph/0501414)
Wills K.A., Tadhunter C.N., Robinson T.G., Morganti R., 2002, MNRAS, 333, 211
Wills K.A., Morganti R., Tadhunter C.N., Robinson T.G., Villar-Martín M., 2004, MNRAS, 347, 771

STARBURSTS IN LOW LUMINOSITY ACTIVE GALACTIC NUCLEI

Rosa M. González Delgado[1] and Roberto Cid Fernandes[2]
[1]*Instituto de Astrofísica de Andalucía (CSIC), Apdo. 3004, 18080 Granada, Spain*
[2]*Depto. de Física, CFM–Universidade Federal de Santa Catarina, C.P. 476, 88040-900, Florianópolis, SC, Brazil*
rosa@iaa.es, cid@astro.ufsc.br

Abstract Low Luminosity Active Galactic Nuclei (LLAGN), which comprise low-ionization nuclear emission-line regions (LINERs) and transition-type objects (TOs), represent the most common type of nuclear activity. Here, we search for spectroscopic signatures of starbursts and post-starbursts in LLAGN, and investigate their relationship to the ionization mechanism in LLAGN. The method used is based on the stellar population synthesis of the circumnuclear optical continuum of these galaxies. We have found that intermediate-age populations ($10^8 - 10^9$ yr) are very common in weak-[OI] LLAGN, but that very young stars ($\leq 10^7$ yr) contribute very little to the central optical continuum of these objects. However, ~ 1 Gyr ago these nuclei harboured starbursts of size ~ 100 pc and masses $10^7 - 10^8$ M$_\odot$. Meanwhile, most of the strong-[OI] LLAGN have predominantly old stellar populations.

1. Introduction

During the past decade, mainly thanks to the high spatial resolution provided by the *Hubble Space Telescope (HST)*, it has been found that stellar clusters are a common phenomenon in starburst galaxies (e.g., Meurer et al. 1995). Young massive clusters are also present in the nuclei of many late-type galaxies (e.g., Böker et al. 2002). However, stellar clusters have also been found in the nuclei of many early-type galaxies (e.g., Carollo et al. 2002). This is remarkable considering that the strong contribution of the bulge component makes the identification of an additional, unresolved cluster in early-type galaxies extremely difficult. However, this result is expected because of the tight correlation between the bulge stellar velocity dispersion and black-hole mass (Ferrarese & Merrit 2000). Thus, in early-type galaxies it is expected that matter accumulates towards the nucleus to feed the black hole, and massive clusters can also form. Therefore, it should be expected that stellar clusters and black holes co-exist in the nuclei of galaxies. Observational evidence in favour of this scenario has been found in the nuclei of Seyfert 2 (Heckman et al. 1997, González Delgado et al. 1998, see also Schmitt, these proceedings). *HST* ultraviolet (UV) spectroscopy has revealed strong resonance lines formed in the winds of massive stars, implying that powerful ($10^{10} - 10^{11}$ L$_\odot$) and compact (~ 100 pc) starbursts dominate the UV nuclear light. These starbursts are resolved into individual young stellar clusters, as their UV images show. Further evidence is found at optical wavelengths through the detection in the galaxies' nuclear spectra of high-order Balmer lines (HOBL) and HeI lines in absorption, suggesting that intermediate-age stars dominate the optical light (González Delgado et al. 2001, Cid Fernandes et al. 2001).

Low Luminosity Active Galactic Nuclei (LLAGN) constitute a sizeable fraction of the nearby AGN population. These include low-luminosity Seyferts, low-ionization nuclear emission-line regions (LINERs), and transition-type objects (TOs) whose properties are in between classical LINERs and HII nuclei. LLAGN comprise about 1/3 of all bright galaxies ($B_T \leq 12.5$) and are the most common type of AGN (Ho, Filippenko & Sargent 1997). LLAGN could constitute a rather mixed phenomenon, as suggested by the several excitation mechanisms (shocks, photo-ionisation by a non-stellar UV/X-ray continuum, and photo-ionisation by hot stars; see, e.g., Filippenko 1996) that have been proposed to explain the origin of their energy source.

To understand the role that starbursts (or stellar clusters) play in the nuclei of early-type galaxies, and to quantify their contribution to the nuclear energy output, we have selected a sample of LLAGN. Here, we summarize the results obtained in several projects that examine the central stellar population in these objects (González Delgado et al. 2004a, Cid Fernandes et al. 2004a,b), and their central morphology (González Delgado et al., in prep.).

2. Sample and Observations

We selected a sample of LLAGN from the Palomar catalogue (Ho et al. 1997) that consists of 24 strong-[OI] ([OI]/H$\alpha \geq 0.25$) and 47 weak-[OI] LLAGN. We have obtained the following observations: (i) Ground-based optical spectra (3400–5500 Å) using ALFOSC on the NOT and the 2.2m telescope at KPNO for 51 objects. These spectra are used to study the nuclear (1×1.1 arcsec^2) stellar population and gradients up to ~ 1 kpc from the nucleus with a resolution of ~ 100 pc (~ 1 arcsec). (ii) *HST*/STIS(G430L) spectra (2900–5700 Å) for 28 LLAGN. We extracted the central 0.2×0.3 arcsec^2, and 0.2×1 arcsec^2. (iii) *HST*/WFPC2 images at optical wavelengths (in at least one of the following filters, F450W, F547M, F555W, F606W, F814W) for 59 LLAGN. We study the central morphology (stellar clusters, dust distribution) and the connection between the central morphology and the stellar populations.

3. Results

Nuclear optical spectra. We have done an empirical stellar population classification of the nuclear spectra by looking for: (i) the presence of the broad WR bump at 4680 Å; (ii) the detection of absorption lines of HeI and HOBL; (iii) the equivalent widths of metallic lines, such as CaII K, G band and MgI. The WR bump probes the presence of very young stars (a few Myr old), while HeI and HOBL probe the young (10–50 Myr) and intermediate-age (100–1000 Myr) stars, and the metallic lines probe the intermediate and old stellar populations. Our main findings are: (i) No features due to Wolf-Rayet stars were convincingly detected in the STIS and ground-based spectra. (ii) Young stars contribute very little to the optical continuum in the ground-based aperture. However, the fraction of light provided by these stars is greater than 10% in most of the weak-[OI] LLAGN STIS spectra. (iii) Intermediate-age stars contribute significantly to the optical continuum of these nuclei. This population is much more important in objects with weak than with strong [OI]. Weak-[OI] LLAGN that have young stars also stand out for their intermediate-age population. (iv) Most of the strong-[OI] LLAGN have a predominantly old stellar population. These results suggest that young and intermediate-age stars do not play a significant role in the ionization of LLAGN with strong [OI]. Introducing a combined stellar population and emission-line classification into four types, young TOs, old TOs, old LINERs and young LINERs, we conclude that this last class is extremely rare.

We have also performed stellar population synthesis to derive the stellar population properties, such as age, mass, size and extinction. This synthesis has been done using the empirical population synthesis code described in Cid Fernandes et al. (2001, 2004c). The code is able to synthesize the entire optical spectra (absorption lines and colors) by means of a spectral grid of different ages and metallicities (e.g., the Bruzual & Charlot [2003] or González Delgado et al. [2005] stellar population models). The output of the code is a population vector, whose components represent flux fractions associated with each population in the grid, plus the extinction. The main result of this analysis is that post-starburst populations ($10^8 - 10^9$ yr) are very common (in 50% of TOs), but young starbursts ($\leq 10^7$ yr), if present, are very weak.

Spatially-resolved spectral properties. For 47 galaxies the spectra are spatially resolved, and we map the stellar population gradients by the radial profiles of absorption-line equivalent widths and the continuum colors along the slit. These variations are also analyzed by determining for each spectrum the contributions of very young ($\leq 10^7$ yr), intermediate-age ($10^8 - 10^9$ yr) and old (10^{10} yr) stellar populations. We have found: (i) The equivalent width profiles can be flat (suggesting spatially uniform stellar populations) and diluted.

Flat profiles occur in galaxies dominated by an old stellar population. Diluted profiles are produced by intermediate-age stars. Since these stars are found mainly in weak-[OI] LLAGN, stellar population gradients are typical of young TOs. (ii) The extinction profiles of young TOs are more complex than those of old LLAGN, which are very often flat. Young TOs have centrally peaked extinction profiles, and are ~ 3 times more extincted than old LLAGN.

Sizes and luminosities of the intermediate-age stellar populations are obtained combining the optical surface brightness and the distribution of the intermediate-age component of the population vector. These intermediate-age populations are highly concentrated close to the nuclei, and their sizes, defined as the half width at half maximum of their surface-brightness profiles, are typically 100 pc or less. The 4020 Å luminosities are of order $\sim 10^{4.3}$ $L_\odot$ Å^{-1} implying masses of order $10^7 - 10^8$ $M_\odot$. This population was 10–100 times more luminous during their formation epoch, at which time a starburst dominated the bulge light.

Central morphology. To find further evidence of the presence of starbursts or post-starbursts in LLAGN, and a possible connection between the properties of the nuclear stellar populations and the central morphology, we retrieved *HST*/WFPC2 optical images for most of the objects of which we have studied their spectral properties. We are performing a surface brightness analysis for each image and producing unsharp masked images to show the possible presence of resolved compact sources (presumably stellar clusters) and the central dust distribution. This preliminary analysis indicates that central compact sources are detected in many of these LLAGN, even in LLAGN that have a nuclear spectrum dominated by an old stellar population. Many of them show also complex dust structures, with many chaotic and/or nuclear spiral dust lanes, and dust disks (Fig. 1). However, most of the images that show a bulge light dominated distribution are found in LLAGN with an old stellar population. This result agrees with the conclusion from the spectral analysis, that young TOs have more complex central dust structures than old LINERs and old TOs.

Comparison with starbursts in other AGN. In a previous study we unambiguously identified nuclear starbursts in $\sim 40\%$ of nearby Seyfert 2 galaxies (González Delgado et al. 2001, Cid Fernandes et al. 2001). These starbursts were identified by means of UV imaging and/or UV-optical spectroscopy. Stellar population synthesis analysis shows that the Seyfert 2s with nuclear starbursts (Seyfert 2 composites) have a dominant intermediate-age ($\leq$ a few 100 Myr) stellar population at optical wavelengths and very young stars ($\leq$ a few 10 Myr) in the UV range. The analogies with the results obtained for LLAGN are evident. While old LLAGN are analogous to pure Seyfert 2 galaxies,

Starbursts in Low Luminosity Active Galactic Nuclei 267

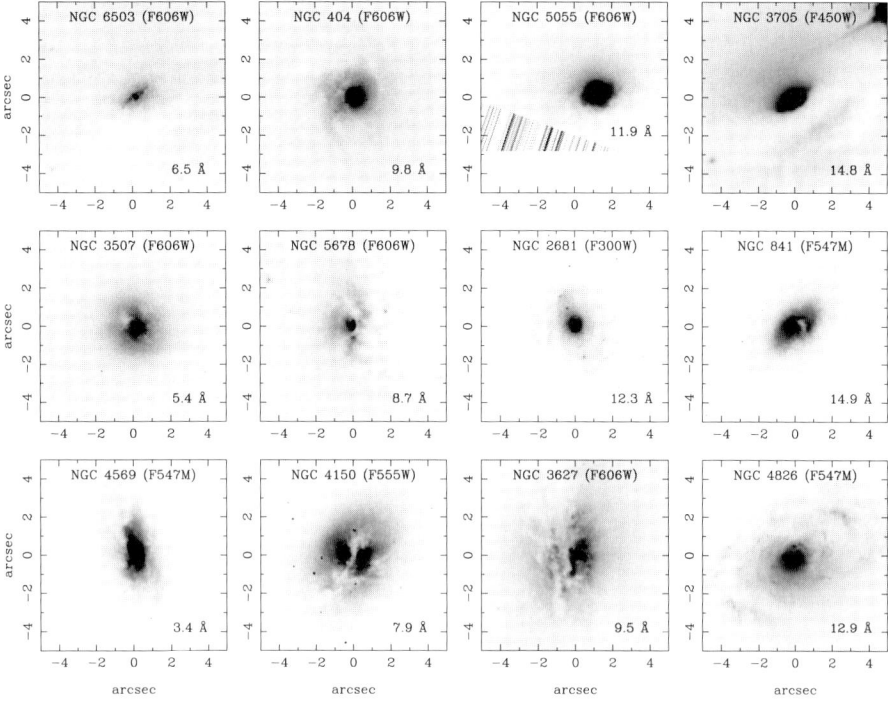

Figure 1. WFPC2 optical images of LLAGN. The objects are sorted from bottom left to top right by increasing equivalent widths of the CaIIK line measured in the nuclear spectra. The values are included at the bottom right of each image. These are some of the LLAGN that have a conspicuous contribution of intermediate-age stars, and thus they are likely young TOs.

young TOs qualitatively resemble evolved Seyfert 2 composites, since the intermediate-age populations in the former are older than in Seyfert 2 galaxies that harbor nuclear starbursts. However, similar post-starburst populations have been found in powerful radio galaxies (Tadhunter et al. 2005, see also Tadhunter, these proceedings). These similarities suggest that some weak-[OI] LLAGN could be the lower-luminosity counterparts of Seyfert 2 composite and/or radio galaxies with post-starbursts. One important consequence of these similarities is that star formation should still be proceeding in the nuclei of young TOs. UV emission has been observed in the core of a few LLAGN (e.g., NGC 4303; Colina et al. 2002). The detection of strong wind lines in the UV spectra clearly suggests that the star formation is still proceeding at some residual level in young TOs (see a few cases in Maoz et al. 1998). However, the UV-optical spectral energy distribution indicates that even though these clusters dominate the UV emission, their contribution at longer wavelengths is low, and the optical continuum is dominated by intermediate-age populations.

Conclusions. Young TOs are clearly separated from old TOs and LINERs in terms of the properties and spatial distribution of the stellar populations. Young TOs have stronger stellar population gradients, a luminous intermediate-age stellar population which is concentrated close to the nucleus (central ~ 100 pc or less) and higher extinction than old TOs and LINERs. These young TOs could be classified as starbursts 1 Gyr ago, or as composite Seyfert 2s (i.e., Seyferts plus nuclear starbursts). Young TOs will become old LLAGN in a few Gyr.

Acknowledgments

We are grateful to our collaborators, Luis Colina, Tim Heckman, Claus Leitherer, Lucimara Martins, Enrique Pérez, Thaisa Storchi-Bergmann, and Henrique Schmitt, with whom we have done the work discussed here.

References

Böker T., Laine S., van der Marel R.P., Sarzi M., Rix H.-W., Ho L.C., Shields J.C., 2002, AJ, 123, 1389
Bruzual A.G., Charlot S., 2003, MNRAS, 344, 1000
Carollo M., Stiavelli M., Seigar M., de Zeeuw P.T., Dejonghe H., 2002, ApJ, 123, 159
Cid Fernandes R., Heckman T., Schmitt H., González Delgado R.M., Storchi-Bergmann T., 2001, ApJ, 558, 81
Cid Fernandes R., et al., 2004a, ApJ, 605, 105
Cid Fernandes R., Gonzíez Delgado R.M., Storchi-Bergmann T., Martins L.P., Schmitt H., 2004b, MNRAS, in press (astro-ph/0410155)
Cid Fernandes R., et al., 2004c, MNRAS, subm.
Colina L., González Delgado R.M., Mas-Hesse J.M., Leitherer C., 2002, ApJ, 579, 545
Ferrarese L., Merrit D., 2000, ApJ, 539, L9
Filippenko A.V., 1996, in: The physics of LINERs in view of recent observations, Eracleous M., Koratkar A., Leitherer C., Ho L., eds., ASP Conf. Series, (ASP: San Francisco), vol. 103, 17
González Delgado R.M., Leitherer C., Heckman T., Lowenthal J.D., Ferguson H.C., Robert C., 1998, ApJ, 495, 698
González Delgado R.M., Heckman T., Leitherer C., 2001, ApJ, 546, 845
González Delgado R.M., Cid Fernandes R., Pérez E., Martins L.P., Storchi-Bergmann T., Schmitt H., Heckman T., Leitherer C., 2004a, ApJ, 605, 127
González Delgado R.M., Cerviño M., Martins L.P., Leitherer C., Hauschildt P.H., 2005, MNRAS, in press (astro-ph/0501204)
Heckman T., González Delgado R.M., Leitherer C., Meurer G.R., Krolik J., Wilson A.S., Koratkar A., Kinney A., 1997, ApJ, 482, 114
Ho L.C., Filippenko A.V., Sargent W.L.W., 1997, ApJS, 112, 315
Maoz D., Koratkar A., Shields J.C., Ho L.C., Filippenko A.V., Sternberg A., 1998, AJ, 116, 55
Meurer G., Heckman T.M., Leitherer C., Kinney A., Robert C., Garnett D.R., 1995, AJ, 110, 2665
Tadhunter C., Robinson T.G., González Delgado R.M., Wills K., Morganti R., 2005, MNRAS, 356, 480

GALEX ULTRAVIOLET SPECTROSCOPY OF LUMINOUS INFRARED GALAXIES

Denis Burgarella[1], Véronique Buat[1], Todd Small[2], Tom A. Barlow[2], and the GALEX Team
[1] *Observatoire Astronomique Marseille Provence, LAM, France*
[2] *California Institute of Technology, Pasadena, California, USA*
denis.burgarella, veronique.buat@oamp.fr; tas, tab@astro.caltech.edu

Abstract   The ELAIS South 1 field was observed by GALEX in spectroscopic and imaging mode. Thanks to the web database built by the ELAIS team, we made use of the multi-wavelength catalogue to select galaxies in common to the two samples. 19 galaxies are found with optical redshifts in the range $0 < z < 1.6$ with $10 < \log(L_{\rm IR}) < 13$ estimated from the 15μm flux. However, the two brightest objects are Seyfert 1 galaxies, while most of the galaxies in the range $0 < z < 0.35$ are normal to Luminous Infrared Galaxies. Using these data, we find that it is not possible to uniquely derive an ultraviolet dust attenuation from the ultraviolet slope, β. These galaxies have a median FUV dust attenuation $A_{\rm FUV} = 2.7 \pm 0.8$.

1. Introduction to GALEX Data

GALEX, the Galaxy Evolution Explorer (Martin et al. 2005) is a 50cm telescope devoted to observing the ultraviolet (UV) sky in two bands centered at about 153μm and 231μm. Several surveys will be done by GALEX, ranging from an All-sky Imaging Survey (AIS) to an Ultra Deep Imaging Survey (UDIS) in imaging and three Spectroscopic Surveys. In addition, a Nearby Galaxy Survey (NGS) will observe about 200 galaxies in the local Universe. The ELAIS South 1 field (ELAIS S1; $\alpha_{\rm J2000.0} = 00^{\rm h}34^{\rm m}44^{\rm s}$ and $\delta_{\rm J2000.0} = -43°28'12''$) was observed by GALEX in spectroscopic (31267s; see Fig. 1) and in imaging mode (12198s; see Fig. 2). We took advantage of the infrared (IR) database of Rowan-Robinson et al. (2004), available from the ELAIS web site, http://astro.imperial.ac.uk/Elais/index.html, to obtain multi-wavelength fluxes and the associated redshifts.

From the 2D spectrum shown in Fig. 2, the GALEX pipeline extracts and co-adds the individual sub-exposures to provide the astronomer with a table containing 1D calibrated and formatted spectra (Fig. 3). The left-hand section

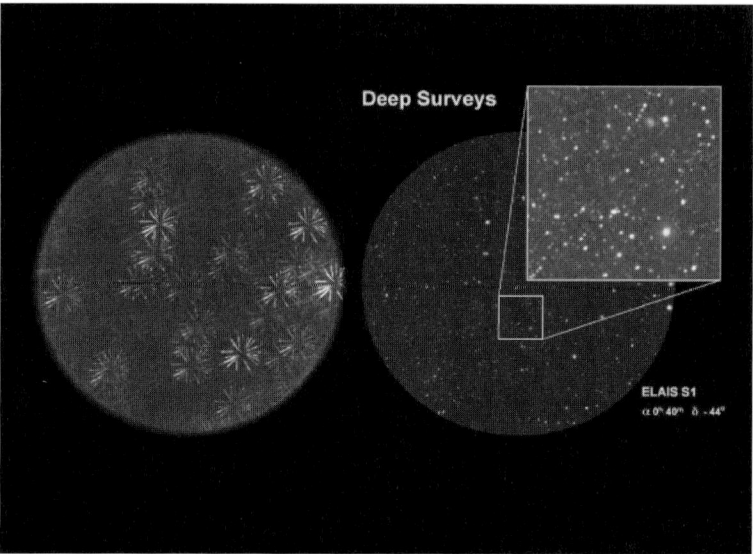

Figure 1. Spectroscopic image of the ELAIS South 1 field over the same area as the image. The shapes resembling fireworks are due to the fact that the grism is rotated for each sub-exposure.

Figure 2. Deep GALEX image of ELAIS South 1. The limiting magnitude reaches $m_{AB} = 25$ mag and the field of view is 1.2 deg in diameter. The central area of the field is zoomed in.

of the spectra corresponds to the far-UV while the right-hand section corresponds to the near-UV, and the full observed spectral range is from approximately 130 to 270μm. The spectral resolution is about $R = 100 - 200$, depending on the wavelength.

2. The Galaxy Sample

The final band-merged ELAIS catalogue has been cross-correlated with GALEX spectroscopic detections; the cross correlation contains 959 sources. The area of the field in common is about 0.5 deg^2, and the number of sources in common is 88, most of which proved to be Galactic stars. We narrowed our selection down to 19 galaxies with spectroscopic redshifts, which we study in more detail in the remainder of this paper.

All of the objects, except two QSOs, are found in the redshift range $0 < z < 0.35$. Their FUV luminosity is $\log(L_{\rm FUV}) = 9.8 \pm 0.6$ (L$_\odot$), and their average IR luminosity, deduced from the 15μm flux, and from Chary & Elbaz (2001), amounts to $\log(L_{\rm IR}) = 11.1 \pm 0.5$ (L$_\odot$). Our sample contains 12 Luminous IR Galaxies (LIRGs; $10^{11} < \log(L_{\rm IR}) < 10^{12}$ [L$_\odot$]; cf. Goldader et al. 2002). The two QSOs, located at $z > 1$, and with $\log(L_{\rm FUV}) > 12$ (L$_\odot$)

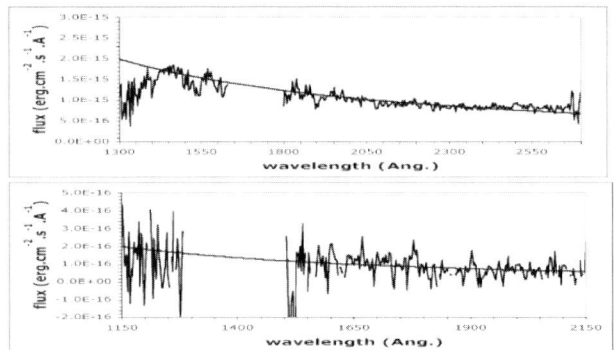

Figure 3. Two spectra extracted from the ELAIS South 1 field: the top spectrum is one of the best-quality spectra of a galaxy at $z = 0.048$; the bottom one is one of the lowest-quality spectra of a galaxy at $z = 0.286$. Power laws are superimposed on both spectra, with ultraviolet slopes of $\beta = -1.58 \pm 0.04$ and -2.04 ± 0.27, respectively.

and $\log(L_{\rm IR}) \sim 10^{13}$ (L$_\odot$) are Seyfert 1 galaxies. One of them, at $z = 1.4$, seems to have a UV slope $\beta = -2$, which is consistent with the hypothesis that some star formation contributes to the UV flux.

3. GALEX UV colors and the UV slope β

The interpretation of UV spectra is mainly related to two parameters: stellar populations (star-formation history, age, metallicity) and the reddening by dust (amount of dust, attenuation laws). Knowing the latter is crucial to get back to the former. To correct for this dust attenuation, Calzetti et al. (1994) proposed to use the UV spectral slope, β, for which they provide a calibration.

We estimated the UV slope of our sample galaxies after visually discarding bad (absorption lines and low-S/N) pixels in the rest-frame range $120-250\mu$m (see Fig. 3). From the IR/FUV flux ratio, we estimated the dust attenuation in our galaxies (calibrated using Buat et al. 2005), and compared these values to the UV slope β. If Calzetti's law is valid, we should find a good correlation between these two parameters, as was shown by Meurer et al. (1999). The diagram presented in Fig. 3 does not show such a correlation; the dispersion amounts to about 2 magnitudes in the FUV dust attenuation $A_{\rm FUV}$. A probable explanation for this behaviour, which is inconsistent with Calzetti et al. (1994), is that our galaxies are not only starburst galaxies, which are expected to scatter around the updated Calzetti curve (see Kong et al. 2004): normal galaxies and LIRGs do not seem to follow Calzetti's law.

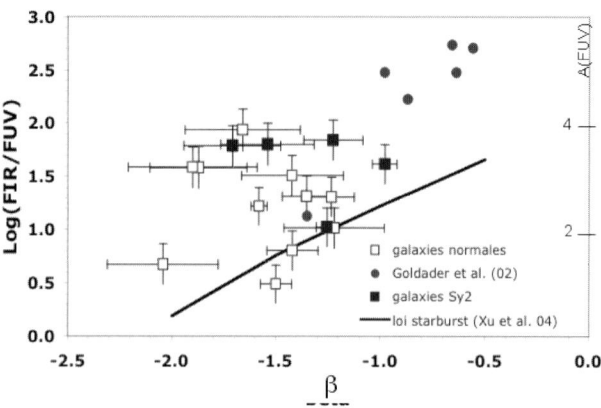

Figure 4. This diagram shows that the UV slope β can hardly be used to estimate the dust attenuation without additional assumptions (e.g., regarding the shape of the attenuation law, star-formation history, etc.

The median FUV dust attenuation for our sample galaxies amounts to $A_{\rm FUV} = 2.7 \pm 0.8$, which is very consistent with the median value for the far-IR-selected sample presented by Buat et al. (2005). On the other hand, the median value for the UV-selected sample of Buat et al. (2005) is $A_{\rm FUV} = 1.1$, i.e., very different from the previous value.

4. Conclusion

This paper presents preliminary results from GALEX spectroscopy (see Burgarella et al. [2005] for a more complete report), which show that the spectroscopic mode of GALEX is performing very well and will provide us with very interesting data. The ability of GALEX to obtain high-quality UV spectroscopic data of a large number of galaxies without a priori selection criteria will prove to be very powerful.

References

Buat V., et al., 2005, ApJ, 619, L51
Burgarella D., et al., 2005, ApJ, 619, L63
Calzetti D., Kinney A.L., Storchi-Bergmann T., 1994, ApJ, 429, 592
Chary R., Elbaz D., 2001, ApJ, 556, 562
Goldader J.D., Meurer G., Heckman T.M., Seibert M., Sanders D.B., Calzetti D., Steidel C.C., 2002, ApJ, 568, 651
Kong X., Charlot S., Brinchmann J., Fall S.M., 2004, MNRAS, 349, 769
Martin D.C., et al., 2005, ApJ, 619, L59
Meurer G.R., Heckman T.M., Calzetti D., 1999, ApJ, 521, 64
Rowan-Robinson M., et al., 2004, MNRAS, 351, 1290

A RECENT REBUILDING OF MOST SPIRALS?

François Hammer[1], Hector Flores[1], Xianzhong Zheng[2], and Yanchun Liang[3]
[1]*Laboratoire Galaxies, Etoiles, Physique et Instrumentation, Observatoire de Paris, 92195 Meudon, France*
[2]*Max-Planck Institut für Astronomie, Germany*
[3]*National Astronomical Observatories, CAS, China*

Abstract Re-examination of the properties of distant galaxies leads to evidence that most present-day spirals built up half of their stellar masses during the last 8 Gyr, mostly during several intense phases of star formation during which they resembled luminous infrared galaxies (LIRGs). Distant galaxy morphologies encompass all of the expected stages of galaxy merging, central core formation and disk growth, while their cores are much bluer than those of present-day bulges. We have tested a spiral rebuilding scenario, in which $75 \pm 25\%$ of spirals experienced their last major merger event less than 8 Gyr ago. It accounts for the simultaneous decreases, during that period, of the cosmic star-formation density, of the merger rate, of the number densities of LIRGs and of compact galaxies, while the densities of ellipticals and large spirals are essentially unaffected.

1. Towards robust evolutionary features from $z = 1$ to $z = 0$

Here, we summarize a study (Hammer et al. 2005, hereafter H05) that is based on a considerable amount of observations done using *HST*, ISO, VLA and VLT. It targets ~ 200 galaxies ($0.4 < z < 1$), mostly from the sample of the Canada France Redshift Survey (CFRS). CFRS is essentially complete up to $z \sim 1$, encompassing all luminous ($M_B < -20$) galaxies with stellar masses ranging from 3 to 30×10^{10} M$_\odot$ – hereafter called intermediate-mass galaxies. Those account for 65 to 80% of the present-day stellar mass (Brinchmann & Ellis 2000, Heavens et al. 2004). Our goals were to collect significant evolutionary features since $z = 1$, by combining: (i) Robust estimates of extinction, star and effective star-formation rates (SFRs and SFR/$M_\star$) and O/H abundances at $z \sim 0.7$. Those quantities were derived from a detailed comparison of mid-IR and extinction-corrected Balmer emission-line fluxes, which provides consistent SFR estimates within a factor of 2 (Flores et al. 2004, Liang et al. 2004a). Notice that Balmer emission lines have been properly corrected for the underlying absorption (based on high-S/N spectra at moderate resolution, see Liang et al. 2004b). Notice also that [OII] or UV fluxes underestimate the SFR by factors averaging 5 to 22 for starbursts and LIRGs, respectively. (ii) A simplified morphological classification (see Zheng et al. 2004) which accounts for E/S0 and spirals as single classes, while another class is assigned to objects barely resolved by *HST* ($r_{\text{half}} < 3.5 h_{70}^{-1}$ kpc luminous compact galaxies, hereafter called LCGs). Our method, also based on color maps, considerably limits the uncertainties related to cosmological dimming, spatial resolution and morphological k correction. It allows for a fair comparison with morphological classification (derived at the same rest-frame wavelengths) of local galaxies in the same mass range (Nakamura et al. 2004).

2. A formation history with violent IR episodes at $z < 1$

H05 found that $\sim 15\%$ of intermediate-mass galaxies at $z > 0.4$ are indeed luminous IR galaxies (LIRGs), a phenomenon far more common than in the local Universe. This is confirmed by preliminary Spitzer results. The high occurrence of LIRGs is easily understood only if they correspond to episodic peaks of star formation, during which galaxies are reddened through short IREs (infrared episodes). We estimate that each galaxy should experience 4 to $5 \times (\tau_{\text{IRE}}/0.1\text{Gyr})^{-1}$ IREs from $z = 1$ to $z = 0.4$. The star formation in LIRGs is sufficient by itself to produce 38% of the total stellar mass of intermediate-mass galaxies and then to account for most of the reported stellar mass formation since $z = 1$. This is not surprising, since integrations of the cosmic star-formation rate density, *if and only if they account for IR emission* (Flores et al. 1999, Elbaz & Cesarsky 2003), match well the evolution of the global stellar mass density since $z = 1$ (e.g., Dickinson et al. 2003, Heavens et al. 2004). It can be considered robust that $45 \pm 15\%$ of the mass locked in present-day stars actually condensed in stars at $z < 1$. This is further supported by the luminosity-metallicity relation of $z \sim 0.7$ emission-line galaxies, which is found to be on average metal deficient by a factor of ~ 2 compared to those of local spirals (Liang et al. 2004a, H05).

A star-formation history with short infrared episodes for most galaxies at intermediate redshifts is consistent with hierarchical galaxy formation scenarios. Observations of LIRG morphologies then reveal the physical processes which are responsible for most of the stellar mass production since $z = 1$. Irregulars, major mergers and compact galaxies represent together about a third of the $z \sim 0.7$ galaxy population, and they have almost disappeared by the present day (see Table 1, and also Lilly et al. 1998). Most of the star formation has occurred in these systems, since they represent almost two-thirds of $z \sim 0.7$ LIRGs (Table 1). We assume in the following that violent infrared episodes are responsible for most morphological changes, which links distant galaxies in a simple way to those of the present-day Hubble sequence.

Table 1. Morphological classification statistics for intermediate-mass galaxies (H05); $z \sim 0.7$ galaxies are compared to those of the SDSS (Nakamura et al. 2004)

Type	$z \sim 0.7$ LIRGs	$z \sim 0.7$ galaxies	local galaxies
E/S0	0%	23%	27%
Spiral	36%	43%	70%
LCG	25%	19%	< 2%
Irregular	22%	9%	3%
Major merger	17%	6%	< 2%

Which present-day galaxy types are related to both morphological changes and episodic violent star formation events? It cannot be present-day ellipticals because: (i) no LIRGs show an E/S0 morphology; (ii) if the fate of all LCGs (or all LIRGs) were to become ellipticals, the density of present-day ellipticals would be much higher than observed (see Table 1). This leads H05 to assume that most of the recent star formation has occurred in progenitors of the numerous present-day spirals (see also Wolf et al. 2004). This is further supported by Zheng et al. (2004a,b), who show that most $z \sim 0.7$ spirals have blue cores (see also Ellis et al. 2001).

3. A scenario accounting for all evolutionary features?

The fraction of major mergers is evolving rapidly (Table 1; see also Conselice et al. 2004). Major mergers in our sample are showing two well-identified nuclei separated by less than a galactic radius (Zheng et al. 2004a,b). Post-mergers are naturally expected to show compact morphologies, and we assume in the following that LCGs are merger remnants. Indeed, this was first claimed by Hammer et al. (2001) on the basis of their detailed spectral properties. Moreover, Östlin et al. (2001) argued that their local counterparts are merger remnants, on the basis of their velocity fields, which are not rotationally supported. Because pair counts evolve very rapidly, the fraction of intermediate-mass galaxies experiencing a major merger event since $z = 1$ is estimated to be 0.75 on the basis of Bundy et al. (2004), a value lower than what was found by Le Fèvre et al. (2000). We now develop a scenario in which most spirals have been rebuilt during the last 8 Gyr (Fig. 1). H05 find that assuming a characteristic time-scale for each phase (major mergers, compact and spirals) from hydrodynamical models (Tissera et al. 2004, see also Cox et al. 2004), one can reproduce the observed fraction of each species if $75\pm25\%$ of spirals were rebuilt since $z = 1$. This simple scenario is consistent with 8 robust evolutionary features, namely the evolution of the global stellar

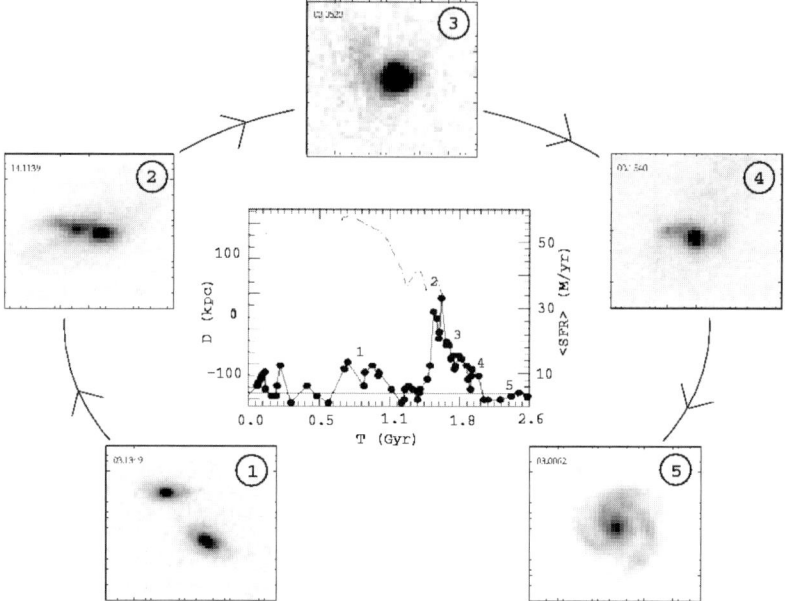

Figure 1. A sequence of $z \sim 0.7$ galaxies illustrating the rebuilding scenario; (1, 2): interaction and merging, (3, 4): post-merger prior to disk re-building and (5): disk growth. *Middle*: the star-formation history derived using hydrodynamical simulations of a major merger (Tissera et al. 2002), which shows the corresponding phases. $75 \pm 25\%$ of present-day spirals have likely experienced such a sequence since $z = 1$.

mass, the L–Z diagram, the pair statistics, the IR light density, the spiral core colours, the number density of E/S0s, spirals, and peculiar galaxies (mergers and compact).

A scenario showing no significant star formation in intermediate-mass galaxies (Cowie et al. 1996, Brinchmann & Ellis 2000) is inconsistent with 6 of these evolutionary features (stellar mass, L–Z, pair statistics, IR light density, spiral core colours and number density of peculiar galaxies). A scenario for which the stellar mass formation is dominated by minor encounters ("collisional starbursts"; Somerville et al. 2001), is better, but still inconsistent with 4 evolutionary features (IR light density, number density of peculiar galaxies, spiral core colours, and difficult to reconcile with pair statistics). In fact, accounting for the IR light evolution leads to our spiral rebuilding scenario. This scenario has the advantage to be predictive, and it can be tested. For example, the formation of a disk after a merger event requires that spiral progenitors have a substantially larger gas fraction than present-day spirals ($\sim$ 10–20% of their baryonic mass). The considerable amount of stars formed during IREs suggests a gas content $\sim$ 5 times higher in $z \sim 0.7$ galaxies. This is consistent with the reported O/H deficiency of the gaseous phases of galaxies at $z \sim 0.7$ (H05, Liang et al. 2004a). An important assumption is that LCGs are merger remnants. Preliminary results from Puech et al. (in prep.) and from Bershady et al. (these proceedings) indicate that most LCGs have velocity fields that are not supported by rotation, as those studied locally by Östlin et al. (2001).

A spiral rebuilding scenario is very efficient in forming large bulges, including those of the numerous early-type spirals (75% of present-day spirals of intermediate mass). Robust conclusions on the past formation history can be derived for only 2 intermediate-mass galaxies, e.g., the Milky Way and M31. Recent studies of stellar populations in the M31 halo are suggestive of a complex formation history, possibly including a recent merger (Brown et al. 2003). On the other hand, the Milky Way shows no trace of such a recent event. It is less bulge-dominated than M31, and we suggest that it could be part of the $\sim 25\%$ of galaxies that escaped a recent major merger at $z < 1$. IR data reveal a strikingly different history from previous studies based on optical/UV data. Indeed, the LIRG number density increases by factors of up to 35 from $z = 0$ to $z = 1$ (Elbaz & Cesarsky 2003), i.e., it evolves much more rapidly than the UV luminosity density. H05 find that the optical properties of LIRGs mimic those of other galaxies, and that besides IR photometry, only detailed spectroscopic studies at moderate resolution can distinguish between them.

References

Brinchmann J., Ellis R.S., 2000, ApJ, 536, L77
Brown T.M., Ferguson H.C., Smith E., Kimble R.A., Sweigart A.V., Renzini A., Rich R.M., VandenBerg D.A., 2003, ApJ, 592, L17
Bundy K., Fukugita M., Ellis R.S., Kodama T., Conselice C.J., 2004, ApJ, 601, 123
Conselice C.J., et al., 2004, ApJ, 600, 139
Cowie L.L., Songaila A., Hu E., Cohen J., 1996, AJ, 112, 839
Cox T.J., Primack J., Jonsson P., Somerville R.S., 2004, ApJ, 607, 87
Dickinson M., Papovich C., Ferguson H.C., Budavári T., 2003, ApJ, 587, 25
Elbaz D., Cesarsky C.J., 2003, Science, 300, 270
Ellis R., Abraham R., Dickinson M., 2001, ApJ, 551, 111
Flores H., et al., 1999, ApJ, 517, 148
Flores H., Hammer F., Elbaz D., Cesarsky C.J., Liang Y.C., Fadda D., Gruel N., 2004, A&A, 415, 885
Hammer F., Gruel N., Thuan T.X., Flores H., Infante, L., 2001, ApJ, 550, 570
Hammer F., Flores H., Elbaz D., Zheng X.Z., Liang Y.C., Cesarsky C., 2005, A&A, 430, 115 (H05)
Heavens A., Panter B., Jímenez R., Dunlop J., 2004, Nature, 428, 625
Le Fèvre O., et al., 2000, MNRAS, 311, 565
Liang Y.C., Hammer F., Flores H., Elbaz D., Cesarsky C.J., 2004a, A&A, 423, 867
Liang Y.C., Hammer F., Flores H., Gruel N., Assemat F., 2004b, A&A, 417, 905
Lilly S.J., et al., 1998, ApJ, 500, 75
Nakamura O., Fukugita M., Brinkmann J., Schneider D.P., 2004, AJ, 127, 2511
Östlin G., Amram P., Bergvall N., Masegosa J., Boulesteix J., Márquez I., 2001, A&A, 374, 800

Somerville R.S., Primack J.R., Faber S.M., 2001, MNRAS, 320, 504
Tissera P.B., Domínguez-Tenreiro R., Scannapieco C., Saiz A., 2002, MNRAS, 333, 327
Wolf C., et al., 2004, ApJ, subm. (astro-ph/0408289)
Zheng X.Z., Hammer F., Flores H., Assemat F., Pelat D., 2004a, A&A, 421, 847
Zheng X.Z., Hammer F., Flores H., Assemat F., Rawat A., 2004b, A&A, subm.

EVOLUTION OF THE IR ENERGY DENSITY AND SFH UP TO $Z \sim 1$: FIRST RESULTS FROM MIPS

E. Le Floc'h[1], C. Papovich[1], H. Dole[2], E. Egami[1], P. Pérez-González[1], G. Rieke[1], M. Rieke[1], E. Bell[3], and the Spitzer/MIPS GTO team[1]
[1] *Steward Observatory, University of Arizona, 933 N. Cherry Avenue, Tucson 85721, AZ, USA*
[2] *Institut d'Astrophysique Spatiale, Université Paris Sud, F-91405 Orsay Cedex, France*
[3] *Max-Planck-Institut für Astronomie, Königstuhl 17, D-69117 Heidelberg, Germany*
(elefloch@as.arizona.edu)

Abstract   Using deep observations of the Chandra Deep Field South obtained with MIPS at 24μm, we present our preliminary estimates on the evolution of the infrared (IR) luminosity density of the Universe from $z = 0$ to $z \sim 1$. We find that a pure density evolution of the IR luminosity function is clearly excluded by the data. The characteristic luminosity $L_{\rm IR}^*$ evolves at least by $(1+z)^{3.5}$ with lookback time, but our monochromatic approach does not allow us to break the degeneracy between a pure evolution in luminosity, or an evolution in both density and luminosity. Our results imply that IR-luminous systems ($L_{\rm IR} \geq 10^{11}$ L$_\odot$) become the dominant population contributing to the co-moving IR energy density beyond $z \sim 0.5 - 0.6$. The uncertainties affecting our measurements are largely dominated by the poor constraints on the spectral energy distributions that are used to translate the observed 24μm flux into luminosities.

1. Introduction

Deep infrared and submillimeter observations with ISO and SCUBA in the late 1990s revealed a very strong evolution of IR-luminous systems with lookback time (e.g., Smail et al. 1997, Elbaz et al. 1999). Since these surveys were only sensitive to the brightest IR sources, the quantification of this evolution has thus far remained under strong debate. In particular, the relative importance of such IR-bright objects compared to less luminous starbursts, and their respective contributions to the total co-moving energy density at high redshift is still unclear. Spitzer, the new infrared facility of NASA, is now providing a unique opportunity to address this issue in more detail.

2. Infrared luminosities of MIPS sources up to $z \sim 1$

We performed deep observations of the Chandra Deep Field South with MIPS at $24\mu m$ down to a sensitivity limit of $80\mu Jy$ (5σ detection). Cross-correlating our data with source catalogs from various optical surveys (i.e., VIMOS VLT Deep Survey, Le Fèvre et al. 2004; GOODS, Vanzella et al. 2004; COMBO-17, Wolf et al. 2004; Chandra source follow-up, Szokoly et al. 2004), we derived the redshifts of 2635 objects detected at $24\mu m$. These MIPS sources are mostly located at $z \leq 1.2$. We believe their redshift identification to be nearly complete up to $z \sim 1$.

Using various libraries of IR luminosity-dependent spectral energy distributions (SEDs; e.g., Dale et al. 2001, Chary & Elbaz 2001), we derived the total IR luminosities of the MIPS sources from their observed $24\mu m$ flux. Given the poor constraint on the true SEDs characterizing these sources at high redshift, the typical uncertainty affecting these estimates could reach a factor of 2 to 3. More importantly, the influence of dust temperature could also add a small systematic bias (Chapman et al. 2003) *that we have not quantified so far*. We will explore this effect more thoroughly in a forthcoming study.

Below $z \sim 0.5$, we find that the MIPS sources are rather modest emitters at infrared wavelengths (median $L_{IR} = 10^{10}$ $L_\odot$). At higher redshifts ($0.5 \lesssim z \lesssim 1$), luminous infrared galaxies (LIRGs, 10^{11} $L_\odot \leq L_{IR} \leq 10^{12}$ $L_\odot$) become the dominant population among the sources detected at $24\mu m$, while we also detect a significant number of more modest starbursts and spirals. The most extreme sources such as the ultraluminous IR galaxies (ULIRGs; $L_{IR} \geq 10^{12}$ $L_\odot$) still remain quite rare up to $z \sim 1.2$.

3. Evolution of the infrared luminosity function

Based on this sample of $24\mu m$-selected sources, and using the libraries of IR SEDs for computing the k corrections, we derived monochromatic luminosity functions (LFs) at different IR wavelengths and in various redshift bins up to $z \sim 1$. Figure 1 shows these LFs plotted at $60\mu m$, and compared to the local $60\mu m$ IRAS LF from Takeuchi et al. (2003). A strong evolution is clearly noticeable. We find that it can not be described by a uniform increase of the density of the local IR galaxy population ("pure density" evolution), but requires a shift of the characteristic luminosity, L^*, by a factor of $(1+z)^{4.0\pm0.5}$. Note that the well-known degeneracy between a pure evolution in luminosity or an evolution combining an increase in both density and luminosity cannot be broken at this stage.

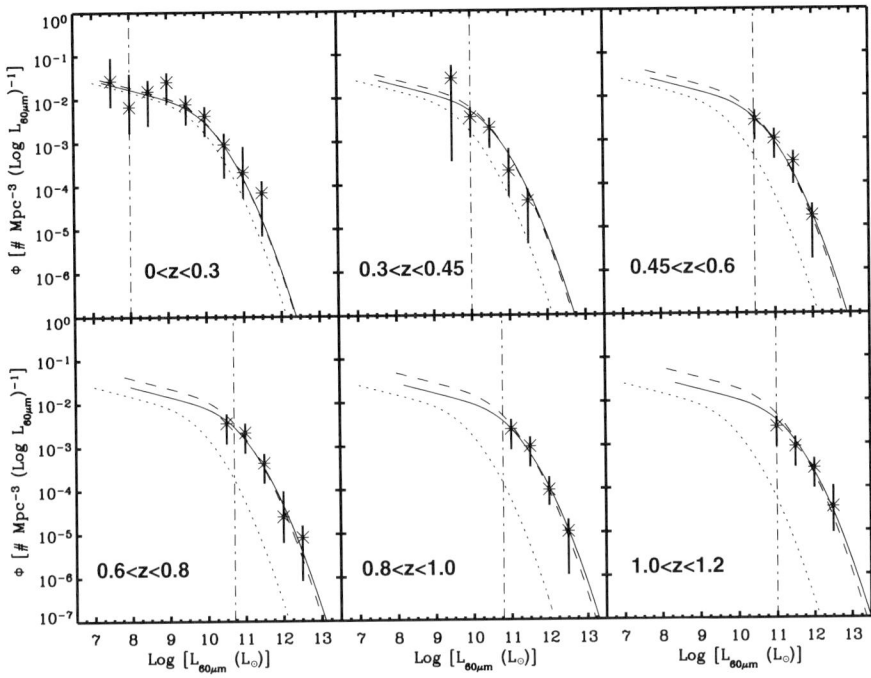

Figure 1. Luminosity functions estimated at 60μm with the $1/V_{\mathrm{max}}$ formalism in different redshift bins between $z = 0$ and $z = 1.2$ (∗ symbols). The 3σ uncertainties are indicated with vertical solid lines. Data points can be fitted by a pure luminosity evolution ($L^\star \propto (1+z)^{4.2}$, solid curve) or a combination of luminosity and density evolution ($L^\star \propto (1+z)^{3.5}, \Phi^\star \propto (1+z)^{1.0}$, dashed curve) to the local 60μm luminosity function (dotted curve). Vertical dashed-dotted lines correspond to the 80% completeness limit in each redshift bin.

4. Star-formation history up to $z \sim 1$

Star-forming galaxies are believed to be the major component of the 24μm source population (Silva et al. 2004). Assuming the calibration between the star-formation rate of galaxies and their infrared luminosities (Kennicutt 1998), we converted the evolution of the luminosity functions into an history of star formation up to $z \sim 1$. Results are shown in Fig. 2, where the relative contribution of normal starbursts, LIRGs and ULIRGs is also illustrated. We find that IR luminous systems ($L_{\mathrm{IR}} \geq 10^{11}$ L$_\odot$) become the dominant population contributing to the co-moving IR energy density beyond $z \sim 0.5 - 0.6$ and represent 70% of the star-forming activity at $z \sim 1$.

References

Blain A., Smail I., Ivison R., Kneib J.-P., Frayer D., 2002, Phys. Rep., 369, 111

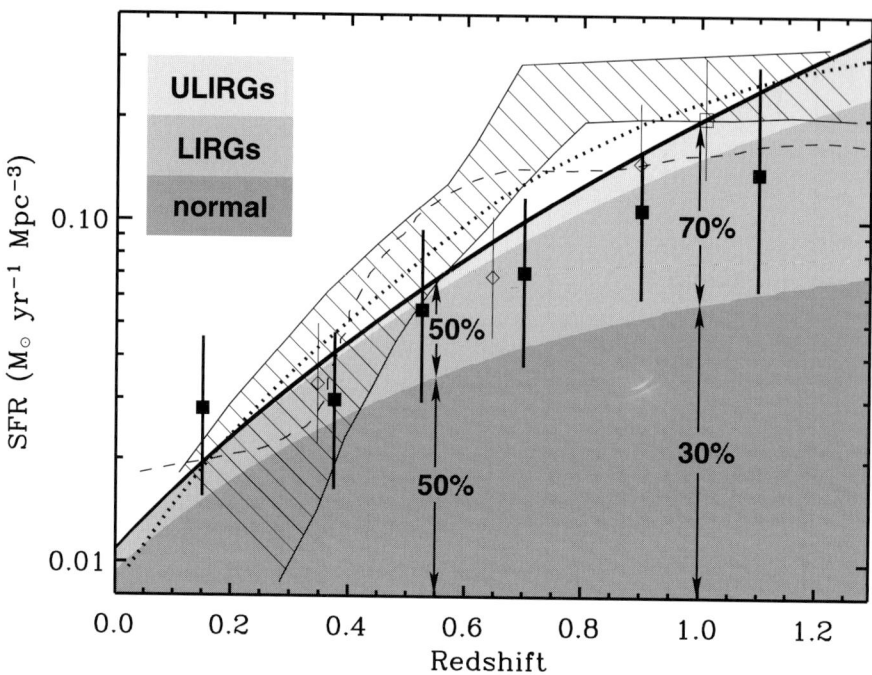

Figure 2. The star-formation history up to $z \sim 1.2$ as seen by MIPS 24μm (thick solid line), decomposed into the contributions of normal galaxies ($L_{IR} \leq 10^{11}$ L$_\odot$), LIRGs and ULIRGs (shaded regions). Filled squares correspond to the contribution of *detected* MIPS sources, while the empty diamonds and the empty square are from Flores et al. (1999) and Thompson et al. (2001), respectively. Model predictions from Chary & Elbaz (2001; cross-hatched region), Blain et al. (2002; dotted line) and Lagache et al. (2004; dashed line) are also plotted.

Chapman S.C., Helou G., Lewis G.F., Dale D.A., 2003, ApJ, 588, 186
Chary R., Elbaz D., 2001, ApJ, 556, 562
Dale D.A., Helou G., Contursi A., Silbermann N.A., Kolhatkar S., 2001, ApJ, 549, 215
Elbaz D., et al., 1999, A&A, 351, L37
Flores H., et al., 1999, A&A, 343, 389
Kennicutt R.C. Jr., 1998, ARA&A, 36, 189
Lagache G., Dole H., Puget J.-L., 2003, MNRAS, 338, 555
Le Fèvre O.L., et al., 2004, A&A, 428, 1043
Silva L., Maiolino R., Granato G.L., 2004, MNRAS, 355, 973
Smail I., Ivison R.J., Blain A.W., 1997, ApJ, 490, L5
Szokoly G.P., et al., 2004, ApJS, 155, 271
Takeuchi T.T., Yoshikawa K., Ishii T.T., 2003, ApJ, 587, L89
Thompson R.I., Weymann R.J., Storrie-Lombardi L.J., 2001, ApJ, 546, 694
Vanzella E., et al., 2004, A&A, in press (astro-ph/0406591)
Wolf C., et al., 2004, A&A, 421, 913

Session VIII

Violent Star Formation and the Properties of Star-Forming Galaxies at High Redshift

UNDERSTANDING INFRARED–LUMINOUS STARBURSTS IN DISTANT GALAXIES

C. Papovich, E. Le Floc'h, H. Dole, E. Egami, P. Pérez-González, G. Rieke, and M. Rieke (for the Spitzer/MIPS GTO team)
Steward Observatory, 933 North Cherry Avenue, Tucson, Arizona 85721, USA
papovich@as.arizona.edu

Abstract New surveys with the Spitzer space telescope identify distant starburst and active galaxies by their strong emission at far-infrared (IR) wavelengths. Using deep Spitzer surveys at 24 and 70μm coupled with *HST* imaging in the Chandra Deep Field South, we study the relation between galaxy morphology and IR-active stages of galaxy evolution. IR-luminous galaxies span a wide range of morphology. At $z \sim 1$, there is a correlation between the relative fraction of galaxies with morphological distortions (multiple nuclei, tidal tails, etc.) and increasing IR luminosity, which suggests that the strong starbursts at high redshift arise from galaxy interactions. However, the majority of IR-luminous galaxies do not have exceptionally asymmetric morphologies, and galaxies with strong asymmetries correspond to a range of galaxy IR activity. We conclude that the relation between galaxy morphology and IR activity is highly complex, and strongly dependent on the initial conditions of galaxy interactions.

1. Introduction

The improvements in far-infrared (IR) sensitivity and survey efficiency now possible with the Spitzer Space Telescope allow major advances in the causes of IR-luminous stages of starbursts, and their impact on galaxy evolution. In Papovich et al. (2004) and Dole et al. (2004), we discuss the Spitzer number counts at 24–160μm from surveys being conducted by the Spitzer Guaranteed Time Observers (GTOs). The Spitzer counts differ strongly from non-evolving models, and from predictions of various contemporary models based on ISO and IRAS results. We interpret the faint 24μm counts as evidence for a substantial population of IR-luminous galaxies at $z > 1$ (Papovich et al. 2004, see also Egami et al. 2004, Le Floc'h et al. 2004 and these proceedings), which is not reflected in contemporary models. For example, hierarchical models that include physical prescriptions for how galaxies form stars currently underproduce the faint IR counts by a factor of 3 (e.g., Balland et al. 2003). In these proceedings, we discuss our efforts to study why such rapid evolution occurs in the IR-luminous galaxy population at high redshifts.

2. Estimating the Total IR Luminosity with Spitzer

Here, we focus on a sample of galaxies in the Chandra Deep Field South (CDF-S), selected in the Spitzer mid-IR 24μm band. The GTO observations of the CDF-S intersect the COMBO-17 photometric redshift survey (Wolf et al. 2003), the *HST*/GEMS survey (Rix et al. 2004), and the VVDS and FORS spectroscopic surveys (Le Févre et al. 2004, Vanzella et al., subm.).

In what follows, we describe the morphological properties of 24μm-selected galaxies as a function of total IR luminosity ($L_{\rm IR} \equiv L[8-1000\mu{\rm m}]$). We used the measured redshifts from 24μm counterparts in the COMBO-17 and spectroscopic catalogs, and converted the 24μm galaxy flux densities to $L_{\rm IR}$ using two techniques. First, we used solely the 24μm data, and calculated k corrections between the rest-frame 24μm luminosity and $L_{\rm IR}$ using the IR templates from Dale et al. (2001) and Chary & Elbaz (2001), assuming that a given rest-frame luminosity corresponds uniquely to a single template (i.e., assuming a direct luminosity-thermal temperature relation in the IR). Second, for 24μm sources with counterparts at 70μm, we used the 24–70μm mid-to-far-IR "color" to select an IR template (akin to selecting a template based on the thermal temperature), scaled this to match the photometry, and then derived $L_{\rm IR}$. Comparing the total IR luminosities from the two methods, we find that they agree to within a factor of 2–3 for galaxies at $z \sim 0.2$–1.2. This scatter is consistent with measurements of the bivariate temperature-luminosity distribution of local IR-selected galaxies (e.g., Chapman et al. 2003).

3. Morphological Indications of IR Activity

The GEMS survey has obtained *HST*/ACS imaging of nearly 0.25 deg^2 in the CDF-S. In Fig. 1, we show the morphological concentration and asymmetry parameters for $\sim$ 400 24μm-selected galaxies at $z \sim 1$ derived from the GEMS *HST* data. Symbol type denotes the level of IR luminosity, including galaxies not detected at 24μm. IR-luminous galaxies with $L_{\rm IR} > 10^{11.5}$ L$_\odot$ span a range of morphology; Figure 2 displays some strongly "interacting" galaxies, and some fairly normal galaxies. Galaxies with $L_{\rm IR} < 10^{11.5}$ L$_\odot$ have very similar concentration and asymmetry distributions, with little change with IR luminosity. There is a trend for galaxies with $L_{\rm IR} > 10^{11.5}$ L$_\odot$ to have higher asymmetry values. While only a few percent of galaxies with $L_{\rm IR} < 10^{11.5}$ L$_\odot$ have $A > 0.35$ (the fiducial value to indicate "interacting/merging" galaxies; e.g., Conselice et al. 2004), more than 10% of galaxies with $L_{\rm IR} > 10^{11.5}$ L$_\odot$ show asymmetries above this value. Qualitatively, we find that the fraction of galaxies that exhibit morphological distortions (mul-

Understanding Infrared–Luminous Starbursts in Distant Galaxies 287

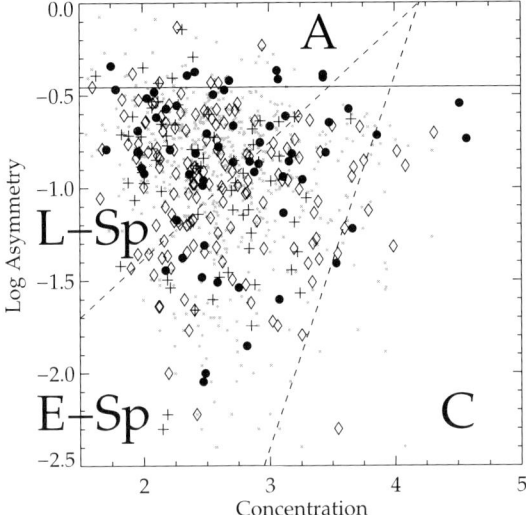

Figure 1. Distribution of morphological concentration and asymmetry of galaxies at $z = 0.9$–1.1 using the ACS z_{850}-band images from GEMS. The symbols distinguish galaxies by total IR luminosity: Gray squares, galaxies not detected at 24μm; Crosses, galaxies with $L_{IR} = 10^{10.5-11}$ $L_\odot$; Diamonds, $L_{IR} = 10^{11-11.5}$ $L_\odot$; Filled circles, $L_{IR} > 10^{11.5}$ $L_\odot$. Fiducial galaxy classifications are delineated by the dashed and solid lines (as labeled). Interacting/merging galaxies typically have high asymmetries ($A \geq 0.35$; Conselice et al. 2003), and occupy the region above the solid line (marked "A"). The dashed lines illustrate regions occupied by irregular and late-type spiral galaxies (L-Sp); early-type spirals (E-Sp); and highly concentrated galaxies (C), including ellipticals and lenticulars.

tiple nuclei, tidal features, etc.) also increases with increasing IR luminosity, from 30% of galaxies with $L_{IR} \sim 10^{10-11.5}$ $L_\odot$ to 50% of galaxies with $L_{IR} > 10^{11.5}$ $L_\odot$ (although the distortions in the majority of these galaxies do not produce excessively large asymmetries).

The quantitative and qualitative morphologies are evidence that galaxy interactions contribute strongly to luminous IR activity in galaxies. However, the period of high IR luminosity and morphological distortions are not simultaneous. We find that if all galaxies with $L_{IR} > 10^{11.5}$ $L_\odot$ arise from interactions, then galaxies are both IR luminous *and* highly asymmetric for only $\sim 10\%$ of the merger sequence. This implies that a range of galaxy-galaxy interactions takes place, each with a set of initial conditions that controls when and if a period of high IR luminosity occurs. Mihos & Hernquist (1996) illustrated that the gas in merging disk galaxies with little or no bulge tends to collapse early in the merger sequence, which gives rise to high star-formation rates (and presumably high IR luminosities) at the same time as the galaxies show substantial morphological distortions. In contrast, adding a bulge to the galaxies stabilizes the gas such that it does not collapse until near the completion of the

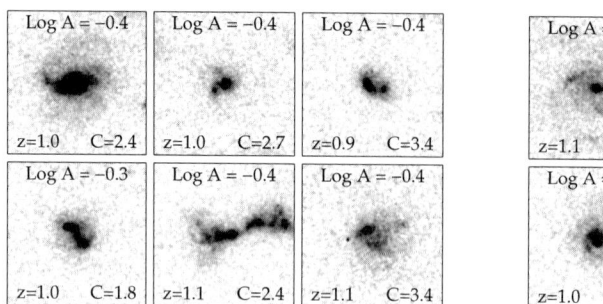
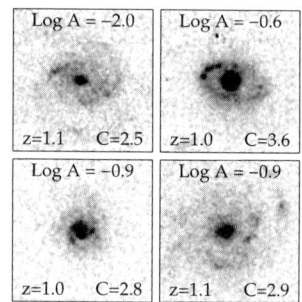

Figure 2. ACS z_{850}-band images of example $z \sim 1$ galaxies with the highest IR luminosities, $L_{\rm IR} > 10^{11.5}$ $L_\odot$. *Left:* Distorted galaxies with asymmetries ≥ 0.35, from the "A" region in Fig. 1. *Right:* Rather normal looking galaxies from the E-Sp region in Fig. 1 (although note the "ring" galaxy). Concentrations, asymmetries, and redshifts of each galaxy are inset.

merger, at which point the merger remnant is fairly compact with little asymmetry. Quantifying the distribution of morphological properties of galaxies as a function of IR luminosity and redshift will allow us to constrain the range of initial conditions in minor mergers, or interactions between galaxies and smaller satellites, and how interactions contribute to the overall star-formation rate and galaxy assembly at high redshifts.

Acknowledgments

I wish to thank the conference organizers for the opportunity to present this material, and for planning such a successful meeting. I am also indebted to the GEMS team, in particular to H.-W. Rix, E. Bell, and D. McIntosh, for their continuous support with the *HST* data and its analysis. Support for this work was provided by NASA through Contract Number 960785 issued by JPL/Caltech, and through a AAS international travel grant.

References

Balland C., Devriendt J.E.G., Silk J., 2003, MNRAS, 343, 107
Chapman S.C., Helou G., Lewis G.F., Dale D.A., 2003, ApJ, 588, 186
Chary R.R., Elbaz D., 2001, ApJ, 556, 562
Conselice C.J., Bershady M.A., Dickinson M., Papovich C., 2003, AJ, 126, 1183
Dale D.A., Helou G., Contursi A., Silbermann N.A., Kolhatkar S., 2001, ApJ, 549, 215
Dole H., et al., 2004, ApJS, 154, 87
Egami E. et al., 2004, ApJS, 154, 130
Le Floc'h E. et al., 2004, ApJS, 154, 170
Mihos C., Hernquist L., 1996, ApJ, 464, 641
Papovich C. et al., 2004, ApJS, 154, 70
Rix H.-W. et al., 2004, ApJS, 152, 163

STARBURSTS IN THE ULTRA DEEP FIELD

Rodger I. Thompson
Steward Observatory, University of Arizona, USA
rthompson@as.arizona.edu

Abstract The Hubble Ultra Deep Field (HUDF) offers a unique opportunity to study high-redshift starbursts with high spatial resolution. The combination of ACS optical and NICMOS near-IR images provide a measure of the redshift, intrinsic SED and extinction. This analysis finds 39 galaxies with SFRs in excess of 50 $M_\odot$ yr^{-1}. The typical starburst galaxy has a very blue SED and relatively high extinction. All of the HUDF starburst galaxies have redshifts higher than 0.9 but no starburst galaxies lie at redshifts greater than 3.5. The highest SFR in the HUDF is 560 $M_\odot$ yr^{-1}. The top ten SFR galaxies, all with SFRs exceeding 100 $M_\odot$ yr^{-1} lie in the redshift range between 1.7 and 3.1. This defines an epoch of massive galaxy assembly similar to that seen in the Northern Hubble Deep Field (NHDF). Unlike the HUDF, however, the NHDF contains starburst galaxies up to the analysis redshift limit of 6. Both fields subtend small areas and any conclusions relative to universal star formation should be viewed with caution.

1.   Introduction

Deep observations with the *Hubble Space Telescope (HST)* provide an opportunity to study star formation in the early Universe (Fernández-Soto et al. 1999, Thompson, Weymann & Storrie-Lombardi 2001, Thompson 2003). These and other studies indicate that star-formation rates (SFRs) at $z > 1$ are significantly higher than present-day rates. There is debate on whether the high SFRs are maintained at $z > 2$, or begin decline at higher reshifts. The small amount of time at high redshifts points to the epoch between $z = 1$ and 3 as the major star-formation period, independent of whether the high redshift SFRs are maintained.

Although the *HST* deep fields give our most sensitive glimpse at high-redshift star formation, their small size makes extension of conclusions derived from them to the Universe as a whole risky. The addition of the HUDF to the deep field suite of observations contributes new data for a region totally uncorrelated to the previous deep fields. Starburst galaxies in the HUDF comprise only 2% of the total number of galaxies observed. As such, extrapolation of starburst results to the Universe in general is particularly risky.

2. Observations and Data Analysis

Detailed descriptions of the observations are given elsewhere (Beckwith et al., Thompson et al., in prep.) so only the basics are given here. The HUDF lies in the Chandra Deep Field South, an area of several recent observations. The *HST* images were taken in the ACS F435W, F606W, F775W, and F850LP and the NICMOS F110W and F160W bands, providing observations between 0.4 and 1.8 μm. The ACS images are from the Director's Discretionary Time observations of the HUDF (Beckwith et al., in prep.). These were degraded in spatial resolution from the 0.03 arcsec drizzled images to 0.09 arcsec through straight 3 by 3 rebinning. We performed standard NICMOS IDT data reductions (Thompson et al. 1999) on the NICMOS images and drizzled the individual images from the 0.2 arcsec pixel size to 0.09 arcsec resolution. The images were aligned and the NICMOS region extracted from the ACS images.

Source extraction was accomplished with SEXTRACTOR operating on the truncated chi-squared image in the two-source mode (Szalay, Connolly & Szokoly 1999, Thompson et al. 2001). The photometric redshift, extinction and SED analysis utilized in the NHDF (Thompson et al. 2001, Thompson 2003) was repeated on the HUDF images, while extending the allowed redshift range to 10. SFRs were determined from the rest-frame 1500 Å flux of the selected SED using the SFR-to-UV flux relation found by Madau, Pozzetti & Dickinson (1998). Use of the intrinsic SED produces an extinction-corrected SFR.

3. Starburst Galaxies, Starburst Properties and Statistics

In this work, we define starbursts as galaxies with SFR $\geq$ 50 $M_\odot$ yr^{-1}. This is a high number for present-day galaxies, but it is probably appropriate at higher redshifts, which is the focus of this presentation. It is 10% of the highest SFRs found it this work. 39 galaxies in the HUDF satisfy this criterion which puts starburst galaxies as 2% of the total number of galaxies as opposed to 3% for the NHDF, using the same criterion. If that is the percentage of time in the starburst phase it would indicate that galaxies with $z = 1-3$ spend about 2–3% of their time in starbursts, either in a sustained or in multiple bursts.

The typical starburst galaxy has a very hot SED with a typical age near 50 Myr, for a low metallicity Salpeter untruncated IMF. The extinction in $E(B-V)$ runs from 0.1 to 1.0 mag, with most starbursts having an $E(B-V)$ value of 0.3 mag or greater. This extinction and luminosities of 10^{11} $L_\odot$ or greater put these galaxies into the LIRG and ULIRG categories. This is partially due to the definition of a starburst galaxy as a galaxy with a SFR of 50 $M_\odot$ yr^{-1} or greater.

The maximum SFR is 560 $M_\odot$ yr^{-1}, and there are 17 galaxies with SFR in excess of 100 $M_\odot$ yr^{-1}. All of the starburst galaxies lie in the redshift range between 0.9 and 3.5 This is in contrast to the NHDF, where starburst galaxies persist up to the maximum analyzed redshift of 6.0. Although one could speculate on physical reasons for the cut-off in the HUDF, it is probably just small-number statistics, since it is not duplicated in the NHDF.

The starburst statistics in the HUDF are dominated by two statistical uncertainties. The first is small-number statistics. Since there is a total of 39 starbursts in 3 redshift bins, the number of starbursts in any bin is uncertain by at least $\sqrt{13}$ or 30% for a typical bin. The other uncertainty is the variance due to large scale structure. For the 5.4 arcmin of the HUDF this is about a factor of 2 for all of the analyzed redshifts.

4. Where the Stars Form

Several speakers at this conference have discussed the question of whether the majority of stars form in starbursts or through quiescent normal star formation. This paper only addresses the question for redshifts beyond 0.5, the lower redshift limit of this analysis. In the unit redshift bins centered on unit redshifts, the top 10 SFR galaxies in each bin produce 50% or more of the total SFR except for $z = 4$ where it is 40%. Although the bins beyond $z = 3$ do not contain any starbursts by our definition, this statistic indicates that roughly half of the stars at $z \geq 0.5$ are produced in the highest-SFR galaxies of the epoch.

An alternative question is what types of regions do stars form in. A region can be characterized by its star-formation intensity x, the SFR per proper kpc^2. Lanzetta et al. (1999) developed a function $h(x)$ that is a measure of the star-formation intensity distribution. The function is defined such that

$$\text{SFR} = \int xh(x)\,\mathrm{d}x \qquad (1)$$

This distribution is plotted in Fig. 1 for $z = 1$. Thompson (2002) has shown that the form of this distribution is essentially independent of redshift. In general, a starburst region is defined as a region with a SFR intensity greater than 0.1 M$_\odot$ yr^{-1} kpc^{-2}, which is marked with a vertical line in Fig. 1. Integration of the distribution indicates that 60% of the SFR is contained in the region with $\log(x)$ greater than -0.25 and 95% in the region with $\log(x)$ greater than -1.25. This indicates that approximately 80% of the stars are created in starburst regions. Within the uncertainties of the measurements it is safe to say that stars are created roughly equally in starburst and non-starburst environments. However, since $h(x)$ is a smooth function, the delineation between starburst and non-starburst environments is rather arbitrary.

Acknowledgments

The NICMOS UDF observations were planned and carried out by the NICMOS UDF Treasury team. This research was funded by NASA Grant HST-GO-09803.01-A from the Space Telescope Science Institute.

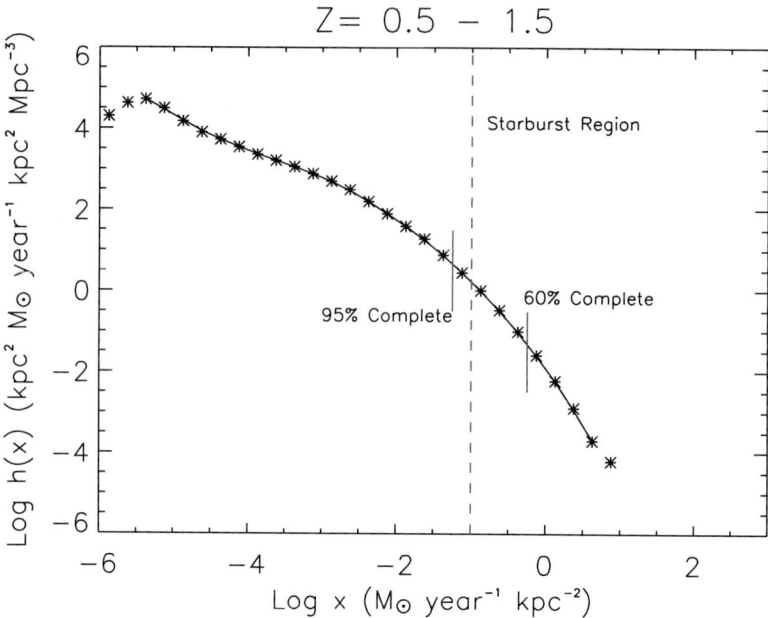

Figure 1. The star-formation intensity distribution function $h(x)$ for redshifts between 0.5 and 1.5. The starburst region is to the right of the dashed lines. The values of x at which the SFR is 60% and 95% complete are marked with vertical lines. The asterisks are the data points and the solid line is the smoothed fit to the data.

References

Fernández-Soto A., Lanzetta K., Yahil A., 1999, ApJ, 513, 34

Lanzetta K., Chen H.-W., Fernández-Soto A., Pascarelle S., Putter R., Yahata N., Yahil A., 1999, in: Photometric Redshifts and High-Redshift Galaxies, Weymann R.J., Storrie-Lombardi L.J., Sawicki M., Brunner R.J., eds., ASP Conf. Ser., (ASP: San Francisco), vol. 191, p. 223

Madau P., Pozzetti L., Dickinson M., 1998, ApJ, 498, 106

Szalay A.S., Connolly A.J., Szokoly G.P., 1999, AJ, 117, 68

Thompson R.I., Storrie-Lombardi L.J., Weymann R.J., Rieke M.J., Schneider G., Stobie E., Lytle D., 1999, AJ, 117, 17

Thompson R.I., Weymann R.J., Storrie-Lombardi L.J., 2001, ApJ, 546, 694

Thompson R.I., 2002, ApJ, 581, L85

Thompson R.I., 2003, ApJ, 596, 748

PROPERTIES OF LYα AND GAMMA RAY BURST-SELECTED STARBURSTS AT HIGH REDSHIFTS

Johan P.U. Fynbo[1], Brian Krog[2,3], Kim Nilsson[1], Gunnlaugur Björnsson[4], Jens Hjorth[1], Páll Jakobsson[1,4], Cédric Ledoux[5], Palle Møller[6], and Bjarne Thomsen[3]
[1]*Niels Bohr Institute, University of Copenhagen, Denmark;* [2]*Nordic Optical Telescope, Santa Cruz de La Palma, Spain;* [3]*Department of Physics & Astronomy, University of Århus, Denmark;* [4]*Science Institute, University of Iceland, Iceland;* [5]*European Southern Observatory, Casilla 19001, Santiago 19, Chile;* [6]*European Southern Observatory, Garching bei München, Germany*

Abstract Selection of starbursts through either deep narrow-band imaging of redshifted Lyα emitters, or localisation of host galaxies of gamma-ray bursts both give access to starburst galaxies that are significantly fainter than what is currently available from selection techniques based on the continuum (such as Lyman-break selection). Here we present results from a survey for Lyα emitters at $z = 3$, conducted with the European Southern Observatory's Very Large Telescope. Furthermore, we briefly describe the properties of host galaxies of gamma-ray bursts at $z > 2$. The majority of both Lyα and gamma-ray burst-selected starbursts are fainter than the flux limit of the Lyman-break galaxy sample, suggesting that a significant fraction of the integrated star formation at $z \approx 3$ is located in galaxies at the faint end of the luminosity function.

1. Introduction

As illustrated by the title of this conference, "From 30 Doradus to Lyman Break galaxies", the term "Lyman Break Galaxy" (LBG) is almost synonymous with "high-redshift starburst galaxy". However, as has been stressed by many authors, including the lead researchers behind the Lyman-break technique, current samples of LBGs consist of starbursts that are extremely luminous in the UV and do not give a complete census of all starbursts at high redshifts. The current magnitude limit in the ground-based surveys for LBGs is $R(\mathrm{AB}) = 25.5$ (Steidel et al. 2003). The luminosity function of LBGs has been extended to $R(\mathrm{AB}) \approx 27$, based on data from the Hubble Deep Fields (e.g., Adelberger & Steidel 2000). The faint end of this LBG luminosity function is very steep, with a slope $\alpha = -1.6$, implying that more than 70% of the light is emitted by galaxies fainter than $R(\mathrm{AB}) = 25.5$. Furthermore, the most vigorous starbursts at high redshifts as, e.g., observed with SCUBA or Spitzer, are often obscured in the rest-frame UV (e.g., Chapman et al. 2004) and hence often do not fulfill the selection criteria for LBGs.

How is it then possible to locate and examine starbursts at high redshifts that are missed by the Lyman-break technique? One other method, not mentioned at this conference, is absorption selection of galaxies. The few galaxy counterparts that thus far have been identified for Damped Lyα Absorbers (DLAs), found in QSO spectra, appear to be starburst galaxies with significantly lower luminosities than LBGs (e.g., Møller et al. 2004, Weatherley et al. 2005). However, the total cross section of DLAs at $z \approx 3$ is much larger than what can be accounted for by LBGs. This implies that most of the neutral gas available for star formation at these redshifts is located in galaxies fainter than the LBG flux limit (Fynbo et al. 1999). Other contributions in these proceedings discuss the dust emission-selected galaxies. Here, we will discuss (i) selection of Lyα emitting starbursts by means of deep narrow-band imaging, and (ii) localisation of gamma-ray burst (GRB) host galaxies.

2. Lyα selection of high-redshift starbursts

The idea to use Lyα to search for primordial galaxies dates back to the 1960s (Partridge & Peebles 1967). The first detection of redshifted Lyα emission from galaxies not powered by active galactic nuclei (AGN) were serendipitous discoveries resulting from searches for galaxy counterparts of QSO absorbers, such as DLAs and Lyman-limit systems (Lowenthal et al. 1991, Møller & Warren 1993, Francis et al. 1996, 2004, Fynbo et al. 2001). Other Lyα emitters were discovered serendipitously in searches for intra-cluster planetary nebulae (Kudritzki et al. 2000). This curiosity reflects the fact that for many years it was thought impossible to locate galaxies by their Lyα emission, as the probability of dust absorption (due to resonant scattering) is much greater for Lyα than for continuum photons. The first dedicated search for Lyα emitters with 8–10m class telescopes was conducted at the Keck telescope (Hu et al. 1998).

The "Building the Bridge" Survey. In 2000, some of the authors started the program "Building the Bridge between Damped Lyα Absorbers and Lyman-Break Galaxies: Lyα Selection of Galaxies" at the European Southern Observatory's Very Large Telescope (VLT). This project is an attempt to use Lyα selection to bridge the gap between absorption and emission-line selected galaxies by characterisation of $z \approx 3$ galaxies, possibly corresponding to the abundant population of faint ($R > 25.5$) galaxies associated with DLAs (Fynbo et al. 1999, Haehnelt et al. 2000). The survey consists of very deep narrow-band observations of three fields at $z = 2.85$, $z = 3.15$, and $z = 3.20$. In each of these fields, we have detected and spectroscopically confirmed ~ 20 Lyα emitters, or LEGOs (Lyα Emitting Galaxy-building Objects). In Fig. 1 we show six examples of LEGOs from the $z = 2.85$ field. Of the total sample, 85%

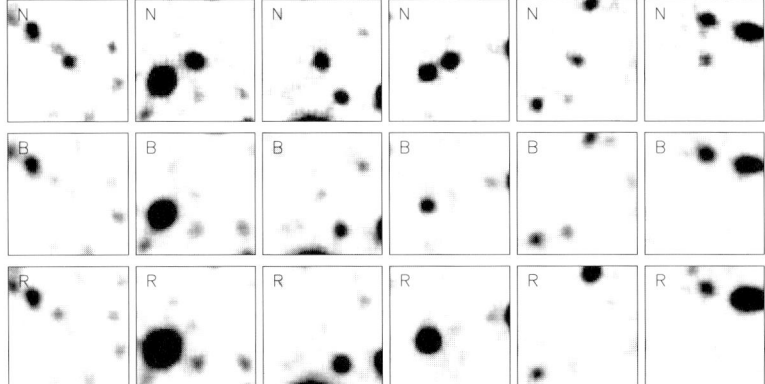

Figure 1. Six examples of spectroscopically confirmed $z = 2.85$ Lyα emitters from the "Building the Bridge Survey" (see Fynbo et al. [2003a] for more information including spectra of the sources). The size of each image is 12×12 arcsec^2. The upper row shows narrow-band images of the galaxies, the middle row the B-band images, and the bottom row the R-band images. About 20% of the LEGOs remain undetected in the broad bands, despite 5σ detection limits of $B(\mathrm{AB}) = 27.0$ and $R(\mathrm{AB}) = 26.4$.

are fainter than the flux limit for LBGs. Furthermore, as only $\approx$25% of the LBGs have Lyα emission lines with equivalent widths large enough to meet our selection criterion for LEGOs (Shapley et al. 2003), it is clear that LBGs and LEGOs are almost unrelated classes of high-redshift starbursts.

LEGOs in the GOODS Field South. Given that most LEGOs are extremely faint, it is very difficult to establish any property beyond the Lyα flux for individual galaxies, even with 8–10m class telescopes. For this reason, we decided to observe a field with existing, very deep broad-band observations covering most of the electromagnetic spectrum, namely the GOODS Field South (Giavalisco et al. 2004). In 2002 we obtained observations of a section of the GOODS Field South. Due to bad weather, we did not reach the same flux limit as for the "Building the Bridge" fields, but nevertheless we detected nearly 20 candidate $z = 3.20$ LEGOs in the field.

The analysis of the broad-band properties of these candidates constitute the thesis work of two of the authors (B. Krog and K. Nilsson). Here we report a few preliminary results. Thus far, the objects have been studied in X-rays (Chandra X-ray Observatory, 1 Ms exposure), near-IR (VLT/ISAAC), and the optical broad bands (*HST*/ACS). The *HST* images (Fig. 2) confirm that the LEGOs have extremely faint continua in the range $V(\mathrm{AB}) = 26$–29. Furthermore, (i) these galaxies are bluer than most LBGs, with spectra that rise toward the blue ($F_\nu \propto \nu^\beta; \beta < 0$) implying younger ages and/or lower dust content than what is typical for LBGs, and (ii) they have extremely compact

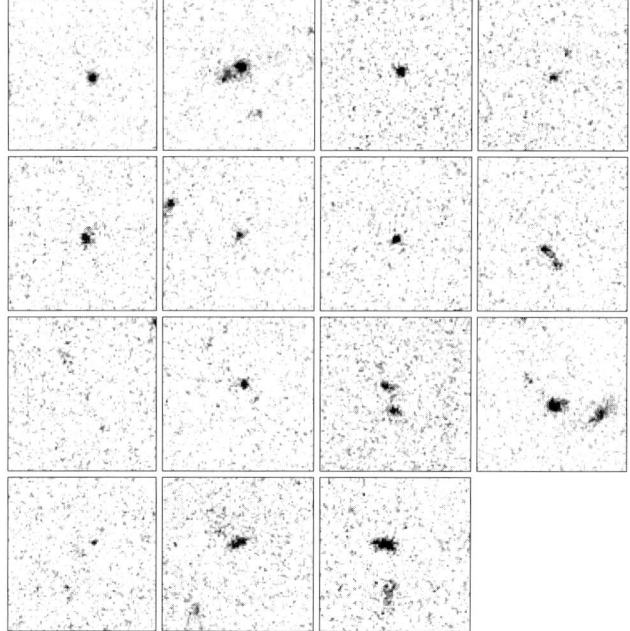

Figure 2. HST/ACS V-band images of size 3×3 arcsec^2 around the positions of 15 of the candidate $z = 3.20$ LEGOs in part of the GOODS Field South. The V(AB) magnitudes range from ~ 26 to > 29.

morphologies and sometimes consist of several knots similar to the LBGs (e.g., Lowenthal et al. 1997). Only upper limits were found in the X-ray and near-IR images, so we can exclude that the galaxies harbour AGN and significant older populations of stars. We are currently working on deriving stronger constraints from stacking the individual sources.

3. GRB selection of starbursts

GRBs are short, extremely energetic bursts of γ-rays associated with energetic core-collapse supernovae (Hjorth et al. 2003). If the GRB rate is directly proportional to the (massive star) formation rate, then the properties of GRB hosts should reflect the diversity of all star-forming galaxies in terms of luminosity, environment, internal extinction and star-formation rate. GRB hosts therefore constitute a central clue for our understanding of galaxy formation and evolution (e.g., Ramírez-Ruiz et al. 2002, Tanvir et al. 2004).

The prompt burst of γ-rays is followed by a so-called afterglow, emitting over a very wide spectral range from radio through the optical/near-IR to X-rays (see van Paradijs et al. 2000 for a review). The afterglow can be extremely

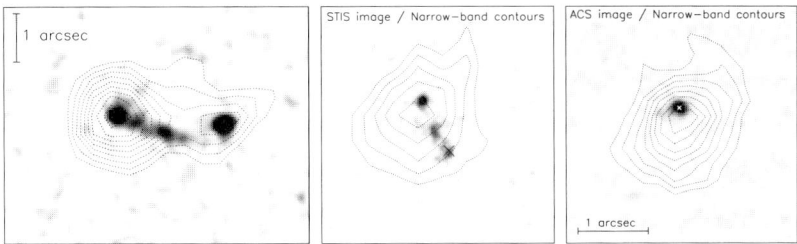

Figure 3. HST images of the host galaxies of GRBs 000926 ($z = 2.04$, $R(AB) = 24.0$), 011211 ($z = 2.14$, $R(AB) = 25.0$), and 021004 ($z = 2.33$, $R(AB) = 24.4$). The contours show the morphology of Lyα emission, as measured by ground-based narrow-band imaging (Fynbo et al. 2002, 2003b, Jakobsson et al., in prep.). GRB 000926 occurred near the centre of the right-most knot. The locations of GRB 011211 and GRB 021004 are marked with crosses. The field sizes for the GRB 011211 and GRB 021004 images are the same.

bright, allowing a precise localisation on the sky. More importantly, spectroscopy of the afterglow can give information about the redshift and the physical conditions within the host, and in intervening absorption systems along the line of sight (see, e.g., Vreeswijk et al. 2004 and Jakobsson et al. 2004a for examples). When the GRB has faded, the host galaxy itself can be observed. The measured redshifts for GRBs cover a very broad range from $z = 0.0085$ to $z = 4.50$ with a median around $z = 1.1$ (e.g., Jakobsson et al. 2004a).

How do GRB hosts compare to LBGs and LEGOs? Thus far, redshifts have been measured for 10 $z > 2$ GRBs. *HST* images of the three brightest of these are shown in Fig. 3. For the remaining seven, the host is either undetected down to faint magnitude limits, or they have magnitudes fainter than $R(AB) = 26$, i.e., the majority of these GRB hosts are fainter than LBGs. Remarkably, all current evidence is consistent with the conjecture that GRB hosts are Lyα emitters (Fynbo et al. 2003b). As shown in Fig. 3, the three brightest high-redshift GRB hosts are Lyα emitters, and for the remaining seven, Lyα emission has either been detected (2 hosts) or has not yet been searched for to sufficient depths to allow detection of even a large equivalent width emission line (5 hosts). Taken at face value, this suggests that faint, LEGO-like galaxies in total account for the majority of the star formation at these redshifts. However, there could be other explanations for why GRBs have, thus far, mainly been localised in such galaxies (Fynbo et al. 2003b). In particular, some of the so-called dark bursts could be located in more massive and dust-obscured galaxies (e.g., Tanvir et al. 2004, Jakobsson et al. 2004b, and references therein). The recently launched Swift satellite (Gehrels et al. 2004) offers a unique chance to resolve this issue.

Summary. In conclusion, surveys for LEGOs and GRB host galaxies reveal that a major fraction of the starburst activity at $z > 2$ may be located in galaxies fainter than the flux limit of the LBG survey. Here we have shown that Lyα emission and GRB selection are two viable methods to probe this population of faint starbursts at high redshifts.

Acknowledgments

We acknowledge benefits from collaboration within the EU FP5 Research Training Network "Gamma-Ray Bursts: An Enigma and a Tool". This work is supported by the Danish Natural Science Research Council (SNF).

References

Adelberger K.L., Steidel C.C., 2000, ApJ, 544, 218
Chapman S., Smail I., Windhorst R., Muxlow T., Ivison R.J., 2004, ApJ, 611, 732
Francis P., et al., 1996, ApJ, 457, 490
Francis P., Palunas P., Teplitz H.I., Williger G.M., Woodgate B.E., 2004, ApJ, 614, 75
Fynbo J.P.U., Møller P., Warren S.J., 1999, MNRAS, 305, 849
Fynbo J.P.U., Møller P., Thomsen B., 2001, A&A, 374, 443
Fynbo J.P.U., et al., 2002, A&A, 388, 425
Fynbo J.P.U., Ledoux C., Møller P., Thomson B., Burud I., 2003a, A&A, 407, 147
Fynbo J.P.U., et al., 2003b, A&A, 406, L63
Gehrels N., et al., 2004, ApJ, 611, 1005
Giavalisco M., et al., 2004, ApJ, 600, L93
Haehnelt M., Steinmetz M., Rauch M., 2000, ApJ, 534, 594
Hjorth J., et al., 2003, Nature, 423, 847
Hu E.M., Cowie L.L., McMahon R.G., 1998, ApJ, 502, L99
Jakobsson P., et al., 2004a, A&A, 427, 785
Jakobsson P., Hjorth J., Fynbo J.P.U., Watson D., Pedersen K., Bjørnsson G., Gorosabel J., 2004b, ApJ, 617, L21
Kudritzki R.-P., et al., 2000, ApJ, 536, 19
Lowenthal J.D., Hogan C.J., Green R.F., Caulet A., Woodgate B.E., Brown L., Foltz C.B., 1991, ApJ, 377, 73
Lowenthal J.D., et al., 1997, ApJ, 481, 673
Møller P., Warren S.J., 1993, A&A, 270, 43
Møller P., Fynbo J.P.U., Fall S.M., 2004, A&A, 422, L33
Partridge R.B., Peebles P.J.E., 1967, ApJ, 147, 868
Ramírez-Ruiz E., Trentham N., Blain A.W., 2002, MNRAS, 329, 465
Shapley A.E., Steidel C.C., Pettini M., Adelberger K.L., 2003, ApJ, 588, 65
Steidel C.C., Adelberger K.L., Shapley A.E., Pettini M., Dickinson M., Giavalisco M., 2003, ApJ, 592, 728
Tanvir N., et al., 2004, MNRAS, 352, 1073
van Paradijs J., Kouveliotou C., Wijers R.A.M.J., 2000, ARA&A, 38, 379
Vreeswijk P., et al., 2004, A&A, 419, 927
Weatherley S., Warren S.J., Møller P., Fall S.M., Fynbo J.P.U., Croom S.M., 2005, MNRAS, in press (astro-ph/0501422)

THE STELLAR POPULATION OF HIGH-Z GALAXIES FROM MEDIUM-RESOLUTION SPECTRA IN THE FORS DEEP FIELD

D. Mehlert[1], C. Tapken[1], I. Appenzeller[1], S. Noll[1,2], D. de Mello[3,4], and T.M. Heckman[5]
[1] *Landessternwarte Heidelberg, Königstuhl 12, 69117 Heidelberg, Germany*
[2] *MPE Garching, Gießenbachstrasse, 85748 Garching, Germany*
[3] *Laboratory for Astronomy and Solar Physics, GSFC, Greenbelt, MD 20771, USA*
[4] *Department of Physics, Catholic University of America, Washington, DC 20064, USA*
[5] *Johns Hopkins University, Department of Physics & Astronomy, Baltimore, MD 21218, USA*

Abstract   We obtained UV rest-frame spectra of 20 galaxies with $2.37 \leq z \leq 3.40$ in the FORS Deep Field, using FORS2 at the ESO VLT. The spectral resolution of $R \approx 2000$ allows us to measure the purely photospheric indices "1370" and "1425" (Leitherer et al. 2001). These indices are good metallicity indicators, unaffected by the starburst age or interstellar absorption lines. We measure an increase of both indices with decreasing redshift, indicating an increase of the average metallicity of bright starburst galaxies with cosmic time.

1. Introduction

Understanding how galaxies formed and evolved is one of the key open questions in astronomy. Therefore, it is important to analyze the properties of the stellar population of the young galaxies at early cosmic epochs. In order to derive information on such galaxies, we obtained high-S/N, low-resolution ($R \approx 200$) spectra of galaxies with $z \leq 5$ (Noll et al. 2004) in the FORS Deep Field (FDF; e.g., Heidt et al. 2003, Bender et al. 2001). Using the equivalent width of the strong stellar-wind line of C IV $\lambda 1550$ in the spectra of $z \leq 3.5$ galaxies as metallicity indicator, we found a significant increase of the average metallicities from about $0.16\,Z_\odot$ at the cosmic epoch corresponding to $z \approx 3.2$ to about $0.42\,Z_\odot$ at $z \approx 2.3$ (Mehlert et al. 2002). These results were subject to two assumptions: (i) The metallicity calibration of C IV based on a local sample is applicable to high-redshift objects. (ii) The contribution of the interstellar absorption is negligible, as observed for local starburst galaxies. In order to verify our early results without these assumptions, we started an investigation of the metallicity of high-redshift galaxies with the two indices "1425" and

"1370" defined by Leitherer et al. (2001), and recently revisited by Rix et al. (2004). These indices are based purely on photospheric stellar lines and vary significantly with metallicity, but only weakly with starburst age. On the other hand, since these two indices rely on weak photospheric stellar lines their measurement requires significantly higher resolution ($R \approx 2000$) spectra. The observations, the spectra, and first results are described in this contribution.

2. Observations and Data

For 20 FDF galaxies with $2.37 \leq z \leq 3.40$ we obtained medium-resolution spectra ($R \approx 2000$), using FORS2 at the ESO VLT with two different setups: 14 objects were observed for 6h each using the 1400V grism. For this setup, the rest-frame $z \approx 2.4$ galaxy spectra are centered around the CIVλ1550 line. 12 objects were observed for 10h using the 1200R grism. For this setup, the rest-frame $z \approx 3.3$ galaxy spectra are centered around the CIVλ1550 line. 6 objects were covered by both setups. The resulting rest-frame resolution element is ≈ 0.75 Å, which closely matches the resolution of the synthetic spectra of Leitherer et al. (1999, 2001), used for comparison. The mean S/N ratio per resolution element in the individual spectra varies between 2 and 10. For 5 galaxies, the S/N ratio of the individual spectra was sufficiently high to measure the 1425 and the 1370 index in the single spectra. In order to evaluate the remaining lower-S/N ratio spectra too, we co-added all spectra to produce composite spectra, which include all galaxies within the two redshift bins, $z \leq 3$ and $z > 3$, respectively. These composite spectra are shown in Figs. 1a and b. In Figs. 1c to e we show enlargements used for spectral regions of the indices 1425 and 1370 and of the two resonance doublets of CI and SiIV.

3. Results

Figure 2 shows the observed values of the two indices as a function of redshift for the single measurements (open squares), as well as the average value of these single measurements in the two redshift bins $z \leq 3$ and $z > 3$ (filled squares). Also included are the index strengths for the two composite spectra described in Section 2 (filled triangles in Fig. 2). Figure 2 shows that both indices increase significantly with decreasing redshift, indicating an increase of the average metallicity with cosmic time. This result is in good agreement with our earlier findings using CIV as a metallicity indicator (Mehlert et al. 2002). Earlier measurements of the 1425 index of high-redshift galaxies exist in the literature for Q1307-BM1163, $z = 1.4$ (Steidel et al. 2004), and the K20 sample, $z = 1.9$ (de Mello et al. 2004). Although both of these measurements correspond to lower redshifts, the values are consistent with the metallicity evolution indicated by Fig. 2. The dependence of the indices on the IMF (Rix

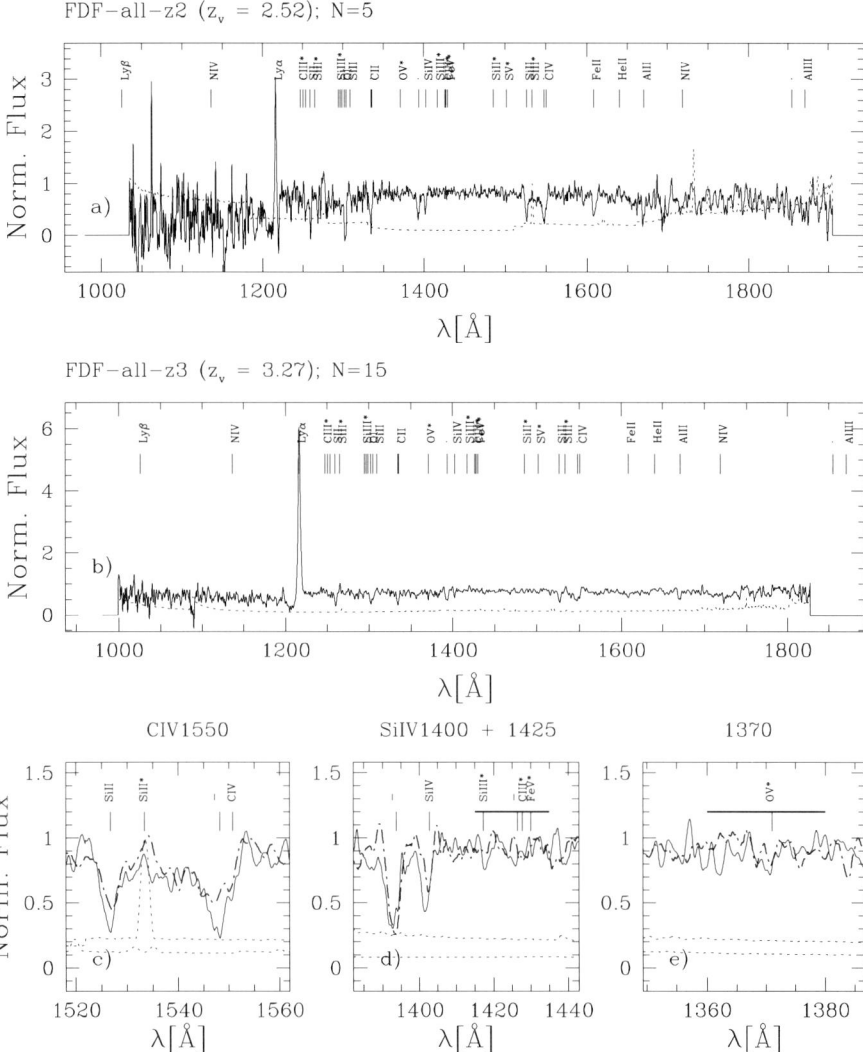

Figure 1. The composite spectrum of *(a)* all $z \leq 3$ galaxies and *(b)* all $z > 3$ galaxies. The dotted line indicates the noise level. The expected positions of some prominent spectral lines are indicated by vertical lines. Purely photospheric lines are indicated by asterisks. Panels *(c)*, *(d)* and *(e)* show enlargements around the C IV wind line, the Si IV wind line plus the 1370 blend, and the 1425 blend, respectively (solid line, composite spectrum of all $z \leq 3$ galaxies; dashed dotted line, composite spectrum of all $z > 3$ galaxies). In panels *(d)* and *(e)*, the position of the 1370 and 1425 blend is indicated by a horizontal bar, respectively.

et al. 2004) is also a factor to be considered and will be further analysed in a forthcoming paper.

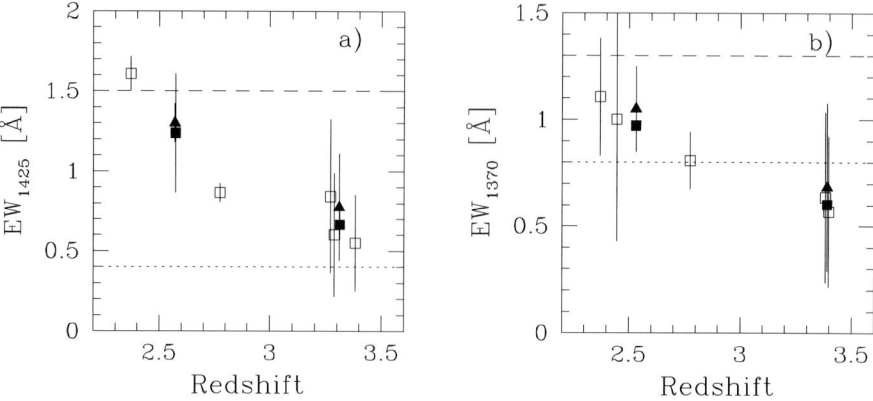

Figure 2. Measured metallicity indices of 1425 and 1370. Open squares: single galaxy measurements. Filled squares: mean values of the indices within the redshift bins $z \leq 3$ and $z > 3$. Filled triangles: measurements of the two composite spectra containing galaxies with $z \leq 3$ and $z > 3$, respectively. The horizontal lines correspond to solar (dashed) and LMC (0.25 $Z_\odot$; dotted) metallicity measured with STARBURST99 (Leitherer et al. 1999) models of continuous star formation with standard IMF and an age of 100 Myr.

Acknowledgments

This research was supported by the German Science Foundation (DFG) (Sonderforschungsbereich 439).

References

Bender R., et al., 2001, in: Deep Fields, ESO/ECF/STScI Workshop, Cristiani S., Renzini A., Williams R., eds., (Springer: New York), p.96

de Mello D.F., Daddi E., Renzini A., Cimatti A., di Serego Alighieri S., Pozzetti L., Zamorani G., 2004, ApJ, 608, L29

Heidt J., et al., 2003, A&A, 398, 49

Leitherer C., Leão J.R.S., Heckman T.M., Lennon D.J., Pettini M., Robert C., 2001, ApJ, 550, 724

Leitherer C., et al., 1999, ApJS, 99, 173

Mehlert D., et al., 2002, A&A, 393, 809

Noll S., et al., 2004, A&A, 418, 885

Rix S.A., Pettini M., Leitherer C., Bresolin F., Kudritzki R.-P., Steidel C.C., 2004, ApJ, 615, 98

Steidel C.C., Shapley A.E., Pettini M., Adelberger K.L., Erb D.K., Reddy N.A., Hunt M.P., 2004, ApJ, 604, 534

STAR-FORMING GALAXIES AT $Z \sim 2$: STELLAR AND DYNAMICAL MASSES

Dawn K. Erb[1], Charles C. Steidel[1], Alice E. Shapley[2], Max Pettini[3], Naveen A. Reddy[1], and Kurt L. Adelberger[4]

[1]*California Institute of Technology, Department of Astronomy, MS 105-24, Pasadena, CA 91125, USA;* [2]*Astronomy Department, 601 Campbell Hall, University of California, Berkeley, CA 84720, USA;* [3]*Institute of Astronomy, Madingley Road, Cambridge CB3 0HA, UK;* [4]*Observatories of the Carnegie Institution of Washington, 813 Santa Barbara Street, Pasadena, CA 91101, USA*

Abstract   We present a comparison of stellar and dynamical masses for 63 UV-selected galaxies with spectroscopic redshifts at $z \sim 2$. Stellar masses are determined by fitting SEDs to $UGRJK$ photometry, with the addition of mid-IR data from the Spitzer Space Telescope in some cases. Dynamical masses are determined using the velocity dispersions from Hα spectra. We find a mean stellar mass of 7×10^{10} M$_\odot$ and a mean dynamical mass of 6×10^{10} M$_\odot$; we discuss the limitations of both mass estimators.

1. Introduction

The epoch of cosmic history formerly known as the "redshift desert," encompassing the range $1.5 \lesssim z \lesssim 2.5$, hosted a large fraction of the star formation we see evidence of in galaxies today (Dickinson et al. 2003), as well as the peak space densities of bright QSOs (Fan et al. 2001) and sub-millimeter selected galaxies (Chapman et al. 2003). In spite of this, until recently relatively little was known about normal galaxies in this redshift range, because of the difficulty of identifying them due to their lack of strong emission lines in the optical window. However, in the past several years surveys have emerged which approach this problem from various wavelengths; the near-IR surveys select galaxies based on their K magnitudes (K20, Cimatti et al. 2002; GDDS, Abraham et al. 2004) or colors (FIRES, Franx et al. 2003), while color selection based on rest-frame UV colors has proven highly successful at identifying galaxies at these redshifts (Steidel et al. 2004).

Here we present preliminary determination of stellar and dynamical masses for a subset of UV-selected galaxies at $z \sim 2$. The total sample includes ~ 1000 galaxies with spectroscopically confirmed redshifts $1.4 \lesssim z \lesssim 2.6$;

we have modeled the stellar populations of 72 of these that have been imaged by the IRAC camera on the Spitzer Space Telescope (Section 2), and have obtained Hα spectra of ~ 100 galaxies in several different fields, from which we estimate dynamical masses and metallicities (Section 3). We use a cosmology with $H_0 = 70$ km s^{-1} Mpc^{-1}, $\Omega_m = 0.3$, and $\Omega_\Lambda = 0.7$ throughout.

2. Rest-frame UV-to-IR SEDs

Using Bruzual & Charlot (2003) model SEDs and $UGRK_s$ and 3.6, 4.5, 5.4 and 8 μm photometry from the IRAC camera on the Spitzer Space Telescope, we have calculated best-fit SEDs for 72 galaxies in the Q1700 field observed during the in-orbit check-out of the IRAC instrument (Barmby et al. 2004). We use a procedure similar to that of Shapley et al. (2001, 2004), fitting for age and extinction and normalizing for star-formation rate and stellar mass; we use a variety of declining star-formation histories, as well as a constant star-formation model. Because of degeneracies between age and extinction, which redden the UV slope similarly, we obtain the strongest constraints on stellar mass. The stellar masses follow a nearly lognormal distribution, with mean and standard deviation $\log M_* = 10.32 \pm 0.51$ M$_\odot$ (Shapley et al., in prep.).

Galaxies at the massive end of the distribution have properties similar to those of bright galaxies found in IR-selected surveys (e.g., Daddi et al. 2004); $\sim 8\%$ of the sample have stellar masses $M_* > 10^{11}$ M$_\odot$ and best-fit ages of 2–3 Gyr. Best-fit SEDs for these galaxies are shown in Fig. 1. With the addition of the IRAC data, we determine stellar masses to $\sim 30\%$, although current star formation may hide significant stellar mass, especially for particularly blue galaxies. We have also obtained $UGRJK$ photometry for ~ 450 galaxies with spectroscopic redshifts $z > 1.4$; modeling the stellar populations of the complete sample will provide a mass function for $z \sim 2$ UV-selected galaxies (Erb et al., in prep.; although not all of these have been imaged by IRAC, a comparison of stellar masses from model SEDs with and without the IRAC data shows no significant systematic bias).

3. Near-IR Spectroscopy

We have also obtained Hα spectra for more than 100 galaxies with redshifts $2 \lesssim z \lesssim 2.5$, using the near-IR spectrograph NIRSPEC on the Keck II telescope. These allow us to measure kinematic properties (velocity dispersions and, in $\sim 15\%$ of the sample, spatially resolved velocity shear), calculate dynamical masses, estimate metallicities, and measure star-formation rates. We focus here on preliminary comparisons of the stellar and dynamical masses.

Figure 2 shows the comparison of dynamical mass from the Hα line width and stellar mass from the SED modeling described above, for 63 galaxies for which we have both Hα spectra and near-IR photometry. In principle, this

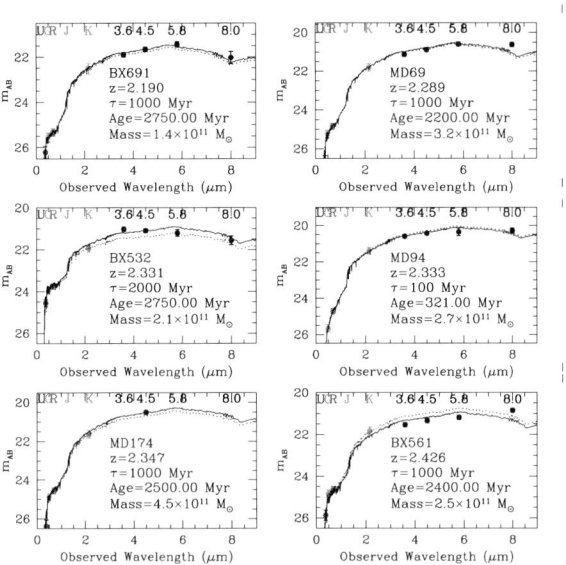

Figure 1. Sample best-fit SEDs from the modeling of rest-frame UV through IR photometry. The galaxies shown here are the most massive in the sample, with $M_* > 10^{11}$ M$_\odot$. The solid line shows the fit including the mid-IR IRAC data, while the dashed line is the fit to only the $UGRK$ photometry. In most cases they agree quite well.

calculation should tell us about the relative contributions of stars and non-luminous matter to the mass of the galaxy; in practice, of course, the comparison is highly imprecise at these redshifts.

For comparison with previous work (Pettini et al. 2001, Erb et al. 2003, 2004) we calculate dynamical masses using the virial theorem and the assumption of a uniform sphere, $M_{\rm dyn} = 5\,r\,\sigma^2/G$. We use $r = 0.3$ arcsec (1 arcsec corresponds to approximately 8 kpc at these redshifts), a typical size from *HST*/ACS images, although the galaxies' morphological irregularity makes appropriate sizes difficult to determine and the variation among individual galaxies is large. The assumption of uniform, spherical density is problematic, but is difficult to improve upon without a more detailed knowledge of the galaxies' structure. The velocity dispersions are also likely to be underestimates, since signal-to-noise issues allow us to detect emission from only the regions of the galaxies with the highest surface brightness (the total dynamical mass should be larger than the stellar mass, but this is generally not what we observe). Given all of these systematic uncertainties, the masses generally agree: we find a mean stellar mass of 7×10^{10} M$_\odot$, and a mean dynamical mass of 6×10^{10} M$_\odot$; the mean dynamical mass varies by a factor of ~ 2 for other reasonable choices of radius.

A further interesting use of the kinematic data may be to investigate the question of evolution; the mean velocity dispersion of the $z \sim 2$ galaxies is $\sim 50\%$ higher than that of the $z \sim 3$ LBGs (a fact that does not appear to be explained by the use of Hα at $z \sim 2$ and [O III] at $z \sim 3$). This may offer a comparison of mass less complicated by issues of sample selection than the comparison of photometry at $z \sim 2$ and $z \sim 3$. Kinematic measurements may also offer insights into star-formation history that are masked by current star formation in the model SEDs (see Erb et al. [2005] for a detailed discussion).

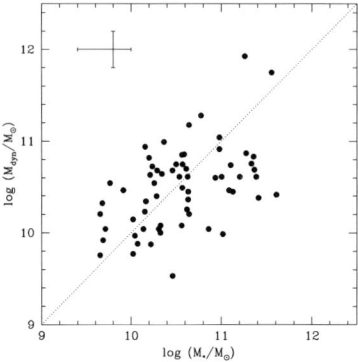

Figure 2. Comparison of stellar and dynamical masses for the 63 galaxies for which we have both Hα spectra and $UGRJK$ photometry. The two galaxies in the upper right have broad Hα lines indicative of an AGN; an AGN continuum may affect their stellar mass estimates as well.

References

Abraham R.G., et al., 2004, AJ, 127, 2455
Barmby P., et al., 2004, ApJS, 154, 97
Bruzual G., Charlot S., 2003, MNRAS, 344, 1000
Chapman S.C., Blain A.W., Ivison R.J., Smail I., 2003, Nature, 422, 695
Cimatti A., et al., 2002, A&A, 392, 395
Daddi E., et al., 2004, ApJ, 600, L127
Dickinson M., Papovich C., Ferguson H.C., Budavári T., 2003, ApJ, 587, 25
Erb D.K., Shapley A.E., Steidel C.C., Pettini M., Adelberger K.L., Hunt M.P., Moorwood A.F.M., Cuby J.-G., 2003, ApJ, 591, 101
Erb D.K., Steidel C.C., Shapley A.E., Pettini M., Adelberger K.L., 2004, ApJ, 612, 122
Fan X., et al., 2001, AJ, 121, 54
Franx M., et al., 2003, ApJ, 587, 79
Pettini M., et al., 2001, ApJ, 554, 981
Shapley A.E., Erb D.K., Pettini M., Steidel C.C., Adelberger K.L., 2004, ApJ, 612, 108
Shapley A.E., Steidel C.C., Adelberger K.L., Dickinson M., Giavalisco M., Pettini M., 2001, ApJ, 562, 95
Steidel C.C., Shapley A.E., Pettini M., Adelberger K.L., Erb D.K., Reddy N.A., Hunt M.P., 2004, ApJ, 604, 534

METALLICITY OF STAR-FORMING GALAXIES

Evolution between $0 < z < 1$

Lisa Kewley[1] and Henry A. Kobulnicky[2]
[1] *Institute for Astronomy, University of Hawaii, USA*
[2] *University of Wyoming, USA*
kewley@ifa.hawaii.edu, chipk@uwyo.edu

Abstract Metallicity is intricately related to star formation because metals are injected into the interstellar medium by stellar mass-loss processes. Theory suggests that metallicity changes less rapidly than star-formation rate as a function of redshift, but until now, there has been no solid observational foundation for the cosmic metallicity history of star-forming galaxies. We present the first results of our new investigation into the metallicity history of star-forming (emission-line) galaxies between redshifts $0 < z < 3$. This analysis provides an initial insight into the evolution of metallicity for star-forming galaxies spanning the redshift range $0 < z < 3$.

1. Introduction

To understand how galaxies in the early Universe evolved into those that we see locally requires an understanding of the chemical and star-formation history of galaxies as a function of redshift. The star-formation history of galaxies has been studied extensively, but our understanding of the chemical history of galaxies as a function of redshift is still largely theoretical. Theory predicts that, as time progresses, the mean stellar metallicity of galaxies increases with chemical enrichment, while the theoretical envelope (spread) of metallicities decreases as the metallicity is diluted by a continual infall of more pristine gas from the interstellar medium (e.g., Nagamine et al. 2001).

Observational investigations into the metallicity history were difficult in the past, because most studies focused on absorption-line systems observed in quasar spectra. Obtaining an absolute galaxy metallicity by this method is non-trivial, because only a single line of sight is available through these systems. Fortunately, alternative metallicity estimates are now possible for star-forming galaxies out to $z \sim 3$ thanks to new efficient spectrographs on 8–10m telescopes and emission-line diagnostics from state-of-the-art stellar evolution and photo-ionization models (e.g., Kobulnicky & Kewley 2004, Lilly, Carollo

& Stockton 2003, Kobulnicky et al. 2003, Shapley et al. 2004). In this paper, we present the first results from our study into the metallicity evolution of star-forming galaxies.

2. Sample Selection and Metallicity Measurements

The aim of this project is to investigate the *relative* change in the mean metallicity and the difference in the spread of metallicities between local galaxies and those at higher redshift in the same mass range. For this preliminary study, we use B-band luminosity as a surrogate for mass.

We use the Nearby Field Galaxy Survey (NFGS) as our local comparison sample. A detailed discussion of the NFGS sample selection is given in Jansen et al. (2000). The 198-galaxy NFGS sample spans the full range of Hubble type and absolute magnitude present in the CfA1 redshift catalog. The NFGS is the only objectively selected sample for which integrated long-slit spectra are available (integrated spectra are essential for avoiding aperture effects as a function of redshift). AGN were removed from the sample using the theoretical classification scheme of Kewley et al. (2001). Balmer emission lines were corrected for underlying stellar absorption as described in Kewley et al. (2002). Emission-line ratios were corrected for extinction using the Balmer decrement and a classical attenuation curve. We calculated metallicities for the NFGS following the prescription outlined in Kewley & Dopita (2002).

Estimates of metallicities for intermediate-redshift ($0 < z < 1$) galaxies have recently been made by Kobulnicky & Kewley (2004), Lilly et al. (2003), Kobulnicky et al. (2003), Kobulnicky & Zaritsky (1999). In Kobulnicky & Kewley (2004), we calculated metallicities for 204 $0.3 < z < 1$ galaxies in the GOODS-North field, using spectra from the Team Keck Redshift Survey (TKRS; Cowie et al. 2004, Wirth et al. 2004). We combined the TKRS metallicities with measurements from the DEEP Groth Strip Survey (Kobulnicky et al. 2003), and additional objects from the Canada-France Redshift Survey (Lilly et al. 2003, Corollo & Lilly 2002). We selected galaxies with [OII], [OIII] and Hβ S/N ratios of at least 8:1. AGN were removed from the samples by searching for broad emission lines or EW([NeIII]λ3826)/EW([OII]λ3727) ratios exceeding 0.4 (e.g., Osterbrock 1989). Note that metallicities for the TKRS and DEEP galaxies are based on equivalent widths, because the survey spectra were not flux calibrated. For galaxies at redshifts between $0.4 < z < 1$, the red emission lines are redshifted out of the optical regime. In this case, metallicities must be obtained using [OII]λ3727, [OIII]λ5007 and Hβ using the R_{23} diagnostic, where R_{23}=([OII]λ3727+[OIII]$\lambda\lambda$5007,4959)/Hβ (see Kewley & Dopita for a review of R_{23}). The major problem with the R_{23} method for our study is that the ratio R_{23} is degenerate – or double-valued – with metallicity, giving both a high and a low metallicity estimate. We assumed that the $0.4 < z < 1$ galaxy metallicities are high, but further observations of additional emission lines are required to reliably constrain the metallicities to either the high or the low value for each galaxy.

Our high-redshift sample consists of emission-line galaxies from Pettini et al. (2001), Kobulnicky & Koo (2000), and Shapley et al. (2004). The metallicities from Shapley et al. (2004) were derived using the [NII]/Hα ratio, which is sensitive to the (unconstrained) ionization parameter of the gas. Metallicities for the remaining galaxies were calculated by Pettini et al. (2001) and Kobulnicky & Koo (2000) using the R_{23} diagnostic.

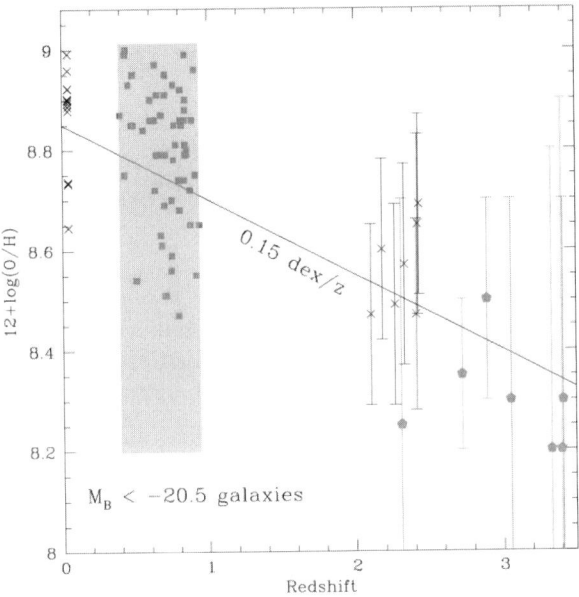

Figure 1. Metallicity evolution for star-forming (emission-line) galaxies between $0 < z < 3$. Crosses, squares, asterisks and hexagons represent the local NFGS sample, the intermediate-redshift sample, the high-redshift [NII]/Hα sample, and the high-redshift R_{23} sample, respectively. The grey square indicates the potential error range of the metallicities in the intermediate-redshift sample.

3. Metallicity History

In Fig. 1 we provide a preliminary view of the metallicity history of star-forming galaxies between $0 < z < 3$, using our three samples. To compare galaxies in the same luminosity class, we plot the $M_B < -20.5$ galaxies in each sample. The grey rectangle indicates the maximum error range in the intermediate-redshift sample metallicities. This error range is defined by the possible metallicities that could be derived using the degenerate R_{23} diagnostic. The metallicities for the high-redshift sample are shown as asterisks and hexagons on Fig. 1. The hexagons indicate that the metallicities were calculated using the R_{23} method, while the asterisks indicate that the metallicities were calculated using the [NII]/Hα line ratio.

Figure 1 shows that the mean metallicity decreases with redshift, as anticipated by chemical evolution models of stellar metallicities (Nagamine et al. 2004). The line of best fit to the data in Fig. 1 gives a mean change in metallicity of 0.15 dex z^{-1}. This estimate should be used with caution: many of the intermediate-redshift galaxy metallicities require confirmation with upcoming observations of a larger number of emission lines. In our future investigations, we will consider emission-line and photometric selection effects which could potentially alter both the mean and spread of the metallicities observed in the intermediate and high-redshift samples.

To summarize, we have provided an initial view of the metallicity history of star-forming galaxies between $0 < z < 1$. In future, we aim to obtain a solid observational understanding of the metallicity history of star-forming galaxies as a function stellar mass.

References

Cowie L.L., Barger A.J., Hu E.M., Capak P., Songaila A., 2004, AJ, 127, 3137
Jansen R.A., Franx M., Fabricant D., Caldwell N., 2000, ApJS, 126, 271
Kewley L., Dopita M., Sutherland R., Heisler C., Trevena J., 2001, ApJ, 556, 121
Kewley L.J., Geller M.J., Jansen R.A., Dopita M.A., 2002, AJ, 124, 3135
Kewley L.J., Dopita M.A., 2002, ApJS, 142, 35
Kobulnicky H.A., Kewley L.J., 2004, ApJ, 617, 240
Kobulnicky H., et al., 2003, ApJ, 599, 1006
Kobulnicky H.A., Koo D.C., 2000, ApJ, 545, 712
Kobulnicky H., Zaritsky D., 1999, ApJ, 511, 908
Lilly S., Carollo C., Stockton A., 2003, ApJ, 597, 730
Nagamine K., Fukugita M., Cen R., Ostriker J.P., 2001, ApJ, 558, 497
Osterbrock D.E., 1989, Astrophysics of Gaseous Nebulae and Active Galactic Nuclei (Mill Valley: University Science Books)
Pettini M., Shapley A.E., Steidel C.C., Cuby J.-G., Dickinson M., Moorwood A.F.M., Adelberger K.L., Giavalisco M., 2001, ApJ, 554, 981
Shapley A.E., Erb D.K., Pettini M., Steidel C.C., Adelberger, K.L., 2004, ApJ, 612, 108
Wirth G.D., et al., 2004, AJ, 127, 3121

NEW METALLICITY DIAGNOSTICS FOR HIGH-REDSHIFT STAR-FORMING GALAXIES

Samantha A. Rix[1,2], Max Pettini[1], Claus Leitherer[3], Fabio Bresolin[4], Rolf-Peter Kudritzki[4], and Chuck C. Steidel[5]
[1] Institute of Astronomy, Madingley Road, Cambridge, CB3 0HA, UK
[2] ING, Apartado de Correos 321, 38700 S/C de La Palma, Spain
[3] STScI, 3700 San Martin Drive, Baltimore, MD 21218, USA
[4] IfA, 2680 Woodlawn Drive, Honolulu HI 96822, USA
[5] Palomar Observatory, Caltech 105–24, Pasadena, CA 91125, USA

Abstract We have modelled the ultraviolet spectra of star-forming regions at metallicities between 1/20 and twice solar, by integrating the output from the model atmosphere code WM-BASIC into the spectral synthesis code STARBURST99. Using our model spectra, we have identified a promising new metallicity diagnostic for high-redshift star-forming galaxies, based upon the strength of a blend of photospheric FeIII absorption lines between 1935 and 2020 Å. We test its validity by applying it to two well-observed high-redshift galaxies.

1. Introduction

Large samples of high-redshift star-forming galaxies, such as the Lyman-break galaxies (LBGs), are now being routinely detected and spectroscopically confirmed over redshifts spanning most of the Hubble time. While many of their global properties are reasonably well characterised, such as their clustering and luminosity functions, there have been fewer studies to date of their individual properties, for instance their stellar populations and chemical compositions. Information on the chemical enrichment of such galaxies will provide important clues as to their evolutionary status, their past history of star formation, their link to today's galaxies and the interplay with their environment.

The limited applicability, at high redshifts, of well-established metallicity indicators has highlighted the need for new diagnostics, particularly ones that can be applied in a "wholesale" manner to large and well-defined samples of galaxies. With this motivation, we have explored the possibility of extracting the information about the chemical enrichment of these galaxies that is encoded in the ultraviolet (UV) spectra of their stellar populations. By bringing together techniques from the fields of massive stars, local starbursts and high-redshift galaxies, we have modelled the UV spectra of star-forming regions over a broad range of metallicities and propose a promising new metallicity diagnostic, based on the strength of FeIII photospheric absorption features near 1978 Å. The results of this work are described in detail in Rix et al. (2004); below is a brief summary.

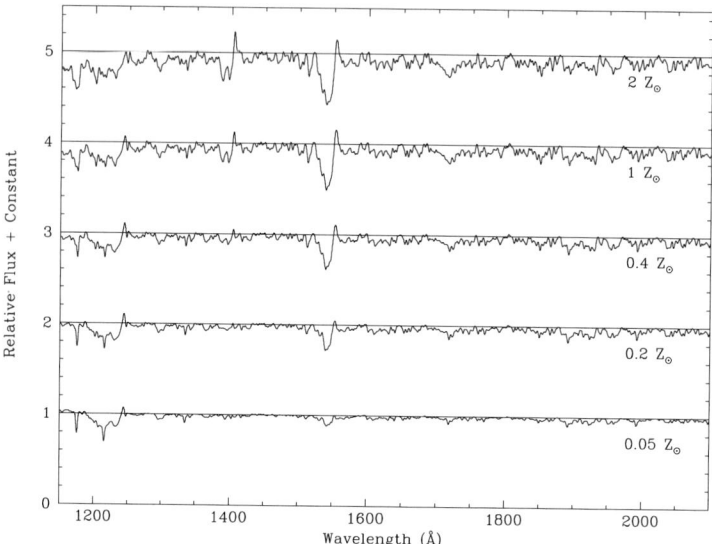

Figure 1. Synthetic UV spectra of star-forming regions with metallicities 2–0.05 $Z_\odot$, at 100 Myr, under the assumption of a continuous star-formation mode with a Salpeter IMF between 1 and 100 $M_\odot$.

2. Marrying Starburst99 with WM-basic

The UV spectra of high-redshift star-forming galaxies, which we can readily observe in the optical regime thanks to the cosmological redshifting of their light, consist of the integrated light of the underlying stellar population, which is dominated by hot and luminous O and B stars. To make progress in interpreting such a complex spectrum, and to extract information about the constituent stellar population and its properties, one can employ an evolutionary spectral synthesis code, such as STARBURST99 (Leitherer et al. 1999, 2001). However, observational constraints limit the empirical libraries of stellar spectra used by STARBURST99 to two metallicity regimes, namely Galactic ($\sim Z_\odot$) and Magellanic Cloud ($\sim 0.25\ Z_\odot$) metallicities. In order to overcome this problem, we synthesized *ab initio* theoretical spectra of stars at the upper end of the Hertzsprung-Russell diagram for a broad range of metallicities, using the non-LTE model atmosphere code WM-BASIC (Pauldrach et al. 2001), and in-

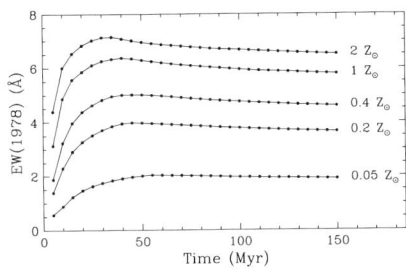

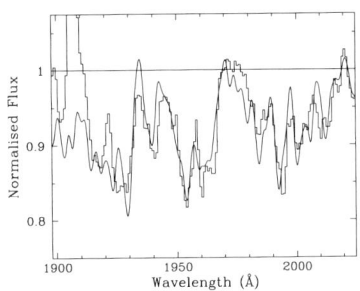

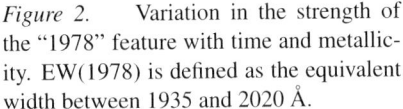

Figure 2. Variation in the strength of the "1978" feature with time and metallicity. EW(1978) is defined as the equivalent width between 1935 and 2020 Å.

Figure 3. Comparison of an LRIS spectrum of the galaxy Q1307-BM1163 ($z = 1.411$; histogram) from a survey by Steidel et al. (2004) with our STARBURST99 + WM-BASIC synthetic spectrum at a metallicity of 1 $Z_\odot$ (line).

corporated the resulting libraries into STARBURST99. Our new updated STARBURST99 code is capable of modelling the UV stellar spectra of star-forming regions as a function of time at metallicities between 1/20 and twice solar, for a range of initial mass functions (IMFs) and for both continuous and instantaneous star-formation modes.

3. Fe III absorption near 1978 Å: A New Metallicity Diagnostic

In Fig. 1 we present spectra synthesized by our code, under the assumption of a continuous star-formation law, with a Salpeter IMF, at the five available metallicities (2, 1, 0.4, 0.2 and 0.05 $Z_\odot$). Although we show the spectra at an age of 100 Myr, the dominant population (and therefore its spectral appearance) reaches an approximate quasi-equilibrium after $\sim$ 50 Myr. Clearly visible is a strong metallicity dependence in both the stellar wind lines (e.g., the C IV λ1549 P Cygni profile) and the photospheric absorption features (e.g., Fe III absorption from 1800 to 2200 Å). In this work, we have concentrated on the photospheric features; the stellar wind lines will be the subject of a future study.

We considered a number of candidate photospheric absorption blends for use as metallicity diagnostics, including the "1370" and "1425" absorption features discussed by Leitherer et al. (2001) and the Fe III absorption from 1935–2020 Å (hereafter the "1978" feature). Under the idealised assumption of continuous star formation, we found all three regions to be only weakly sensitive to time after $\sim$ 50 Myr, while much more strongly dependent upon metallicity (see, e.g., Fig. 2). The "1978" feature proves to be the most promising metallicity diagnostic given (i) its broader wavelength range and stronger absorption than the "1370" and "1425" features, (ii) the lack of other "contaminating" spectral features in star-forming galaxies at these wavelengths, (ii) its accessibility to optical observations over the redshift range $z \sim 1-3$ (where recent surveys have been prolific), and (iv) its sensitivity to the abundance of mainly one element, namely iron.

4. Comparison of our Models with Spectral Observations of High-Redshift Galaxies

Before we can apply our new metallicity diagnostic with confidence to the study of high-redshift star-forming galaxies, we must first test its validity against more established metallicity measures. For this reason, we compared our model spectra to the rest-frame UV spectra of two bright and well-studied high-redshift galaxies, MS1512-cB58 ($z = 2.73$) and Q1307-BM1163 ($z = 1.41$), both of which have previous metallicity estimates. We find that our completely theoretical models do a remarkably good job at reproducing the shape of the line profile of the "1978" feature (see, e.g., Fig. 3), and that the metallicities we deduce from the strength of the absorption agree, to within a factor of ~ 2, with those derived from analyses of interstellar absorption and emission lines.

5. Conclusions

The synergy of the model atmosphere code WM-BASIC and the spectral synthesis code STARBURST99 has enabled us to model the UV spectra of star-forming regions at metallicities between 1/20 and twice solar. From our models, we propose using the strength of the "1978" photospheric absorption feature as a metallicity diagnostic for distant star-forming galaxies. Our new technique seems to be a promising alternative, or at least complement, to established methods, which have only a limited applicability at high redshifts.

References

Leitherer C., et al., 1999, ApJS, 123, 3
Leitherer C., Leão J.R.S., Heckman T.M., Lennon D.J., Pettini M., Robert C., 2001, ApJ, 550, 724
Pauldrach A.W.A., Hoffmann T.L., Lennon M., 2001, A&A, 375, 161
Rix S.A., Pettini M., Leitherer C., Bresolin F., Kudritzki R.-P., Steidel C.C., 2004, ApJ, 615, 98
Steidel C.C., Shapley A.E., Pettini M., Adelberger K.L., Erb D.K., Reddy N.A., Hunt M.P., 2004, ApJ, 604, 534

UV LUMINOSITY FUNCTION AT $Z \sim 4, 3,$ AND 2

Marcin Sawicki[1] and David Thompson[2]
[1]*Dominion Astrophysical Observatory, Herzberg Institute of Astrophysics, National Research Council, 5071 West Saanich Road, Victoria, BC, V9E 2E7, Canada*
[2]*Caltech Optical Observatories, California Institute of Technology, MS 320-47, Pasadena, California, 91125, USA*
marcin.sawicki@nrc.ca, djt@irastro.caltech.edu

Abstract   We use very deep ($R_{\rm lim} = 27$) $U_n GRI$ imaging to study the evolution of the faint end of the UV-selected galaxy luminosity function from $z \sim 4$ to $z \sim 2$. We find that the number of sub-L^* galaxies increases from $z \sim 4$ to $z \sim 3$, while the number of bright galaxies appears to remain constant. We find no evidence for continued evolution to lower redshift, $z \sim 2$. If real, this differential evolution of the luminosity function suggests that *differentially* comparing key diagnostics of dust, stellar populations, etc., as a function of z and L may allow us to isolate the key mechanisms that drive galaxy evolution at high redshift, and we describe several such studies currently underway.

1. The Keck Deep Fields

The shape of the galaxy luminosity function bears the imprint of galaxy formation and evolutionary processes, and suggests that galaxies below L^* differ substantially from those above it in more than just luminosity. Our understanding of galaxy formation may profit from studying the evolution of not just the bright but also the faint component of the galaxy population at high redshift.

To study the evolution of the galaxy luminosity function at high redshift, we have carried out a very deep imaging survey that uses the very same $U_n GRI$ filter set and color-color selection technique as used in the work of Steidel et al. (1999, 2003, 2004), but that reaches to $R = 27 - 1.5$ magnitudes deeper and significantly below L^* at $z = 2$–4 (Sawicki & Thompson 2005). These Keck Deep Fields (KDF) were obtained with the LRIS imaging spectrograph on Keck I and represent 71 hours of integration, split into five fields that are grouped into 3 spatially-independent patches, to allow us to monitor the effects of cosmic variance. We use the $U_n GRI$ filter set and spectroscopically-confirmed and optimized color-color selection techniques developed by Steidel et al. (1999, 2003, 2004). Consequently, we can confidently select sub-L^* star-forming galaxies at $z \sim 4, 3, 2.2$, and 1.7, without the need for what at

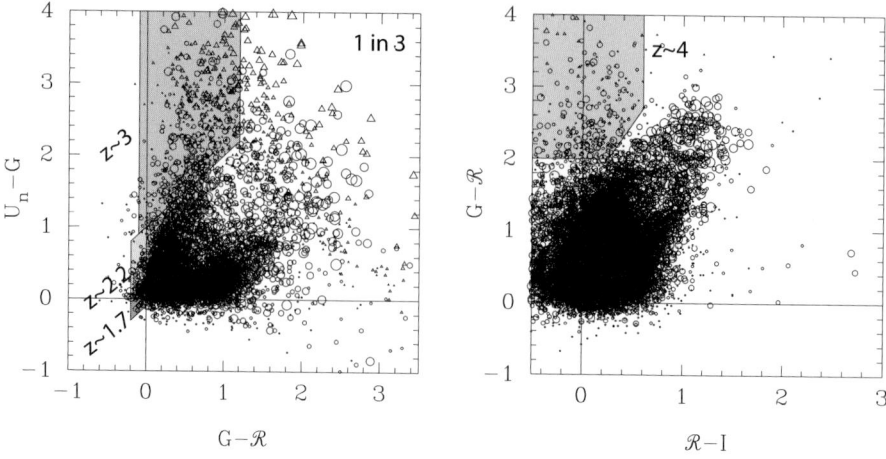

Figure 1. Color-color diagrams showing the selection regions for selecting high-z samples in the KDF. The left-hand panel illustrates selection of $z \sim 3, 2.2$, and 1.7 samples; only a third of the objects in our catalog are plotted. The right-hand panel shows the selection of $z \sim 4$ galaxies. The filters we used and our color selection are both *identical* to those used by the Steidel team to select galaxies at $z \sim 1.7$–4. However, our data reach to $R = 27$, or 1.5 magnitudes deeper and thus significantly below L^* at these redshifts.

the magnitudes we probe would be extremely expensive spectroscopic characterization of the sample. To $R = 27$, the KDF contains 427, 1481, 2417, and 2043, $U_n GRI$-selected star-forming galaxies at $z \sim 4, 3, 2.2$, and 1.7, respectively (Fig. 1). A detailed description of the Keck Deep Field observations, data reductions, and the high-z galaxy selection can be found in Sawicki & Thompson (2005).

2. Luminosity function at high redshift

Figure 2 shows the luminosity function of UV-selected star-forming galaxies at $z \sim 4, 3$, and 2.2. At $z \sim 4$ and 3, we augment our KDF data with the identically-selected and similarly-computed Steidel et al. (1999) LF measurements at bright magnitudes, $R < 25.5$. As in Steidel et al. (1999), our LF calculation uses the effective volume technique, calculating V_{eff} using recovery tests of artificial high-z galaxies implanted into the imaging data. We do not present the results for $z \sim 1.7$ here, because that analysis is still ongoing as the narrow selection window in the $(U_n - G)$ color makes it necessary to carry out a more sophisticated treatment of completeness and effective volume at $z \sim 1.7$ than in the higher-redshift bins. Our data reach down to very faint luminosities, which correspond to star-formation rates (not adjusted for dust) of only 1–2 $M_\odot$ yr^{-1} (Fig. 2).

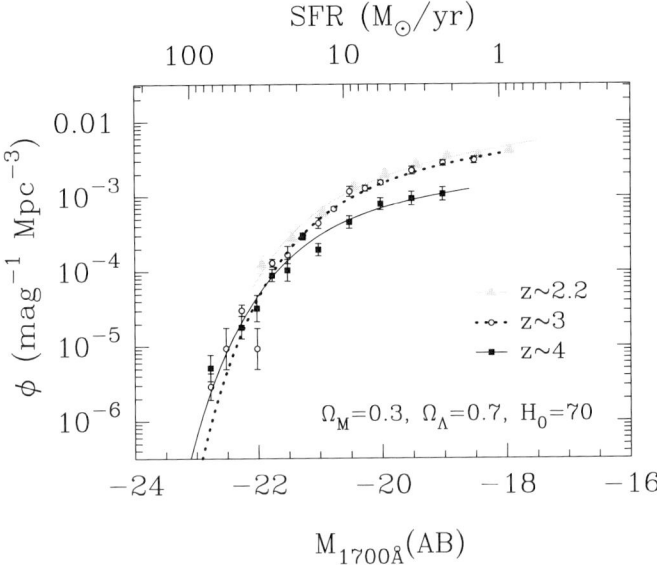

Figure 2. The luminosity function of high-z rest-frame UV-selected galaxies at $z \sim 4, 3$, and 2.2. No dust correction has been applied; error bars include both $\sqrt{N}$ counting statistics and field-to-field scatter. There is a clear deficit of faint LBGs at $z \sim 4$ compared to $z \sim 3$ and $z \sim 2.2$.

As Fig. 2 shows, we find a factor of ~ 3 evolution in the number counts (or, alternatively, 2 mag in luminosity) of *faint* Lyman Break Galaxies from $z \sim 4$ to $z \sim 3$, while at the same time there is no evidence for evolution from $z \sim 4$ to $z \sim 3$ at the *bright* end (Steidel et al. 1999, and Fig. 2). Thus, it seems that the luminosity function of high-z galaxies evolves in a *differential* way, suggesting that different mechanisms drive the evolution of the faint and of the luminous galaxies. It is unlikely that the observed evolution is due to a selection bias because (i) if it were, we would expect the deficit to be present at bright *and* faint magnitudes, and (ii) to make up the deficit would require an unreasonably enormous expansion of the $z \sim 4$ color selection box (Fig. 2). We therefore conclude that the evolution from $z \sim 4$ to $z \sim 3$ is likely a reflection of a true differential, luminosity-dependent evolutionary effect. At the same time, we see no evidence for evolution at the the faint end from $z \sim 3$ to $z \sim 2.2$ (the bright end remains unconstrained at present), suggesting that the mechanism responsible for the evolution at earlier epochs saturates at lower redshifts.

3. What is behind the evolution of the LF?

It is presently not clear what is responsible for the observed differential evolution of the LF. One straightforward possibility is that the number of faint (but not bright) LBGs simply increases over the 500 Myr from $z \sim 4$ to $z \sim 3$ (but not beyond). Another possibility is that dust properties – its amount or covering fraction – may be decreasing in faint LBGs, thus making them brighter. Alternatively, if star formation in individual faint LBGs is time-variable, then they may brighten and fade (and thus move into and out of a given magnitude bin in the LF) with duty cycle properties that change with redshift. However, whichever mechanism is responsible, it appears to saturate by $z \sim 2.2$.

The differential appearance of the LF evolution suggests that different evolutionary mechanisms are at play as a function of UV luminosity. Studies that *differentially* compare key galaxy properties as a function of luminosity and redshift should help us isolate the mechanisms that are responsible for this evolution. For example: (i) The KDF is designed to measure galaxy clustering as a function of both luminosity and redshift, and doing so will let us relate the potentially time-varying UV luminosity to the more stable dark matter halo mass. (ii) We will also use a high-quality ~ 1000-hour 80-object *composite* spectrum of a faint $z \sim 3$ LBG (Gemini GMOS observations are underway) to compare key diagnostics of dust, superwinds, and stellar populations in a faint and a luminous (e.g., Shapley et al. 2003) composite LBG. (iii) Broadband rest-frame UV-optical colors constrain the stellar population age and the amount of dust in LBGs (e.g., Sawicki et al. 1998, Papovich et al. 2001), and we can use this approach to look for systematic differences in age and dust content. Significantly, all such studies will be making *differential* comparisons, thereby reducing our exposure to systematic biases in models or low-z analogs.

A key point is that we have identified luminosity and redshift as important variables in galaxy evolution at high z. While LBG follow-up studies to date have primarily focused on luminous galaxies at $z \sim 3$, extending such studies as a function of z and L should yield valuable insights into how galaxies form and evolve: studying how key diagnostic properties of high-z galaxies vary will help us constrain the drivers of galaxy evolution in the early Universe.

References

Papovich C., Dickinson M., Ferguson H.C., 2001, ApJ, 559, 620
Sawicki M., Thompson D., ApJ, 2005, in press
Sawicki M., Yee H.K.C., 1998, AJ, 115, 1329
Shapley A.E., Steidel C.C., Pettini M., Adelberger K.L., 2003, ApJ, 588, 65
Steidel C.C., Adelberger K.L., Giavalisco M., Dickinson M., Pettini M., 1999, ApJ, 519, 1
Steidel C.C., et al., 2003, ApJ, 592, 728
Steidel C.C., et al., 2004, ApJ, 604, 534

MASSIVE GALAXIES AT $Z = 2$ IN COSMOLOGICAL HYDRODYNAMIC SIMULATIONS

Kentaro Nagamine[1], Renyue Cen[2], Lars Hernquist[3], Jeremiah P. Ostriker[2], and Volker Springel[4]

[1]*University of California, San Diego, USA;* [2]*Princeton University, USA;* [3]*Harvard University, USA;* [4]*Max-Planck-Institut für Astrophysik, Germany*

Abstract   We study the properties of galaxies at redshift $z = 2$ in a Λ cold dark matter (ΛCDM) Universe, using two different types of hydrodynamic simulation methods – Eulerian TVD and smoothed particle hydrodynamics (SPH) – and a spectrophotometric analysis in the U_n, G, R filter set. The simulated galaxies at $z = 2$ satisfy the color-selection criteria proposed by Adelberger et al. (2004), when we assume Calzetti-type extinction with $E(B-V) = 0.15$. We find that the number density of simulated galaxies brighter than $R = 25.5$ at $z = 2$ is about $1 \times 10^{-2} h^3$ Mpc^{-3} for $E(B-V) = 0.15$, which is roughly twice that of the number density found by Erb et al. (these proceedings) for the ultraviolet (UV) bright sample. This suggests that roughly half of the massive galaxies with $M_\star > 10^{10} h^{-1}$ M$_\odot$ at $z = 2$ are part of the UV-bright population, and the other half is bright at infrared (IR) wavelengths. The most massive galaxies at $z = 2$ have stellar masses $\geq 10^{11-12}$ M$_\odot$. They typically have been forming stars continuously, at a rate exceeding 30 M$_\odot$ yr^{-1} over a few Gyr from $z = 10$ to $z = 2$, together with a significant contribution from starbursts reaching up to 1000 M$_\odot$ yr^{-1}, which lie on top of the continuous component. TVD simulations indicate a more sporadic star-formation history than the SPH simulations. Our results do *not* imply that hierarchical galaxy formation fails to account for the observed massive galaxies at $z \geq 1$. The global star-formation rate density in our simulations peaks at $z \geq 5$, a much higher redshift than predicted by the semi-analytic models. This star-formation history suggests an early build-up of the stellar mass density, and predicts that 70 (50, 30)% of the total stellar mass at $z = 0$ had already been formed by $z = 1\,(2, 3)$. Upcoming observations by Spitzer and Swift might help to better constrain the star-formation history at high redshift.

1. Introduction

A number of recent observational studies have revealed a new population of red, massive galaxies at redshift $z \sim 2$ (e.g., Chen et al. 2003, Daddi et al. 2004, Franx et al. 2003, Glazebrook et al. 2004, McCarthy et al. 2004, van Dokkum et al. 2004), utilizing near-IR wavelengths which are less affected by

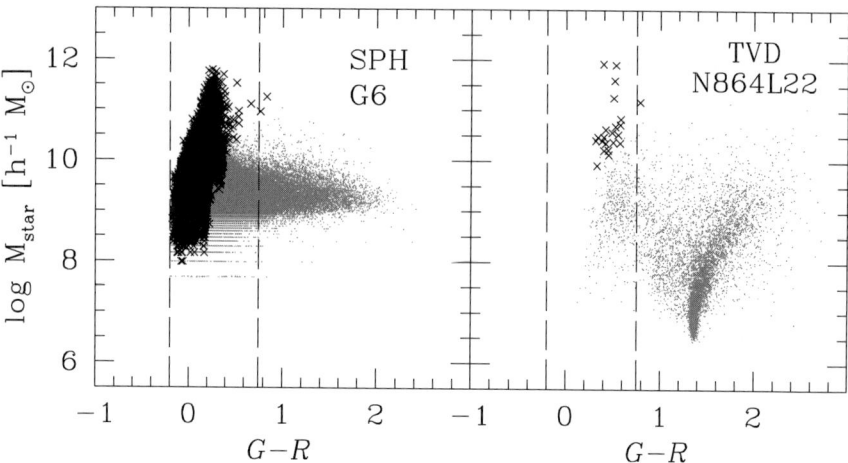

Figure 1. Stellar mass vs. $(G-R)$ color of simulated galaxies at $z=2$. Gray points show the total galaxy population, and the black crosses indicate the galaxies brighter than $R=25.5$. The vertical long-dashed lines roughly indicate the UV color selection range of Adelberger et al. (2004).

dust extinction. At the same time, some studies focused on the assembly of the stellar mass density at high redshift by comparing the observational data and semi-analytic models of galaxy formation (e.g., Dickinson et al. 2003, Fontana et al. 2003, Poli et al. 2003). These studies have suggested that the hierarchical structure formation theory may have difficulty to account for sufficient early star formation. The concern grew with the mounting evidence for high-redshift galaxy formation, including the discovery of Extremely Red Objects (EROs) at $z \geq 1$, sub-millimeter galaxies at $z \geq 2$, Lyman-break galaxies (LBGs) at $z \geq 3$, and Lyα emitters at $z \geq 4$. We now face the important question as to whether this evidence for high-redshift galaxy formation is consistent with the concordance ΛCDM model. See Nagamine et al. (2004, 2005) for the details of the simulations and the present work.

2. Results

Figure 1 shows the stellar mass vs. $(G-R)$ color of the simulated galaxies at $z=2$. The most massive galaxies have stellar masses $M_\star > 10^{10} h^{-1}\,\mathrm{M}_\odot$, and UV colors $-1.2 < (G-R) < 0.8$, consistent with the UV color selection criteria of Adelberger et al. (2004). The differences in the distribution of the points can be understood in terms of the different box sizes and the randomness of the initial condition of the simulations. On the right-hand side of the panels, there are a few red, passive systems that have stellar masses $M_\star > 10^{10} h^{-1}\,\mathrm{M}_\odot$. The near-IR properties of these passive systems will be reported in future papers. Figure 2 shows the star-formation histories of the most massive and

Massive galaxies at $z = 2$ 321

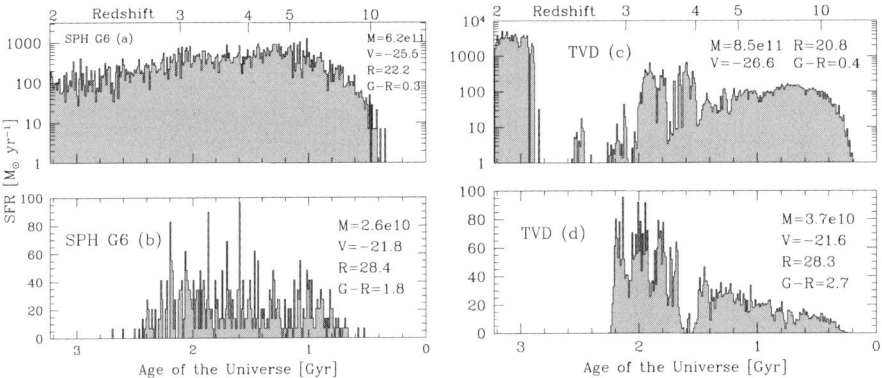

Figure 2. Star-formation history of the most massive galaxies (a and c), and the reddest galaxies (b and d) in the simulation. The two left-hand panels are for the SPH G6 run, and the right-hand ones are for the TVD run.

reddest galaxies in the simulations. For the most massive systems, the star-formation rate has a continuous component of ≥ 30 M$_\odot$ yr^{-1} over a few Gyr, and a starburst component that reaches up to 1000 M$_\odot$ yr^{-1} exists on top of the continuous component. Such extreme star-formation histories allow these galaxies to build up stellar masses greater than a few $\times 10^{11}$ M$_\odot$ by $z = 2$. The TVD simulation indicates a somewhat more sporadic star-formation history, which is perhaps due to the differences in the details of the star-formation recipe and the numerical resolution.

Figure 3 shows the star-formation history of the entire simulation box. In panel (a), all models – including the analytic model of Hernquist & Springel (2003) – show that the SFR density peaks at $z \geq 5$. These SFR histories lead to an early build-up of the stellar mass density compared to both the current observational estimates and the results from the semi-analytic models (see Nagamine et al. 2005 for a direct comparison), and we predict that 70 (50, 30)% of the total stellar mass at $z = 0$ had already been formed by $z = 1\,(2, 3)$ based on these theoretical models.

In summary, we have shown that the simulations based on the hierarchical ΛCDM model can in fact account for the masses and the co-moving number densities of the massive galaxies at $z = 2$ that are found from recent observations. Our simulations indicate that the properties (i.e., stellar mass, color, SF history, clustering) of the UV-bright LBGs at $z \geq 2$ can be understood if they are identified with galaxies that reside in massive dark matter halos.

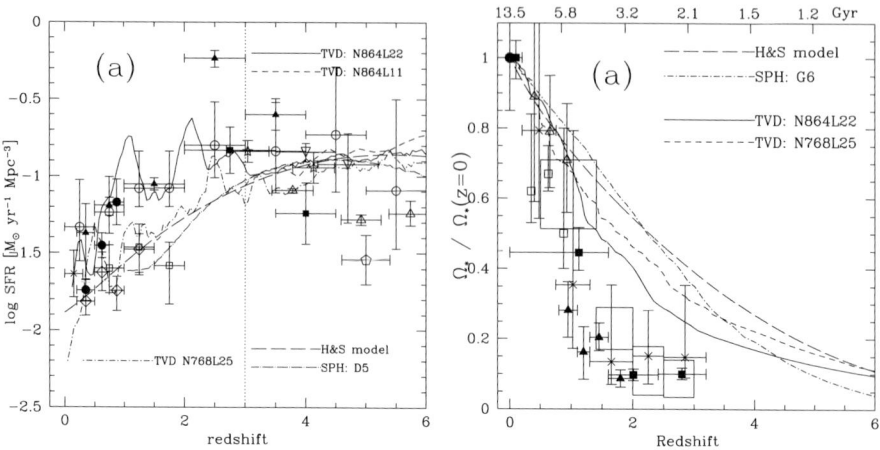

Figure 3. (Left) Star-formation rate density as a function of redshift for simulations and for the analytic model by Hernquist & Springel (2003). The sources of the extinction-corrected data points are described in Nagamine et al. (2004). (Right) Stellar mass density as a function of redshift.

Acknowledgments

This work was supported by NSF grants ACI 96-19019, AST 00-71019, AST 02-06299 and AST 03-07690, and NASA ATP grants NAG5-12140, NAG 5-13292, and NAG5-13381.

References

Adelberger K.L., Steidel C.C., Shapley A.E., Hunt M.P., Erb D.K., Reddy N.A., Pettini M., 2004, ApJ, 607, 226
Chen H.-W., et al., 2003, ApJ, 586, 745
Daddi E., et al., 2004, ApJ, 600, L127
Dickinson M., Papovich C., Ferguson H., Budavári T., 2003, ApJ, 587, 25
Fontana A., et al., 2003, ApJ, 594, L9
Franx M., et al., 2003, ApJ, 587, L79
Glazebrook K., et al. 2004, Nature, 430, 181
Hernquist L., Springel V., 2003, MNRAS, 341, 1253
McCarthy P.J., et al., 2004, ApJ, 614, L9
Nagamine K., Cen R., Hernquist L, Ostriker J.P., Springel V., 2004, ApJ, 610, 45
Nagamine K., Cen R., Hernquist L., Ostriker J.P., Springer V., 2005, ApJ, 618, 23
Poli F., et al., 2003, ApJ, 593, L1
van Dokkum P. et al., 2004, 611, 703

K-LUMINOUS GALAXIES AT $Z \sim 2$

Metallicity and B Stars

Duília de Mello[1,2] and the K20 Team
[1] *Laboratory for Astronomy and Solar Physics, GSFC/NASA – Catholic University of America*
[2] *Johns Hopkins University, Baltimore, MD21218, USA*
duilia@ipanema.gsfc.nasa.gov

Abstract   We present the results from the analysis of the composite spectrum of five near-IR luminous ($K < 20$) galaxies at $z \sim 2$. Several of the strongest absorption lines are present in the UV spectrum of the merging galaxy NGC 6090 and not in the spectra of Lyman Break Galaxies at $z \sim 3$. They were identified as SiIIIλ1296, CIIIλ1428, SiIIIλ1485, and Fe $\sim \lambda$1380 Å, which are photospheric lines typical of B stars. A metallicity higher than solar is suggested by comparing the pure photospheric lines known as the 1425 Å index (SiIII, CIII, FeV) with STARBURST99 models. The evidence of high metallicity, together with the high masses, high star-formation rates, and possibly strong clustering, suggest that these galaxies are candidates to become massive spheroids.

1.   Introduction – The K20 Survey

A key open question in galaxy evolution is the epoch of formation of massive spheroidal galaxies. As the rest-frame optical–near-infrared traces the galaxy mass, the K_s band allows a fair selection of galaxies based on their masses up to $z \sim 2$. Based on this, a VLT spectroscopic survey of about 500 galaxies with $K_s < 20$ in the GOODS southern field was conducted (Cimatti et al. 2002). In this contribution (see also de Mello et al. 2004), we analyze the average spectrum of five K20 galaxies at $1.7 < z < 2.3$ with the highest S/N ratios among the ones presented in Daddi et al. (2004). These $K < 20$ galaxies at $z \sim 2$ appear to be massive ($\geq 10^{11}$ M$_\odot$) and have high star-formation rates (SFR 100–500 M$_\odot$ yr^{-1}; Daddi et al. 2004), thus qualifying as good candidates for assembling/forming massive early-type galaxies.

2.   Local starbursts and B stars

We have compared the K20 average spectrum with the local starburst galaxies NGC 1705, NGC 1741, NGC 4214, and NGC 6090. The best match is ob-

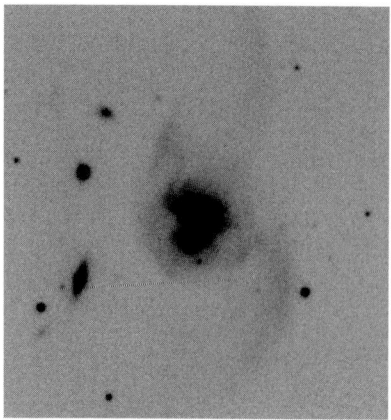

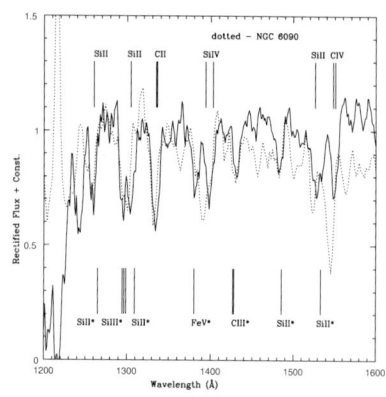

Figure 1. Sloan r-band image of NGC 6090, extracted from the National Virtual Observatory (~ 4 arcmin^2).

Figure 2. Spectrum of NGC 6090 (dotted lines) and the K20 composite spectrum. Photospheric lines are marked with ∗.

tained with NGC 6090 which is an interacting system at $v \sim 9062$ km s^{-1}, and in the process of merging. It is a luminous infrared galaxy ($\log L_{\rm IR} = 11.51$ (L$_\odot$; Scoville et al. 2000), containing a number of luminous clusters triggered by the galaxy-galaxy interaction. The Sloan Digital Sky Survey r-band image (Fig. 1) [1] shows tidal tails that extend several arcminutes from the two merging objects. *HST/NICMOS* images of NGC 6090 (Scoville et al. 2000) show the inner site of the interaction in more detail, where a less massive galaxy seems to be merging with a disk. The spectrum of NGC 6090 taken by the HUT during the Astro-2 mission (González Delgado et al. 1998) is shown in Fig. 2. It has several absorption lines which are similar in strength to the K20 composite spectrum, such as the photospheric lines Si IIIλ1295 Å, C IIIλ 1430 Å and Si IIIλ1485 Å, and a marginally weaker Fe Vλ1380 Å line.

In order to search for the stellar population from which these photospheric lines originate, we examined a far-UV (IUE) library of Milky Way OB stars (de Mello et al. 2000). The similarity between the spectra of B stars and the K20 composite spectrum is remarkable. In Figs. 3 and 4, we show the average spectrum of two main sequence stars (B0v and B8v) and a supergiant (B3I), where the main photospheric lines are identified. B stars live longer than the more luminous short-lived O stars and become a major source of the UV flux in the integrated spectrum of starbursts. The photospheric lines found in the spectrum of K20 galaxies, and in NGC 6090, are stellar features of B stars.

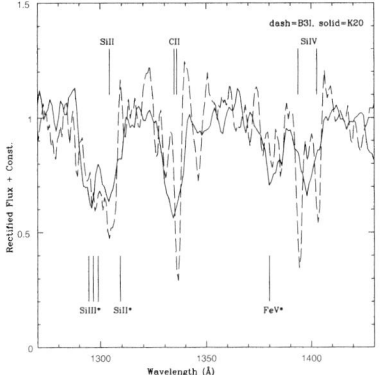
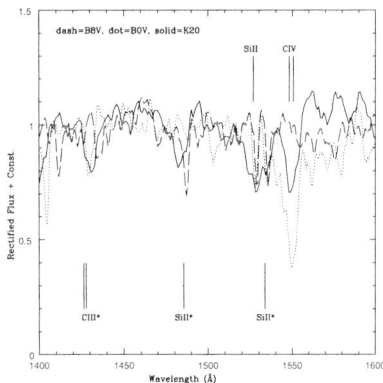

Figure 3. The spectrum of a B3I star (dashed) and the K20 composite spectrum.

Figure 4. The spectrum of a B8V (dashed) star and a B0V (dotted) star and the K20 composite spectrum. Photospheric lines are marked with ∗.

3. Lyman Break Galaxies and Metallicity

We also compared the K20 average spectrum with the average spectrum of Lyman Break Galaxies (LBGs) at $z \sim 3$ (Shapley et al. 2003). The best match is obtained with the LBGs without Lyα emission (Fig. 5). The most striking differences between the LBG composite spectra and the K20 average spectrum are the photospheric lines described above. One caveat that one has to keep in mind, before further interpreting this comparison, is the fact that the average LBG spectrum contains several hundred spectra, whereas the spectrum we present here is the average of only 5 objects. Therefore, a few peculiar objects could be present and co-addition of a larger number of spectra is desirable in the future, in order to smooth out the contribution of individual objects. Nevertheless, they comprise an interesting class of objects that might be important in galaxy evolution scenarios.

We have used the pure photospheric lines known as the 1425 Å index (SiIII, CIII, FeV; Leitherer et al. 2001) to estimate metallicity (Fig. 6), since this index does not strongly depend on age. The equivalent width of the index is 2.3 ± 0.4 Å, a value much larger than given in STARBURST99 models (Leitherer et al. 1999) for solar (1.5 Å) and LMC metallicity (0.4 Å). It corresponds to models with continuous star formation, fewer massive stars, and metallicities greater than solar (see Fig. 11 in Rix et al. 2004). Recently, Shapley et al. (2004) also suggested that $K < 20$ galaxies at $2.1 < z < 2.5$ have at least solar metallicity, based on near-IR spectroscopic measurements of seven galaxies. Near-IR spectroscopy of a larger sample of K20 galaxies is needed to confirm the metallicities and estimate the relative importance of O and B stars in these objects.

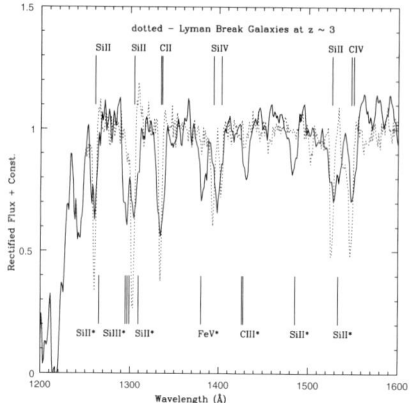

Figure 5. The spectrum of Lyman Break Galaxies (dotted) and the K20 composite spectrum.

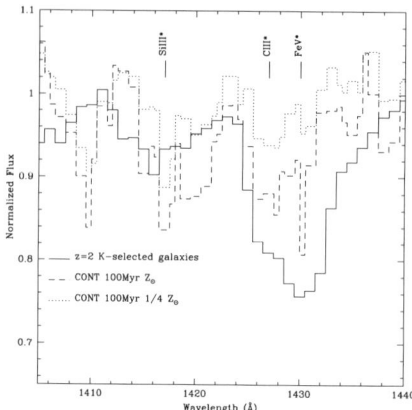

Figure 6. The 1425 Å index region of the K20 composite spectrum. STARBURST99 models for solar metallicity (dashed) and $0.25\,Z_\odot$ (dotted).

Notes

1. We acknowledge use of the National Virtual Observatory, which is funded by the National Science Foundation under Cooperative Agreement AST0122449 with The Johns Hopkins University.

References

Cimatti A., et al., 2002, A&A, 392, 395
de Mello D.F., Leitherer C., Heckman T., 2000, ApJ, 530, 251
de Mello D.F., Daddi E., Renzini A., Cimatti A., di Serego Alighieri S., Pozzetti L., Zamorani G., 2004, ApJ, 608, L29
Daddi E., et al., 2004, ApJ, 600, L127
González Delgado R.M., Leitherer C., Heckman T., Lowenthal J.D., Ferguson H.C., Robert C., 1998, ApJ, 495, 698
Leitherer C., Leão J.R.S., Heckman T.M., Lennon D.J., Pettini M., Robert C., 2001, ApJ, 550, 724
Leitherer C., et al., 1999, ApJS, 123, 3
Rix S.A., Pettini M., Leitherer C., Bresolin F., Kudritzki R.-P., Steidel C.C., 2004, ApJ, 615, 98
Scoville N.Z., et al., 2000, AJ, 119, 991
Shapley A.E., Steidel C.C., Pettini M., Adelberger K.L., 2003, ApJ, 588, 65
Shapley A.E., Erb D.K., Pettini M., Steidel C.C., Adelberger K.L., 2004, ApJ, 612, 108

RESOLVED MOLECULAR GAS EMISSION IN A QSO HOST GALAXY AT $Z = 6.4$

Fabian Walter
Max Planck Institut für Astronomie, Heidelberg, Germany

Abstract We present high-resolution VLA observations of the molecular gas in the host galaxy of the highest-redshift quasar currently known, SDSS J1148+5251 ($z = 6.42$). Our VLA data of the CO(3–2) emission have a maximum resolution of 0.17×0.13 arcsec^2 (≤ 1 kpc^2), and enable us to resolve the molecular gas emission both spatially and in velocity. The molecular gas in J1148+5251 is extended to a radius of 2.5 kpc, and the central region shows 2 peaks, separated by 0.3 arcsec (1.7 kpc). Assuming that the molecular gas is gravitationally bound, we estimate a dynamical mass of $\sim 4.5 \times 10^{10}$ M$_\odot$ within a radius of 2.5 kpc. This dynamical mass estimate leaves little room for matter other than the detected molecular gas, and – in particular – the data are inconsistent with a $\sim 10^{12}$ M$_\odot$ stellar bulge, which would be predicted based on the $M_{\rm BH} - \sigma_{\rm bulge}$ relation.

1. Introduction

More and more objects have been found at the highest redshifts in recent years, back to the "Dark Ages" (the epoch of formation of the first luminous structures), at $z > 6$ when the Universe was less than a Gigayear old (e.g., Fan et al. 2002, 2003, 2004, Hu et al. 2002, Stanway et al. 2003). Optical spectra of the brightest quasars at $z > 6$ show a clear Gunn & Peterson (1965) effect, demonstrating that these objects are located at the end of cosmic re-ionization (Becker et al. 2001, Fan et al. 2003, White et al. 2003). It is of paramount importance to study these sources in detail, to measure the stellar and gaseous constituents, chemical abundances and dynamical masses of the host galaxies, which in turn constrain galaxy evolution models in the very early Universe.

The highest-redshift quasar currently known is SDSS J114816.64+525150.3 (hereafter: J1148+5251) at $z = 6.42$ (Fan et al. 2003). J1148+5251 is a highly luminous object, which is thought to be powered by mass accretion onto a supermassive black hole of mass $(1-5) \times 10^9$ M$_\odot$ (which accretes at the Eddington limit; Willot et al. 2003). Thermal emission from warm dust has been detected at millimetre wavelengths, implying a far-infrared (FIR) luminosity

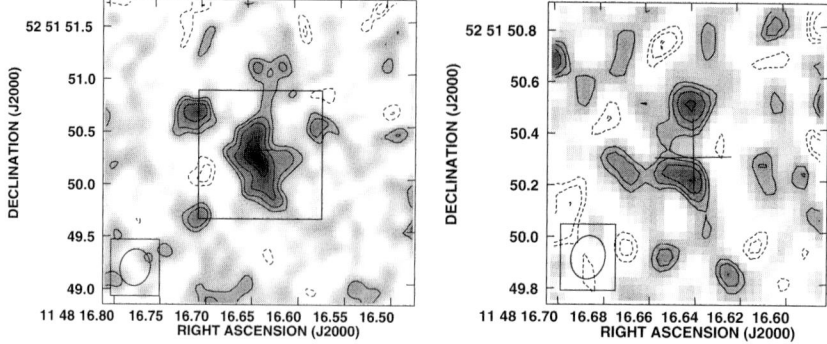

Figure 1. Left: CO(3–2) map of J1148+5251 of the combined B and C-array data set (covering the total bandwidth, 37.5 MHz or 240 km s^{-1}). Contours are shown at $-2, -1.4, 1.4, 2, 2.8, 4\sigma$ ($1\sigma = 43\mu$Jy beam^{-1}). The beam size (0.35×0.30 arcsec^2) is shown in the bottom left corner. *Right:* CO(3–2) map at ~ 1 kpc resolution of the central region, shown as a box to the left. Contours are shown at $-2.8, -2, -1.4, 1.4, 2, 2.8, 4\sigma$ ($\sigma = 45\mu$Jy beam^{-1}). The beam size (0.17×0.13 arcsec^2) is shown in the bottom left corner. The positional uncertainty of the SDSS position is of order ± 0.1 arcsec, or about the size of the cross in both panels

of 1.3×10^{13} L$_\odot$ (Bertoldi et al. 2003a), corresponding to about 10% of the bolometric luminosity of the system. Using the Very Large Array (VLA) and the Plateau de Bure interferometer (PdBI), we have detected various rotational lines of carbon monoxide (CO), the most common tracer of molecular gas, in J1148+5251 (Walter et al. 2003, Bertoldi et al. 2003b).

Here we present follow-up high-resolution observations of the molecular gas distribution of J1148+5251, obtained with the NRAO VLA. For details on the observations and more discussion on this subject, the reader is referred to Walter et al. (2004).

2. Results

Global Distribution of Molecular Gas

In Fig. 1 we present the CO(3–2) emission in J1148+5251 over the total measured bandwidth (37.5 MHz, 240 km s^{-1}) at 0.35×0.30 arcsec^2 (1.9×1.7 kpc^2) resolution. The emission is clearly extended and Gaussian fitting in the map plane gives a deconvolved major axis (FWHM) of 0.65 ± 0.12 arcsec (~ 3.6 kpc), a marginally resolved minor axis of 0.25 ± 0.12 arcsec (~ 1.4 kpc), and a position angle of $15 \pm 10°$ (measured east from north). Molecular gas can be seen out to distances of 0.42 arcsec from the centre (~ 2.5 kpc; numbers are deconvolved for the beam size). The source is possibly extended towards the north but more sensitive observations are needed to confirm this extension. If we assume that the main molecular gas concentration forms an inclined disk, this gives an inclination of $\sim 65°$ (where $0°$ corresponds to face-on), which implies that any inclination correction to the measured rotational velocities is minimal. The fitted peak to the central distribution is 0.21 mJy beam^{-1}, and the integral intensity is 0.55 mJy. This corresponds to an H$_2$ mass of $M(\text{H}_2) \sim 1.6 \times 10^{10}$ M$_\odot$. Given our beam size of 0.35×0.30 arcsec^2, the peak brightness of 0.21 mJy beam^{-1} corresponds to a surface brightness of 1.1 K, or a beam-smoothed brightness temperature of $1.1 \times (1+z) = 8.3$ K at $z = 6.42$.

Dynamics of the Molecular Gas

Assuming that the molecular gas is gravitationally bound and forms an inclined disk with a radius of 2.5 kpc (see previous section), and that the gas seen in the PdBI spectrum emerges from the same region (full width at zero intensity: 560 km s^{-1} or $v_{\text{rot}} \sim 280$ km s^{-1}; Bertoldi et al. 2003b), we derive an approximate dynamical mass for J1148+5251 of $\sim 4.5 \times 10^{10}$ M$_\odot$ (or $\sim 5.5 \times 10^{10}$ M$_\odot$ if we correct for an inclination of $\sim 65°$), with an error of order 50%. Within the large uncertainties, this number is compatible with the derived molecular gas mass and the mass of the central black hole, but does not leave much room for additional matter (e.g., a massive stellar bulge, see discussion below).

Distribution of CO at 1 kpc scale

The distribution of molecular gas at even higher resolution reveals that the situation is likely more complex than described by the simple disk model above. In Fig. 1, we show our highest resolution CO(3–2) map derived from B-array data only. Here, the resolution is only 0.17×0.13 arcsec^2, or $\sim 0.95 \times 0.72$ kpc^2. This map shows that the emission breaks up into two regions, a northern source at $11^\text{h}48^\text{m}16.640^\text{s}$, $52°51'50.51''$ (peak flux density: $192 \pm 45 \mu$Jy), and a southern source at $11^\text{h}48^\text{m}16.641^\text{s}$, $52°51'50.21''$ (flux density: $180 \pm 45 \mu$Jy), i.e., separated by ~ 0.3 arcsec (1.7 kpc). On a Kelvin scale, both sources are at ~ 4.5 K, which corresponds to a lower limit for the beam-averaged brightness temperature of ~ 35 K at $z = 6.42$.

3. Discussion

We present the first resolved maps of a system located at the end of cosmic re-ionization at $z = 6.4$. The molecular gas distribution in J1148+5251 is extended out to radii of 2.5 kpc. The central region is resolved and shows 2 peaks, separated by 1.7 kpc; they account for about half of the total emission, with the other half present in the more extended molecular gas distribution. We have assumed that the gas is gravitationally bound and is situated in a disk, although the data do not rule out the presence of an unbound system, such as an ongoing merger. Based on the extent of the molecular gas distribution and the line-width measured from the higher CO transitions, we derive a dynamical mass of $\sim 4.5 \times 10^{10}\,M_\odot$ ($\sim 5.5 \times 10^{10}\,M_\odot$ if we correct for an inclination of $i \sim 65°$). This dynamical mass estimate can account for the detected molecular gas mass within this radius, but leaves little room for other matter. In particular, given a black-hole mass of $\sim 1 - 5 \times 10^9\,M_\odot$ (Willot et al. 2003), this dynamical mass could not accommodate an order a few $\times 10^{12}\,M_\odot$ stellar bulge, which is predicted by the present-day $M_{\rm BH} - \sigma_{\rm bulge}$ relation (Ferrarese & Merritt 2000, Gebhardt et al. 2000), if this relation were to hold at these high redshifts. Our finding therefore suggests that black holes may assemble before the stellar bulges.

Acknowledgments

I would like to thank my collaborators on this project: C. Carilli, F. Bertoldi, P. Cox, K. Menten, K.Y. Lo, X. Fan and M. Strauss.

References

Becker R.H., et al., 2001, AJ, 122, 2850
Bertoldi F., et al., 2003a, A&A, 409, L47
Bertoldi F., Carilli C.L., Cox P., Fan X., Strauss M.A., Beelen A., Omont A., Zylka R., 2003b, A&A, 406, L55
Carilli C.L. et al., 2002, AJ, 123, 1838
Fan X., Narayanan V.K., Strauss M.A., White R.L., Becker R.H., Pentericci L., Rix H., 2002, AJ, 123, 1247
Fan X., et al., 2003, AJ, 125, 1649
Fan X., et al., 2004, AJ, 128, 515
Ferrarese L., Merritt D., 2000, ApJ, 539, L9
Gebhardt K., et al., 2000, ApJ, 539, L13
Gunn J.E., Peterson B.A., 1965, ApJ, 142, 1633
Hu E.M., Cowie L.L., McMahon R.G., Capak P., Iwamuro F., Kneib J.-P., Maihara T., Motohara K., 2002, ApJ, 568, L75
Stanway E.R., et al., 2004, ApJ, 604, L13
Walter F., et al., 2003, Nature, 424, 406
Walter F., Carilli C., Bertoldi F., Menten K., Cox P., Lo K.Y., Fan X., Strauss M.A., 2004, ApJ, 615, L17
White R.L., Becker R.H., Fan X., Strauss M.A., 2003, AJ, 126, 1
Willott C.J., McLure R.J., Jarvis, M.J., 2003, ApJ, 587, L15

Session IX

Conference Summary

CONFERENCE SUMMARY: STARBURSTS AND GALAXY EVOLUTION

Robert W. O'Connell
Astronomy Department, University of Virginia, Charlottesville, VA 22903-0818, USA
rwo@virginia.edu

Abstract   Starbursts are extreme concentrations of star-forming activity with mass conversion surface densities reaching over 1000 times higher than normal for disk galaxies. They are responsible for a large fraction of all cosmic star formation. They have shaped the cosmic landscape not just in individual galaxies, but though the effects of "superwinds" that enrich the intergalactic gas, constrain supermassive black hole growth, and perhaps facilitate cosmic re-ionization. This conference provided vivid testimony to the importance of starbursts in galaxy evolution and to the speed with which our understanding of them is being transformed by a flood of new pan-spectral data, especially at high redshifts. A key result is that it appears possible to scale the physics of smaller star-forming events upward to encompass starbursts.

1.   Introduction: Solved and Unsolved Problems

"Galaxy formation is a solved problem." That was the most memorable quote from the last meeting I attended at Cambridge, the 1996 conference on the Hubble Deep Field. The prominent astronomer who offered this sentiment was perhaps a little premature, but his excess of enthusiasm was forgivable considering the stream of beautiful data on the high-redshift Universe that had just begun to emerge. That stream has exponentiated over the last 8 years, and it may well be that we can solve the problem of galaxy formation within the next couple of decades.

Before that is possible, however, we need to understand star formation. We have already solved the basic problem of stellar structure and evolution. That can fairly be said to be the primary accomplishment of astrophysics in the 20th century (especially because it had been several million years since humans had first wondered about the stars!). There are only a few remaining dark corners of the evolutionary process. But one is crucial: star formation. This is central to galaxy astrophysics, but the deficiencies in our understanding are obvious. For instance, faulty prescriptions for star formation are thought to be the culprits in discrepancies between predictions of CDM models for galaxy formation and the observations.

A great deal of observational firepower will be directed at the problem of star formation in the coming years, but we already know one essential fact: star formation is a *collective* process. Most stars (perhaps nearly 100% in our Galaxy) form in clusters. There are strong interactions with the surrounding environment and among protostars. Quantifying feedback processes, both positive and negative, is a key to understanding star formation. All this means that star formation is a more difficult problem than was the astrophysics of isolated stars. Progress will be importantly informed by observations of other galaxies and a wider range of environments than are found in our Galaxy.

Starbursts are important because they are clearly a collective phenomenon and represent one extreme of the star-formation process. Furthermore, they are bright enough to be detected throughout the observable Universe and can serve as tracers of the cosmic history of star formation. This conference provided vivid testimony to the importance of starbursts as keys to galaxy evolution and to the speed with which our understanding of them is being transformed by the flood of new data, especially at high redshifts.

2. What Are Starbursts?

The term *starburst* conveys the dual notions of intensity and limited duration. There is no strict definition, however, so the term has been used (or abused) to encompass a huge variety of star-formation events. Several speakers proposed useful definitions, and I will follow their lead.

Significant star formation is a hallmark of about half the galaxies in the local Universe. There are several convenient proxies for active star formation: blue optical/UV colors, emission lines, or strong infrared output. Although there was a general awareness of the statistics, the Sloan Digital Sky Survey has recently brought home the unmistakable *bimodality* of optical colors: galaxies fall into either a red or blue sequence (separated by about 0.4 mag in $(B-V)$ color), with few systems in between. This means they are either active star formers (blue) or have not hosted significant star formation for $\gtrsim 1$ Gyr. Interphase types are rare because once star formation ceases, color evolution from the blue to the red sequences occurs in only ~ 500 Myr (as long as the active galaxy is itself old).

The blue systems are mainly disk-dominated. The normal structure and dynamics of disks favor relatively slow conversion of gas into stars. Global *self-regulation* within disks is evidently effective over long time-scales because there are good correlations between ionizing populations (lifetimes $\lesssim 10$ Myr) and broad-band optical colors (characteristic times of $\gtrsim 1000$ Myr). The range of what might be called "quiescent" star formation encountered along the normal Hubble sequence is about 4 orders of magnitude in both star-formation rate ($\dot{s}$, measured in $M_\odot$ yr^{-1}) and star formation surface density (Σ_{SFR}, measured in $M_\odot$ yr^{-1} kpc^{-2}). A significant number of local galaxies, mostly dwarfs, exhibit elevated activity, ranging above $\Sigma_{SFR} \sim 0.1$, which might be called "enthusiastic" star formation. This accounts for only $\sim 15\%$ of all local star formation.

The most interesting cases, naturally, are the "psychopathic" ones at the extremes, which are the starbursts. These are often, though not always, associated with a large disturbance to normal disk kinematics. The central feature of a starburst is the concentration of star-forming activity and especially the large feedback it produces on its surroundings, often driving a "superwind" out of the host galaxy. For definiteness, I will define a "major starburst" as an episode where such effects are important. This requires $\Sigma_{SFR} > 1\ M_\odot$ yr^{-1} kpc^{-2}, equivalent to $\Sigma_{L_{bol}} > 10^{10}\ L_\odot$ kpc^{-2}, or about 1000× normal values for disk galaxies. The relevant quantities must be averaged over a finite cell size in area and time (say, 1 kpc^2, 10 Myr), and the initial mass function (IMF) must contain massive stars capable of producing ionization, winds, and core-collapse supernovae.

The fact that this definition is arbitrary is actually significant: the properties of star-formation regions appear to be *continuous* across the range of amplitudes observed.

Starbursts constitute an important fraction of all detected high-redshift galaxies. Because of the fierce dimming effects in distance modulus and surface brightness at large z, there is a powerful selection effect operating here. Nonetheless, statistics based on co-moving volume densities have shown that while major starbursts are rare locally ($\sim 0.5\%$ of nearby systems), they were much more common at earlier times (perhaps 15% in number and 70× in luminosity density at $z = 1$). IR-bright starbursts have been responsible for as much as 80% of the local stellar mass.

Although we do not yet have a definitive understanding of starbursts and their effects, the wealth of new data is making progress rapid. Along the way, there are two central difficulties: (i) major starbursts are rare in the local Universe, and we are forced to *scale up* our understanding of physical processes from local samples and conditions; and (ii) starbursts are notoriously complex 3D systems, made especially difficult to probe by often severe differential internal extinction.

3. Are Starbursts Important?

Starbursts are certainly fascinating, but how important are they? An interesting way to frame the question is to ask: if we did not know that major starbursts existed from direct observations, would we have difficulties explaining what we see in the Universe? That is, are starbursts a *necessary inference* from other phenomena? The answer is an emphatic "yes," and here is a tentative and incomplete list of the essential fingerprints of starbursts based on issues raised in this conference:

- Super star clusters: These very massive but compact systems, ranging from very young clusters still in dust cocoons to classic globular clusters, can evidently form only in a high-pressure medium, with $P/k \sim 10^{8-9}$, over $10^4 \times$ higher than normal for disk galaxies. This requires abnormal, non-equilibrium conditions such as prevail in starbursts. Young SSCs are found to be mainly associated with interaction-induced starbursts.

- Massive bulges and E galaxies: The high stellar densities found in the centers of nearby early-type galaxies (ETGs) imply conversion of large amounts of gas into stars at rates equivalent to the most extreme starbursts known, $\dot{s} \sim 1000$ $M_\odot$ yr^{-1}, if only a single event was involved. There is good statistical evidence that many, if not all, ETGs originate from gas-rich mergers, which are well known to produce violent starbursts. A few large mergers rather than a series of minor mergers are favored. (The best direct evidence for bulge-producing starbursts at early times is probably the high-redshift sub-mm sample – e.g., from SCUBA – with $\dot{s}$ up to ~ 1000 $M_\odot$ yr^{-1}.)

- Cosmic mass deposition in stars: For a decade, we have been able to estimate the conversion rate of gas into long-lived stellar mass at redshifts $z \lesssim 4$. The present-day mean mass density cannot by itself place very good constraints on the range of $\Sigma_{\rm SFR}$, since over 13 Gyr have elapsed since the Big Bang (although, as noted above, direct statistics on distant starburst progenitors can do so). However, recent deep probes to $z \sim 2$, such as GDDS, reveal that a considerable fraction of all stars then are in "dead and red" systems comparable to local gE galaxies, with little star formation in the preceding 1.5 Gyr. The data imply that massive galaxy assembly begins early ($z_{\rm f} \gtrsim 3$–5) in dense regions and, since there is so little time for this to happen, massive starbursts with $\dot{s} \gtrsim 300$ $M_\odot$ yr^{-1} must be involved, possibly through gas-rich mergers.

The remaining items on the list involve *superwinds* generated by starbursts:

- Chemical enrichment of the ICM and IGM: Although metal abundances at higher redshifts are generally lower than prevailing local values, much of the gas to the highest z's yet probed has been processed through stars. The primordial generations of stars responsible can be explored only theoretically now, but it is clear that the natural mechanism for dispersing new metals from the halos in which they form is a superwind.

- The mass-metallicity relation in galaxies: The correlation between larger galaxy masses and higher metal abundances has been known for about 40 years. Optical, UV, and IR observations are currently providing much better information on metallicites and dust abundances in both local and high-redshift samples. Again, the natural explanation for the mass dependence is that starburst superwinds evacuate gas preferentially from lower-mass systems.

- Absence of super-supermassive black holes: Supermassive black holes (SMBHs) are now thought to exist in all spheroidal systems, and their masses are linked to the surrounding stellar population. Their growth by gas accretion is self-limited to the Eddington rate. Nonetheless, in the absence of another inhibiting mechanism, SMBH masses would exponentiate with an e-folding time of about 100 Myr. Star formation in nuclei is often found associated with SMBHs, and it may be that starburst winds act to regulate SMBH growth in the same way they limit star formation itself.

- Cosmic re-ionization: It is likely that massive stars, rather than AGN, are responsible for cosmic re-ionization at redshifts $z \gtrsim 7$. But the optical depth in typical nearby galaxies is such that only 3–10% of ionizing photons can escape. Superwinds in protogalaxies may be necessary to clear out channels for ionizing radiation.

4. We're All Pan-Spectral Now

In the past 5 years, the necessity as well as the opportunities to attack the problem of starbursts using a multi-wavelength approach have become manifest. No single band suffices, and the full EM spectrum from radio to gamma rays is now enthusiastically embraced in starburst research. Some examples: (i) stellar ages and abundances are best deduced from UV–optical–near-IR observations; (ii) the best $\dot{s}$ estimator is $L(\mathrm{UV}) + L(\mathrm{IR})$, meaning that different instruments are always necessary; (iii) a long-wavelength baseline is essential to overcome distortions by extinction of statistical samples and of physical inferences from any given band; (iv) starburst regions can be opaque, even at mid-IR wavelengths; radio/mm observations are needed for the youngest (~ 2 Myr) embedded sources; (v) mid-IR photometry and spectroscopy, now just coming into their own with the Spitzer Space Telescope, show great promise as dust/gas tracers within starbursts.

Understanding the physical coupling mechanisms between wavelength domains is essential: (i) there has recently been good progress in modeling the UV through IR spectral energy distributions of starbursts taking all three major components (stars, gas, dust) into account, but this remains a key area for additional effort; (ii) the long-recognized relation between radio continuum and far-

IR dust emission in star-forming systems is sometimes said to be the best correlation known in extragalactic astronomy, yet we do not fully understand its origins or implications for the star-formation process. A less well-established radio/X-ray correlation in young systems is also important to understand.

The fastest increments in observational insight into starbursts are currently coming from Spitzer and GALEX (IR and UV). Probably the fastest increments for the coming decade will be from ALMA (mm-wave).

Although this conference emphasized observations, we cannot forget that theoretical and computational astrophysics have to be part of a "pan"-discipline approach to starbursts.

5. The Limits of Spatial Resolution

A hard lesson in the study of starbursts has been that their scales and complexities push the limits of instrumental spatial resolution even in nearby systems. For instance, it is difficult: (i) to measure the diameters of SSCs (~ 2–10 pc), in order to obtain reliable mass and IMF inferences; (ii) to study superwind substructures in nearby starbursts and to determine host morphologies in distant ones; and (iii) to obtain kinematics of starbursts on the appropriate physical scales.

The *Hubble Space Telescope (HST)* has been the mainstay of high-resolution (~ 0.05 arcsec) imaging and spectroscopy for 14 years. An informal count shows that over half the contributions in this conference relied in some way on *HST* data. But *HST* will not last much more than another 6 years even if NASA can find a safe way of servicing it. In the foreseeable future, we will have the EVLA and ALMA for high-resolution radio/mm observations and JWST and ground-based AO systems for high-resolution near and mid-IR observations. However, it is doubtful that AO systems will operate well for $\lambda \lesssim 1\mu$m. Unfortunately, there are no current plans to replace or improve (to ~ 0.01 arcsec?) high-resolution optical/UV capability in space. It is vital to remedy that situation.

6. Scalability

"Scalability" was a major theme of the conference. It arises from two main concerns: To what extent can we scale local starburst systems to cosmically distant ones? And to what extent can we scale the physics of modest to extreme star-formation amplitudes? The evidence, fortunately, is that scalability is *good*, implying modest rather than fundamental adjustments with changes in environment and scale.

The premier example of scalability is the Schmidt-Kennicutt "law," under which $\Sigma_{\rm SFR} \propto \Sigma_{\rm gas}^{1.4}$. The quantities refer to global averages over the surfaces of individual galaxies. The relation applies over a remarkable 6 decades. As noted above, a similar degree of scalability applies to the radio/far-IR correlation for star-forming systems.

Other encouraging, if less firmly established, examples of scalability include:

- Congruences in EM spectral shape for starbursts across a wide range of environments and amplitudes.

- The continuity of starburst properties across a large range of amplitudes. It is possible to define a scaling sequence between the nearby (3.5 Mpc) archetypal starburst M82 ($L \sim 2 \times 10^{10}$ $L_\odot$), more distant ULIRGs (10^{12} $L_\odot$), and high-redshift SCUBA sub-mm sources (10^{13} $L_\odot$).

- The smooth increase of starburst activity with lookback time exhibits no evidence of a *transition* point where starbursts suddenly become more important.

- Continuity of Lyman-break galaxies (LBGs) at $z \gtrsim 3$ with more local systems. Careful studies, lately including GALEX data, show that properties (sizes, surface brightnesses, masses, kinematics) of LBGs are continuous with those of lower-redshift luminous blue compact galaxies, some of which may be the progenitors of local dE galaxies (i.e., dynamically hot systems).

- The mass-metallicity-extinction relation, which changes slowly with redshift and has no transition points. The abundance scale seems to decrease smoothly with redshift.

- The duration of starburst episodes is $\delta t \sim 100$ Myr and seems similar at all redshifts. Individual galaxies may experience a number of such episodes.

- The IMF for star formation on the scales of star clusters or galaxies now appears to be *universal*, except in a small number of SSCs where there may be changes in $M_{\rm low}$. The massive star IMF appears universal, which is very important for analyzing feedback processes. (Progress here has been excellent despite many complications, e.g., limited spatial resolution, large differential extinction effects, and mass segregation.)

- Star-formation histories of nearby galaxies may all be similar for a given gas density, despite the presence of "noise," which gives rise to "mini-bursts." It is important to understand the disk self-regulation mechanism.

7. Conclusion

Recent progress in understanding starbursts and placing them in the context of galaxy evolution has been outstanding and is healthily accelerating. We are fortunate to be riding a tidal wave of marvelous new data highlighted by unprecedented large sample sky surveys, *HST* high-resolution imaging, sensitive new infrared and sub-mm instrumentation, and the inauguration of the Spitzer and GALEX observatories.

To close, let me mention some critical aspects of starburst physics that deserve special attention. How does feedback operate in young starbursts to regulate processes like saturation, quenching, and outflows? In particular regarding the latter, the largest effects of starbursts are related to galactic superwinds, but there are numerous uncertainties regarding their underlying physics. Nearby systems are the benchmarks for detailed scrutiny of superwinds. A crucial open question is the mechanism of starburst triggers: for a given trigger, there is apparently a large variation in the resulting star-formation amplitude, which remains poorly understood as yet. A final important problem concerns the drivers and time-scales for dust shroud dissipation, which transforms an IR-bright galaxy into a UV/optical-bright one. All of these areas will benefit from a combined observational/theoretical attack.

Acknowledgments

All the conference participants wanted to extend their genuine gratitude to the organizing committees for a very productive and enjoyable conference, but most couldn't do so because of page limits. This work has been supported in part by *HST* grants GO-09117 and GO-09455.

MEMORABLE QUOTES

- Max (Pettini) told me I had to be dynamic and harsh.
  Linda Smith (on chairing a session)

- You can plot whatever model you like.
  Almudena Alonso Herrero

- Same data, different authors, different results.
  Marco Sirianni

- What did you ask me?
  Andrea Gilbert (halfway through answering a question)

- No man is an island, even if he is British.
  Mark McCaughrean (on collaborators)

- It's easy to get a good fit, which means that your fit doesn't mean much...
  Ariane Lançon

- Blue now means redder colours and red bluer colours.
  Eva Schinnerer

- The question mark here summarises everything, and I have nothing to add...
  Claus Leitherer

- Everything I say is going to be reasonably uncertain. It's going to have caveats, which are probably also going to have caveats.
  Neil Trentham

- Maybe burstiness is an illusion.
  Curtis Struck

- We are talking about M31 compressed into NGC 205, with a starburst added on.
  Matthew Bershady

- Is there a population of galaxies that essentially form stars like popcorn?
  Rob Kennicutt

- The reason for that is that we cheated – and they didn't.
  Bob Abraham

- You have to be really careful when you are going to put up tomb stones at talks!
  Bob Abraham

- Surprisingly good agreement, although it looks different...
  Jarle Brinchmann

- Because it's the brightest cluster in region A, "A1" seemed quite a good name for it...
  Linda Smith

- Everybody has got gas in that group; it's the Saudi Arabia group.
  Jay Gallagher (on the Cen A group)

- Of course, I cannot discuss everything about the high-redshift Universe.
  Daniel Schaerer

- I am a Madau plotter.
  Rodger Thompson

- Did you help organise this meeting?
  (Name withheld), to Richard de Grijs (Chair of the SOC), upon leaving at the end of the meeting

- It's hard to comment on other people's work.
  Jerry Ostriker

- She is going to talk about – hey, that's different!
  Evan Skillman (chairing a session)

- Theorists have a tendency to plot things that observers find hard to understand.
  Duília de Mello

- This is small-number statistics, with only one source.
  Fabian Walter

- Some people claim to understand it, but I have never understood they are understanding it.
  Bob O'Connell

Author Index

Aalto, S.	251	Calzetti, D.	**97**, 215, P5, P70
Abraham, R.C.	**195**	Cannon, J.M.	**P9**, P73
ACS Science Team	41, P7, P24	Carlberg, R.G.	195
Adelberger, K.L.	303	Carney, B.W.	P61
Akiyama, M.	187, P29, P37	Carollo, M.	P43
Alexander, D.	P25	Castander, F.J.	P59
Alexander, P.	P14	Castangia, P.	**P10**
Allard, E.L.	**P1**	Castellanos, M.	P60
Alloin, D.	P28	Cen, R.	319
Aloisi, A.	**P2**, P4, P35	Cerviño, M.	P30
Alonso-Herrero, A.	**35**, P28	Cesarsky, C.	P40, P45
Amram, P.	P47	Charlot, S.	P45
Anders, P.	**P3**	Chen, H.-W.	195
Ando, M.	P29	Christensen, L.	**P11**
Angeretti, L.	**P4**	Chyży, K.T.	**P12**
Aoki, K.	P29	Ciardi, B.	P34
Appenzeller, I.	299, P53	Cid Fernandes, R.	263, P20, P71
Armus, L.	P5, P70	Cimatti, A.	P34
Arribas, S.	173, P19	Ciroi, S.	P78
Ashby, M.	P8	Clampin, M.	41, P24
		Clark, J.S.	**P13**
Balard, P.	P47	Colina, L.	173, P19
Barlow, T.A.	269	Comastri, A.	P56, P64
Barmby, P.	P8	Combes, F.	**167**
Bauer, F.	P25	Condon, J.	223
Beck, R.	223, 251	Conselice, C.	215, P63, P79
Bell, E.	279	Contini, T.	P39
Bendo, G.J.	**P5**	Crampton, D.	195
Benvenuti, P.	P10	Crowther, P.A.	**21**, P13
Bergvall, N.	**103**, P47, P86	Cuisinier, F.	P32
Bershady, M.	17, **177**, P26	Cullen, H.	**P14**
Bertone, S.	**P6**		
Björnsson, G.	293	Daddi, E.	P34
Blakeslee, J.P.	P24	Dale, D.A.	P5
Boily, C.	71	Deep Team	P52
Bonnell, I.A.	**65**	de Grijs, R.	**157**, 227, P3, P41
Bordalo, V.	P77	Deharveng, J.-M.	P27
Borys, C.	P63	Demarco, R.	P24
Boulesteix, J.	P47	de Mello, D.F.	299, **323**, P79
Bouwens, R.	**P7**, P24	Deng, L.	P40
Bradley, L.D.	P24	Dennefeld, M.	P68
Brandl, B.R.	**49**, P8	D'Ercole, A.	P46
Bremer, M.	P16	De Rossi, M.E.	**P15**
Bresolin, F.	311	Désert, J.-D.	P35
Brodie, J.P.	219	Devereux, N.	P8
Brown, T.M.	P79	Díaz, A.I.	P26, P60
Buat, V.	**133**, 269, P27	Dickinson, M.	P63
Buckalew, B.	**P8**	di Serego Alighieri, S.	P34
Bunker, A.	P74	Dole, H.	279, 285, P68
Burgarella, D.	**269**, P27	Dopita, M.A.	**137**

343

Dorman, B.	P61	Gorosabel, J.	P11
Douglas, L.	**P16**	Goto, T.	P24
do Vale Asari, N.	P20	Graham, J.R.	45, 75
Downes, D.	P84	Grauer, A.D.	P5
Draine, B.T.	P5	Green, E.M.	P61
Dunlop, J.S.	**121**	Greggio, L.	P4
Durham, R.N.	17	Grimes, J.	**P21**
		Gronwall, C.	P24
Egami, E.	285	Guseva, N.G.	P58
Elbaz, D.	**241**, P40, P45	Guzmán, R.	17, 177, P26, P59
Ellis, R.	P74		
Elmegreen, B.G.	**57**	Hadfield, L.J.	21, P13
Engelbracht, C.W.	P5, P8	Hammer, F.	**273**, P40, P45
Erb, D.	**303**, P65	Harford, A.G.	**P22**
Evans, C.J.	**P17**	Harris, J.	**215**
		Hartig, G.	P24
Fazio, G.	P8	Hébrard, G.	P35
FDF Team	P53	Heckman, T.	**3**, 299, P2, P21, P25
Fellhauer, M.	**P18**		P46, P70, P71, P80
Ferguson, H.C.	P79	Helou, G.	223, P5
Ferrara, A.	P34	Henkel, C.	P57, P84
Fletcher, A.	251	Hernquist, L.	319
Flores, H.	273, P40, P45	Hidalgo-Gámez, A.	**P23**
Forbes, D.	P51	Hinz, J.	P8
Ford, H.C.	P24	Hjorth, J.	293, P11
Förster Schreiber, N.M.	**233**	Holden, B.	P24
Fricke, K.J.	P58	Hollenbach, D.J.	P5
Fritze–v. Alvensleben, U.	**209**, P41	Homeier, N.L.	41, **P24**
Fuentes, J.G.	187, P37	Hoopes, C.G.	P2
Fynbo, J.P.U.	**293**	Hornschemeier, A.E.	**P25**
		Hoyos, C.	177, **P26**
Gach, J.-L.	P47	Humphreys, R.	P8
GALEX Team	269, P69	Hunter, D.A.	219
Gallagher, III, J.S.	**11**, 215, 227, P85	Hüttemeister, S.	P50
Gallego, J.	17		
García-Marín, M.	**173**, **P19**	Iglesias-Páramo, J.	133, P66
García-Rissman, A.	**P20**	Illingworth, G.	P7, P24
Gardner, J.P.	P79	Inoue, A.K.	**P27**
Garland, C.A.	P59	Iovino, A.	P78
Gehrz, R.	P8	Ivanov, V.D.	**P28**
Giavalisco, M.	P42, P79	Iwata, I.	P27, **P29**
Gieles, M.	P3	Izotov, Y.I.	P58
Gilbert, A.M.	**45**		
Gil de Paz, A.	P52	Jakobsson, P.	293
Glazebrook, K.	195	James, P.	**P51**
Gnedin, N.Y.	P22	Jamet, L.	**P30**
González Delgado, R.M.	**263**, P20, P30, P71	Jappsen, A.-K.	**P31**
Goodwin, S.	P13	Jarret, T.H.	P5
Gordon, K.D.	P5, P8, P70, P82	Jee, M.J.	P24

Author Index 345

Jørgensen, I.	195	Lilly, S.	P43
Juneau, S.	195	Lilly, T.	**P41**
		Loiseau, N.	P51
K20 Team	323	Lotz, J.	**P42**
Kehrig, C.	**P32**	Lowenthal, J.	**17**
Kennicutt, R.C.	31, **187**, 251, P5, P37, P70, P80	Lyons, B.J.	17
Kewley, L.J.	**307**, P5	Mac Low, M.-M.	P31
Kimble, R.	41	Madau, P.	P42
Kissler-Patig, M.	**P33**	Maier, C.	**P43**
Klein, U.	P50	Maiolino, R.	P56
Klessen, R.S.	P31	Maíz-Apellániz, J.	**P44**
Knapen, J.H.	**181**, P1, P72	Malhotra, S.	P5
Kneib, J.-P.	P39	Marcillac, D.	P40, **P45**
Knödlseder, J.	**81**	Marcolini, A.	**P46**
Kobulnicky, H.	307, P8	Marquart, T.	103, **P47**, P86
Koo, D.	17, 177, P26, P52, P62	Martel, A.R.	P24
		Martínez-Valpuesta, I.	**P48**
Krog, B.	293	Marzke, R.	195
Kroupa, P.	P18, P83	Masegosa, J.	P47
Kudritzki, R.-P.	311	Mas-Hesse, J.M.	P44, P73
Kunth, D.	247, P35, P73	Matteucci, F.	P67
Kurk, J.	**P34**	Mattson, L.	P86
		Mazzuca, L.	P72
Lacy, M.	P68	McCarthy, P.J.	195
Lagache, G.	P68	McCrady, N.	**75**
Landsman, W.	P61	McMahon, R.	P74
Lançon, A.	**71**	Mehlert, D.	**299**, P53
Larsen, S.	**219**	Mei, S.	P24
Larson, R.B.	P31	Melbourne, J.	**27**, P38, P52, P62
Le Borgne, D.	195	Menanteau, F.	P24, P79
Le Borgne, J.-F.	P39	Meyer, M.	P5
Lebouteiller, V.	**247**, **P35**	Meurer, G.R.	41, P70
Lecavelier des Etangs, A.	P35	Meurs, E.J.A.	P54
Ledoux, C.	293	Mihos, C.	**153**
Lee, H.	**P36**	Misselt, K.	P8
Lee, J.C.	187, **P37**, **P38**	Mobasher, B.	P63
Le Floc'h, E.	**279**, 285	Møller, P.	293
Lehnert, M.	P16	Monreal, A.	173
Leitherer, C.	**89**, 311, P2, P4, P5, P70, P71, P73	Moorwood, A.	P81
		Moss, C.	**P49**
Lemoine-Busserolle, M.	**P39**	Mühle, S.	P50
Lennon, D.J.	P17	Mundell, C.G.	**P51**
Lequeux, J.	P35	Murowinski, R.	195
Li, A.	P5	Murphy, E.	P5
Li, Y.	P31		
Liang, Y.	273, **P40**, P45	Nagamine, K.	**319**
Lidman, C.	P39	Negueruela, I.	P13
Liebert, J.	P61	Nikolic, B.	P14

Nilsson, K.	293	Romano, D.	**P67**
Noeske, K.	**P52**, P58, P62	Rood, R.T.	P61
Noll, S.	299, **P53**	Rosati, P.	P24
Norci, L.	**P54**	Roth, K.	195
		Roussel, H.	**223**, P5
O'Connell, R.W.	227, **333**, P61	Rupen, M.P.	251
O'Halloran, B.	**P55**	Ryder, S.	P72
Ohta, K.	P29		
Origlia, L.	**P56**	Sabbi, E.	P4
Östlin, G.	103, P73, P86	Sajina, A.	**P68**
Ostriker, J.P.	319	Sakai, S.	187, P37
Ott, J.	**P57**	Salim, S.	**P69**
		Salzer, J.	P38
Pahre, M.	P8	Savaglio, S.	195, P2
Papaderos, P.	P52, **P58**	Sawicki, M.	**315**
Papovich, C.	279, **285**	Scannapieco, C.	P15
Parry, I.	P72	Schaerer, D.	CD-ROM: **T1**
Pasquali, A.	P10	Schiavon, R.P.	P61
Peletier, R.F.	P1	Schild, H.	21
Pelló, R.	P39	Schinnerer, E.	**251**, P51
Pérez, E.	P30, P44	Schmitt, H.R.	P20, **P70**, **P71**
Pérez Gallego, J.	**P59**	Schmutz, W.	21
Pérez-González, P.	279, P8	Schweizer, F.	**143**
Pérez-Montero, E.	**P60**	Scott, D.	P63, P68
Peterson, R.C.	**P61**	Scoville, N.Z.	251
Petrosian, A.	P73	Sembach, K.R.	P2
Pettini, M.	303, 311	Setti, G.	P64
Phillips, A.	P52, **P62**	Shapley, A.E.	303
Pisano, D.J.	P59	Sharp, R.	**P72**
Poggianti, B.	P25	Sheth, K.	P5
Polomski, E.	P8	Shlosman, I.	P48
Pompei, E.	P78	Silk, J.	**201**
Pope, A.	**P63**	SINGS Team	31
Postman, M.	P24	Sirianni, M.	**41**
Primack, J.	P42	Skillman, E.	P36, **P73**
Ptak, A.	P21	Small, T.	269
		Smith, D.A.	215
Radovich, M.	P78	Smith, J.-D.T.	**31**, P5
Rafanelli, P.	P78	Smith, L.J.	**227**, P85
Ranalli, P.	P56, **P64**	Snijders, L.	109
Reddy, N.	303, **P65**	Spitzer/MIPS Team	279, 285
Regan, M.W.	P5	Springel, V.	319
Reverte-Payá, D.	**P66**	Stanway, E.	**P74**
Richard, J.	P39	Stasińska, G.	P30
Richtler, T.	219	Steidel, C.	303, 311, P65
Rieke, G.H.	279, 285, P5, P8	Stoehr, F.	P6
Rieke, M.J.	279, 285, P5, P28	Storchi-Bergmann, T.	P71
Rix, S.A.	**311**, P17	Strickland, D.	P21, P46
Roellig, T.	P8	Struck, C.	**163, T2**

Tadhunter, C.	**257**	Zirm, A.	P24
Tamura, N.	P29	Zucker, D.	**P87**
Tanvir, N.	**P75**		
Tapken, C.	299		
Taylor, G.	**P76**		
Telles, E.	P32, **P77**		
Temporin, S.	**P78**		
Teplitz, H.I.	**P79**		
Thompson, D.	315		
Thompson, R.I.	**289**, P7		
Thomsen, B.	293		
Thornley, M.D.	P5		
Thuan, T.X.	P58		
Tissera, P.B.	P15		
Tosi, M.	P4, P67		
Tremonti, C.	P25, **P80**		
Trentham, N.	**115**		
Trundle, C.	P17		
UDF NICMOS Team	P7		
Ulvestad, J.S.	**127**		
Vacca, W.D.	75		
van der Werf, P.	**109**, P81		
van Loon, J.	P8		
van Starkenburg, L.	**P81**		
Vega, L.R.	P20		
Vernet, J.	P34		
Vidal-Madjar, A.	P35		
Vijh, J.P.	**P82**		
Vílchez, J.M.	P30, P32, P66		
Vils, M.	177		
Walborn, N.R.	P17		
Walter, F.	**327**, P5, P57, P84		
Weidner, C.	**P83**		
Weiß, A.	251, P57, **P84**		
Westmoquette, M.S.	227, **P85**		
White, S.D.M.	P6		
Wilcots, E.M.	P50		
Willmer, C.	P52		
Willner, S.	P8		
Witt, A.N.	P82		
Woodward, C.E.	P8		
Yan, L.	P81		
Zackrisson, E.	103, **P86**		
Zheng, X.	273		

Object Index

ω Cen	157, 201, P17	AzV 104	P17
		AzV 170	P17
2MASX J08380769 +6508579	P73	AzV 210	P17
		AzV 215	P17
30 Doradus	3, 45, 49, 57, 71, 115, P9, P10, P26	AzV 216	P17
		AzV 220	P17, P21, P70
3C277.3	257	AzV 327	P17
3C293	257, P76	AzV 362	P17
3C305	257	AzV 423	P17
3C459	257		
		BMB B-28	P28
Abell 262	P49	BX 418	21
Abell 347	P49	BX 531	303
Abell 370	T1	BX 532	303
Abell 400	P49	BX 691	303
Abell 426	P49		
Abell 569	P49	Cam 0840+1044	P77
Abell 634	P66	Cam 0840+1201	P32
Abell 779	P49	Cas A	127
Abell 1367	P49	cB58	see MS 1512-cB58
Abell 1656	P49	Centaurus A	see NGC 5128
Abell 1689	P24	Cetus dSph	177
Abell 1835-IR 1916	T1	Circinus	P57
Abell 2218	T1	CG J1720−67.8	P78
Abell 2390	P24	CGCG 127.056	P49
AC 114	P24, P39	CGCG 127.071	P49
AC 114-A2	P39	CGCG 287.053	P66
AC 114-S2	P39	CL0023	P24
AM 1238-362	P78	CL0024+17	P24
Antennae System	21, 35, 45, 49, 57, 109, 127, 143, 209, P3, P12, P18, P33, P45, P55	CL0152−1357	P24
		CL1054−1245	P16
		CL1604+4304	P24
		CL1604+4321	P24
Antennae-WS80	109	CL2244−02	P39
Apples 1 dSph	177	Cloverleaf	P84
Arches	P13	Coma	P25, P37
Arcturus Stream	201	CTS1008	P32
Arp 94	see NGC 3226/7		
Arp 220	109, 115, P57	DDO 154	P36
Arp 284	163	DDO 165	P83
Arp 299	81, 127	DSF 2237+116 C2	P27
Arp 299-A/B1	127		
AzV 15	P17	ELAIS N1-01	P68
AzV 18	P17	ELAIS N1-02	P68
AzV 22	P17	ELAIS N1-04	P68
AzV 49	P17	ELAIS N1-07	P68
AzV 69	P17	ELAIS N1-09	P68
AzV 75	P17	ELAIS N1-10	P68
AzV 80	P17	ELAIS N1-12	P68
AzV 95	P17	ELAIS N1-13	P68

ELAIS N1-15	P68	HD 91969	P17
ELAIS N1-16	P68	HD 96248	P17
ELAIS N1-24	P68	HD 111973	P17
ELAIS N1-29	P68	HD 148688	P17
ELAIS N1-45	P68	HDF-N	17, 89, 289, T1, P7, P29, P79
ELAIS N1-68	P68		
ELAIS N1-101	P68	HDF-S	P7, P27
ELAIS South 1	269	HDFS 85	P27
ESO 185-G13	P86	HDFS 1825	P27
ESO 194-G13	P78	He 2-10	45, 89, 127, P21, P86
ESO 311-G012	P48	HERC1-13088	17
ESO 338-G04	P86	HERC1-14739	17
ESO 338-IG04B	P73	HV 888	P28
ESO 381-G006	P78	HV 957	P28
ESO 381-G009	P78	HV 11417	P28
ESO 400-G43	P70, P86	HV 2084	P28
ESO 421-G02	P86		
ESO 462-G20	P86	IC 10	11, 21, P12, P87
ESO 480-G12	P86	IC 342	109, P4
		IC 1613	167, P87
FLY99-824	P27	IC 2560	P20
FLY99-825	P27	IC 4710	P5
FLY99-957	P27	IC 4767	P48
Fornax cluster	P18	II Zw 40	P32, P77
Fornax dSph	157	IRAS 05189−2524	P21
		IRAS 08339+6517	P73
G1	see M31-G1	IRAS 16007+3743	173, P19
GOODS(-N/S)	293, 307, 323, P7, P29, P42, P62, P63, P65, P74	IRAS 20551−4250	P21
		IRAS 23128−5919	P21
		IRASF 10214	P84
GRB 000926	293	I Zw 18	247, P2, P35, P73, P58
GRB 010222	115	I Zw 36	247
GRB 011211	293		
GRB 021004	293	J0053+1234	P29
GRB 030115	P75	J0738+0507b	P81
GRB 030323	115	J1023+1952	P51
GRB 030329	115	J114816.64+525150.3	see SDSS J1148+5251
GRB 970508	P55	J1720−67.8	see CG J1720-67.8
GRB 981226	P11		
		LMC	11, 21, 49, 57, 157, 233, 247, P9, P10, P18, P26, P53
H1-13088	P26		
H1-13385	P26		
Haro 3	127	LCBG 313088	177
Haro 11	P86	Leo I	167
KCS 1166	P34	Lockman 850.2	121
KCS 1173	P34	Lockman 850.5	121
HCM 6A	T1	Lockman Hole	121
HD 14818	P17	Lynx2-1635	17
HD 58350	P17		

Object Index 351

M31	21, 157, 177	Mrk 309	21
M31-G1	157	Mrk 332	P70
M33	21, 49, 57, 89, 127, 157, 247, P8, P10, P30, P35, P44	Mrk 334	P71
		Mrk 493	P71
		Mrk 538	P59
M51	31, 97, 157, 251, P5, P60	Mrk 573	P71
		Mrk 710	P77
M81	11, 157, 277, P5	Mrk 900	P86
M82	11, 31, 35, 45, 57, 71, 75, 81, 89, 109, 127, 157, 163, 277, 233, 333, P13, P55, P56, P85	Mrk 1318	P77
		MS 1512-cB58	89, 115, 311, P17
		NGC 87	P78
		NGC 88	P78
M82-A/A1	277	NGC 89	P78
M82-B	157	NGC 92	P78
M82-C	277	NGC 205	177, P52
M82-D	277	NGC 253	81, 109, 127, 233, P33, P56, P57
M82-E	277		
M82-F	57, 71, 75, 277, P13	NGC 278	181
M82-L	277	NGC 303	P71
M82 MGG-9	57, 75, P13	NGC 330	P17
M82 MGG-11	57, 75, P13	–A01,A02,B22,B37	P17
M83	see NGC 5236	NGC 346 – 12,113, 324,355,368,487	P17
M83-74	21		
M100	P1	NGC 404	263
MBR H013	P28	NGC 588	P30
MBR H037	P28	NGC 604	49, 57, 247, P30, P35, P44
MBR H044	P28		
MBR H071	P28	NGC 625	P9, P36
MBR H091	P28	NGC 841	263
MBR P106	P28	NGC 891	163
MBR P123	P28	NGC 1019	P20
MD 69	303	NGC 1275	11
MD 94	303	NGC 1313	21
MD 174	303	NGC 1365	P57
Mice, The	143, 157, 201, P51	NGC 1672	P71
Milky Way	3, 11, 21, 57, 81, 115, 143, 157, 167, 201, 209, 233, 251, 323, P3, P13, P17, P18,P25, P52, P80, P82, P86	NGC 1377	223
		NGC 1399	P18, P41
		NGC 1530	35
		NGC 1560-30	P4
		NGC 1569	21, 41, 209, P4, P13, P21, P50, P67, P83
Mrk 42	P71	NGC 1569-A/A1/A2	41, 57
Mrk 36	P77	NGC 1569-B	41
Mrk 54	89	NGC 1614	57
Mrk 59	247	NGC 1705	11, 41, 89, 323, P2, P13, P17, P21, P36, P67, P83
Mrk 231	P21, P76		
Mrk 273	P21, P76		
Mrk 297	P55	NGC 1705-I/1	41, 57

NGC 1705-2	41	NGC 5236-502	219
NGC 1741	323	NGC 5236-805	219
NGC 1808	109, P33, P57	NGC 5253	35, 45, 57, 89, 127, 215,P21
NGC 1960	57		
NGC 2070	49	NGC 5253-C1	35
NGC 2194	57	NGC 5253-C2	35
NGC 2403	163	NGC 5256	P71
NGC 2681	263	NGC 5674	P71
NGC 2903	P60	NGC 5678	263
NGC 3077	215, P21	NGC 5866	P5, P69
NGC 3125	21	NGC 5940	P71
NGC 3125-1	21	NGC 6090	323
NGC 3256	35, 57	NGC 6240	11, P21
NGC 3226/7	P51	NGC 6300	P71
NGC 3227	P71	NGC 6503	263
NGC 3310	11, 89, P60	NGC 6722	P48
NGC 3351	P60	NGC 6814	P71
NGC 3393	P71	NGC 6946	57, 81, 127
NGC 3504	P60	NGC 6946-1447	219
NGC 3507	263	NGC 6951	P71
NGC 3603	P13	NGC 7023	31
NGC 3627	263	NGC 7130	P71
NGC 3628	P21	NGC 7252	143, 209, P18, P33, P83
NGC 3705	263	NGC 7252-W3	P18
NGC 3921	143	NGC 7331	31, P5
NGC 3982	P71	NGC 7469	35, P71
NGC 4038/39	see Antennae	NGC 7469-C1	35
NGC 4150	263	NGC 7479	P71
NGC 4214	89, 323	NGC 7496	P71
NGC 4214-10	219	NGC 7552	109
NGC 4214-13	219	NGC 7673	11, 177, P10, P52
NGC 4253	P71	NGC 7674	P71
NGC 4303	263	NGC 7677	P10
NGC 4321	see M100	NGC 7714	P26, P60
NGC 4449	219, P12, P21		
NGC 4449-27	219	Ophiuchus	57, 65
NGC 4449-47	219	Orion Region	11, 65, 115
NGC 4569	263	OSK SP 29-13	P28
NGC 4579	P5	OSK SP 29-14	P28
NGC 4676	see Mice		
NGC 4826	263	Perseus	57
NGC 4945	P57	– h and χ Persei	57
NGC 5033	P5	Pox 186	P58
NGC 5055	263	PKS 1345+12	257
NGC 5128	57, 153, P57	PKS 1549−79	257
NGC 5135	P71	PKS 1932−46	257
NGC 5194	157		
NGC 5236	11, 81, 89, 215, P57, P72	Q0000-263 D6	P27
		Q1307-BM1163	299, 311

Object Index

Quintuplet	P13	SN 1982aa	P55
		SN 1986J	P54
R136	45, 49, 57, 71, P13	SN 1987A	P54
		SN 1988Z	P54
SA57-1501	17	SN 1993J	P54
SA57-5482	17, P26	SN 1994I	P54
SA57-7042	17, P26	SN 1994W	P54
SA57-10601	17, P26	SN 1995N	P54
SA68-1067	17	SN 1998bw	P54, P55
SA68-3307	17	SN 1998s	P54
SA68-6134	P26	SN 1999em	P54
SA68-8846	17	SN 1999gi	P54
SA68-17169	17	SN 2002ap	P54
SA68-17418	17	SN 2003bg	P54
SBM03-1	P74	SN 2003dh	115
SBS 0335−052	247, T1, P58	SN 2004am	81
SBS 0807+580B	P66	SNR 41.95+575	81
SBS 0808+581A	P66	SNR 43.31+592	81
SBS 0808+581B	P66	Subaru Deep Field	T1
SBS 0809+582	P66	SXDF	121
SBS 0810+581	P66		
SCG 0018−4854	P78	Tadpole, The	157, 209
Sco-Cen OB assoc.	57	Taurus	57
Sculptor group	P36	Tol 0226−390	P32
SDSS J0124+0050	P59	Tol 0341−407	P86
SDSS J0834+0139	P59	Tol 0538−416	P32
SDSS J1148+5251	327	Tol 1214−277	P58
Sk 191	P17	Tol 1223−359	P77
SL 0119+1452	P59	Tol 1238−36.4	P78
SL 0218+0757	P59	Tol 1345−420	P77
SL 0222-0830	P59	Tol 1924−416	P32, P73, P70
SL 0745+2826	P59	Tol 2019−405	P77
SL 0834+0139	P59	Tol 65	P58
SL 0904+5136	P59	Tucana dSph	177
SL 0934+0014	P59	TXS 2226−184	P76
SL 0936+0106	P59		
SL 0943−0215	P59	UCM 0014+1829	17, P3
SL 1231+0435	P59	UCM 0019+2201	17
SL 1234+0319	P59	UCM 0040+0220	17
SL 1241−0007	P59	UCM 0135+2242	17
SL 1354+0205	P59	UCM 0148+2123	17
SL 1402+0955	P59	UCM 0159+2354	17
SL 1507+5511	P59	UCM 1253+2756	17
SMC	21, 57, 89, 157, 167, 247, P17, P53, P82	UCM 1302+2853	17
		UCM 1324+2926	17
SMMJ 14011	P84	UCM 1656+2744	17
SN 1978K	P54	UCM 2304+1640	17
SN 1979C	P54, P55	UCM 2351+2321	17
SN 1980K	P54		

353

UDF	289, T1, P7, P16, P42, P52, P74, P79
UDF 0206	P52
UDF 0900	P52
UDF 0901	P52
UDF 4142	P52
UDF 4445	P52
UDF 7559	P52
UDSR 23	P40
UGC 5296	P23
UGC 8387	57
UGC 10214	see Tadpole
UM422	P47
UM423	P47
UM439	P47
UM446	P47
UM452	P47
UM456	P47
UM461	P47, P77
UM462	P47
UM463	P47
UM465	P47
UM477	P47, P77
UM483	P47, P77
UM491	P47
UM499	P47
UM500	P47
UM501	P47
UM504	P47
UM523	P47
UM533	P47
UM538	P47
UM559	P47
Virgo	187, P37
W49A	127
W51	57
Westerlund 1	21, P13
WR6	31

Astrophysics and Space Science Library

Volume 324: *Kristian Birkeland – The First Space Scientist,* by A. Egeland, W.J. Burke. Hardbound ISBN 1-4020-3293-5, April 2005

Volume 323: *Recollections of Tucson Operations,* by M.A. Gordon. Hardbound ISBN 1-4020-3235-8, December 2004

Volume 322: *Light Pollution Handbook,* by K. Narisada, D. Schreuder Hardbound ISBN 1-4020-2665-X, November 2004

Volume 321: *Nonequilibrium Phenomena in Plasmas,* edited by A.S. Shrama, P.K. Kaw. Hardbound ISBN 1-4020-3108-4, December 2004

Volume 320: *Solar Magnetic Phenomena,* edited by A. Hanslmeier, A. Veronig, M. Messerotti. Hardbound ISBN 1-4020-2961-6, December 2004

Volume 319: *Penetrating Bars through Masks of Cosmic Dust,* edited by D.L. Block, I. Puerari, K.C. Freeman, R. Groess, E.K. Block. Hardbound ISBN 1-4020-2861-X, December 2004

Volume 318: *Transfer of Polarized light in Planetary Atmospheres,* by J.W. Hovenier, J.W. Domke, C. van der Mee. Hardbound ISBN 1-4020-2855-5. Softcover ISBN 1-4020-2889-X, November 2004

Volume 317: *The Sun and the Heliosphere as an Integrated System,* edited by G. Poletto, S.T. Suess. Hardbound ISBN 1-4020-2830-X, November 2004

Volume 316: *Civic Astronomy - Albany's Dudley Observatory, 1852-2002,* by G. Wise
Hardbound ISBN 1-4020-2677-3, October 2004

Volume 315: *How does the Galaxy Work - A Galactic Tertulia with Don Cox and Ron Reynolds,* edited by E. J. Alfaro, E. Pérez, J. Franco Hardbound ISBN 1-4020-2619-6, September 2004

Volume 314: *Solar and Space Weather Radiophysics - Current Status and Future Developments,* edited by D.E. Gary and C.U. Keller
Hardbound ISBN 1-4020-2813-X, August 2004

Volume 313: *Adventures in Order and Chaos,* by G. Contopoulos. Hardbound ISBN 1-4020-3039-8, January 2005

Volume 312: *High-Velocity Clouds*, edited by H. van Woerden, U. Schwarz, B. Wakker
Hardbound ISBN 1-4020-2813-X, September 2004

Volume 311: *The New ROSETTA Targets- Observations, Simulations and Instrument Performances*, edited by L. Colangeli, E. Mazzotta Epifani, P. Palumbo
Hardbound ISBN 1-4020-2572-6, September 2004

Volume 310: *Organizations and Strategies in Astronomy 5*, edited by A. Heck
Hardbound ISBN 1-4020-2570-X, September 2004

Volume 309: *Soft X-ray Emission from Clusters of Galaxies and Related Phenomena*, edited by R. Lieu and J. Mittaz
Hardbound ISBN 1-4020-2563-7, September 2004

Volume 308: *Supermassive Black Holes in the Distant Universe,* edited by A.J. Barger
Hardbound ISBN 1-4020-2470-3, August 2004

Volume 307: *Polarization in Spectral Lines*, by E. Landi Degl'Innocenti and M. Landolfi
Hardbound ISBN 1-4020-2414-2, August 2004

Volume 306: *Polytropes – Applications in Astrophysics and Related Fields,* by G.P. Horedt
Hardbound ISBN 1-4020-2350-2, September 2004

Volume 305: *Astrobiology: Future Perspectives,* edited by P. Ehrenfreund, W.M. Irvine, T. Owen, L. Becker, J. Blank, J.R. Brucato, L. Colangeli, S. Derenne, A. Dutrey, D. Despois, A. Lazcano, F. Robert
Hardbound ISBN 1-4020-2304-9, July 2004
Paperback ISBN 1-4020-2587-4, July 2004

Volume 304: *Cosmic Gammy-ray Sources,* edited by K.S. Cheng and G.E. Romero
Hardbound ISBN 1-4020-2255-7, September 2004

Volume 303: *Cosmic rays in the Earth's Atmosphere and Underground*, by L.I, Dorman
Hardbound ISBN 1-4020-2071-6, August 2004

Volume 302: *Stellar Collapse,* edited by Chris L. Fryer
Hardbound, ISBN 1-4020-1992-0, April 2004

Volume 301: *Multiwavelength Cosmology*, edited by Manolis Plionis
Hardbound, ISBN 1-4020-1971-8, March 2004

Volume 300: *Scientific Detectors for Astronomy*, edited by Paola Amico, James W. Beletic, Jenna E. Beletic
Hardbound, ISBN 1-4020-1788-X, February 2004

Volume 299: *Open Issues in Local Star Fomation,* edited by Jacques Lépine, Jane Gregorio-Hetem
Hardbound, ISBN 1-4020-1755-3, December 2003

Volume 298: *Stellar Astrophysics - A Tribute to Helmut A. Abt,* edited by K.S. Cheng, Kam Ching Leung, T.P. Li
Hardbound, ISBN 1-4020-1683-2, November 2003

Volume 297: *Radiation Hazard in Space,* by Leonty I. Miroshnichenko
Hardbound, ISBN 1-4020-1538-0, September 2003

Volume 296: *Organizations and Strategies in Astronomy, volume 4,* edited by André Heck
Hardbound, ISBN 1-4020-1526-7, October 2003

Volume 295: *Integrable Problems of Celestial Mechanics in Spaces of Constant Curvature*, by T.G. Vozmischeva
Hardbound, ISBN 1-4020-1521-6, October 2003

Volume 294: *An Introduction to Plasma Astrophysics and Magnetohydrodynamics,* by Marcel Goossens
Hardbound, ISBN 1-4020-1429-5, August 2003
Paperback, ISBN 1-4020-1433-3, August 2003

Volume 293: *Physics of the Solar System,* by Bruno Bertotti, Paolo Farinella, David Vokrouhlický
Hardbound, ISBN 1-4020-1428-7, August 2003
Paperback, ISBN 1-4020-1509-7, August 2003

Volume 292: *Whatever Shines Should Be Observed,* by Susan M.P. McKenna-Lawlor
Hardbound, ISBN 1-4020-1424-4, September 2003

Volume 291: *Dynamical Systems and Cosmology*, by Alan Coley
Hardbound, ISBN 1-4020-1403-1, November 2003

Volume 290: *Astronomy Communication*, edited by André Heck, Claus Madsen
Hardbound, ISBN 1-4020-1345-0, July 2003

Volume 287/8/9: *The Future of Small Telescopes in the New Millennium*, edited by Terry D. Oswalt
Hardbound Set only of 3 volumes, ISBN 1-4020-0951-8, July 2003

Volume 286: *Searching the Heavens and the Earth: The History of Jesuit Observatories*, by Agustín Udías
Hardbound, ISBN 1-4020-1189-X, October 2003

Volume 285: *Information Handling in Astronomy - Historical Vistas*, edited by André Heck
Hardbound, ISBN 1-4020-1178-4, March 2003

Volume 284: *Light Pollution: The Global View*, edited by Hugo E. Schwarz
Hardbound, ISBN 1-4020-1174-1, April 2003

Volume 283: *Mass-Losing Pulsating Stars and Their Circumstellar Matter*, edited by Y. Nakada, M. Honma, M. Seki
Hardbound, ISBN 1-4020-1162-8, March 2003

Volume 282: *Radio Recombination Lines*, by M.A. Gordon, R.L. Sorochenko
Hardbound, ISBN 1-4020-1016-8, November 2002

Volume 281: *The IGM/Galaxy Connection*, edited by Jessica L. Rosenberg, Mary E. Putman
Hardbound, ISBN 1-4020-1289-6, April 2003

Volume 280: *Organizations and Strategies in Astronomy III*, edited by André Heck
Hardbound, ISBN 1-4020-0812-0, September 2002

Volume 279: *Plasma Astrophysics, Second Edition*, by Arnold O. Benz
Hardbound, ISBN 1-4020-0695-0, July 2002

Volume 278: *Exploring the Secrets of the Aurora*, by Syun-Ichi Akasofu
Hardbound, ISBN 1-4020-0685-3, August 2002

Volume 277: *The Sun and Space Weather*, by Arnold Hanslmeier
Hardbound, ISBN 1-4020-0684-5, July 2002

Volume 276: *Modern Theoretical and Observational Cosmology*, edited by Manolis Plionis, Spiros Cotsakis
Hardbound, ISBN 1-4020-0808-2, September 2002

Volume 275: *History of Oriental Astronomy*, edited by S.M. Razaullah Ansari
Hardbound, ISBN 1-4020-0657-8, December 2002

Volume 274: *New Quests in Stellar Astrophysics: The Link Between Stars and Cosmology*, edited by Miguel Chávez, Alessandro Bressan, Alberto Buzzoni, Divakara Mayya
Hardbound, ISBN 1-4020-0644-6, June 2002

Volume 273: *Lunar Gravimetry*, by Rune Floberghagen
Hardbound, ISBN 1-4020-0544-X, May 2002

Volume 272: *Merging Processes in Galaxy Clusters*, edited by L. Feretti, I.M. Gioia, G. Giovannini
Hardbound, ISBN 1-4020-0531-8, May 2002

Volume 271: *Astronomy-inspired Atomic and Molecular Physics*, by A.R.P. Rau
Hardbound, ISBN 1-4020-0467-2, March 2002

Volume 270: *Dayside and Polar Cap Aurora*, by Per Even Sandholt, Herbert C. Carlson, Alv Egeland
Hardbound, ISBN 1-4020-0447-8, July 2002

Volume 269: *Mechanics of Turbulence of Multicomponent Gases*, by Mikhail Ya. Marov, Aleksander V. Kolesnichenko
Hardbound, ISBN 1-4020-0103-7, December 2001

Volume 268: *Multielement System Design in Astronomy and Radio Science*, by Lazarus E. Kopilovich, Leonid G. Sodin
Hardbound, ISBN 1-4020-0069-3, November 2001

Volume 267: *The Nature of Unidentified Galactic High-Energy Gamma-Ray Sources,* edited by Alberto Carramiñana, Olaf Reimer, David J. Thompson
Hardbound, ISBN 1-4020-0010-3, October 2001

Volume 266: *Organizations and Strategies in Astronomy II*, edited by André Heck
Hardbound, ISBN 0-7923-7172-0, October 2001

Volume 265: *Post-AGB Objects as a Phase of Stellar Evolution*, edited by R. Szczerba, S.K. Górny
Hardbound, ISBN 0-7923-7145-3, July 2001

Volume 264: *The Influence of Binaries on Stellar Population Studies*, edited by Dany Vanbeveren
Hardbound, ISBN 0-7923-7104-6, July 2001

Volume 262: *Whistler Phenomena - Short Impulse Propagation*, by Csaba Ferencz, Orsolya E. Ferencz, Dániel Hamar, János Lichtenberger
Hardbound, ISBN 0-7923-6995-5, June 2001

Volume 261: *Collisional Processes in the Solar System*, edited by Mikhail Ya. Marov, Hans Rickman
Hardbound, ISBN 0-7923-6946-7, May 2001

Volume 260: *Solar Cosmic Rays*, by Leonty I. Miroshnichenko
Hardbound, ISBN 0-7923-6928-9, May 2001

Volume 259: *The Dynamic Sun*, edited by Arnold Hanslmeier, Mauro Messerotti, Astrid Veronig
Hardbound, ISBN 0-7923-6915-7, May 2001

Volume 258: *Electrohydrodynamics in Dusty and Dirty Plasmas-Gravito-Electrodynamics and EHD*, by Hiroshi Kikuchi
Hardbound, ISBN 0-7923-6822-3, June 2001

Volume 257: *Stellar Pulsation - Nonlinear Studies*, edited by Mine Takeuti, Dimitar D. Sasselov
Hardbound, ISBN 0-7923-6818-5, March 2001

Volume 256: *Organizations and Strategies in Astronomy*, edited by André Heck
Hardbound, ISBN 0-7923-6671-9, November 2000

Volume 255: *The Evolution of the Milky Way- Stars versus Clusters*, edited by Francesca Matteucci, Franco Giovannelli
Hardbound, ISBN 0-7923-6679-4, January 2001

Volume 254: *Stellar Astrophysics*, edited by K.S. Cheng, Hoi Fung Chau, Kwing Lam Chan, Kam Ching Leung
Hardbound, ISBN 0-7923-6659-X, November 2000

Volume 253: *The Chemical Evolution of the Galaxy*, by Francesca Matteucci
Paperback, ISBN 1-4020-1652-2, October 2003
Hardbound, ISBN 0-7923-6552-6, June 2001

Volume 252: *Optical Detectors for Astronomy II*, edited by Paola Amico, James W. Beletic
Hardbound, ISBN 0-7923-6536-4, December 2000

Volume 251: *Cosmic Plasma Physics*, by Boris V. Somov
Hardbound, ISBN 0-7923-6512-7, September 2000

Volume 250: *Information Handling in Astronomy*, edited by André Heck
Hardbound, ISBN 0-7923-6494-5, October 2000

Volume 249: *The Neutral Upper Atmosphere*, by S.N. Ghosh
Hardbound, ISBN 0-7923-6434-1, July 2002

Volume 247: *Large Scale Structure Formation*, edited by Reza Mansouri, Robert Brandenberger
Hardbound, ISBN 0-7923-6411-2, August 2000

Volume 246: *The Legacy of J.C. Kapteyn*, edited by Piet C. van der Kruit, Klaas van Berkel
Paperback, ISBN 1-4020-0374-9, November 2001
Hardbound, ISBN 0-7923-6393-0, August 2000

Volume 245: *Waves in Dusty Space Plasmas*, by Frank Verheest
Paperback, ISBN 1-4020-0373-0, November 2001
Hardbound, ISBN 0-7923-6232-2, April 2000

Volume 244: *The Universe*, edited by Naresh Dadhich, Ajit Kembhavi
Hardbound, ISBN 0-7923-6210-1, August 2000

Volume 243: *Solar Polarization*, edited by K.N. Nagendra, Jan Olof Stenflo
Hardbound, ISBN 0-7923-5814-7, July 1999

Volume 242: *Cosmic Perspectives in Space Physics*, by Sukumar Biswas
Hardbound, ISBN 0-7923-5813-9, June 2000

Volume 241: *Millimeter-Wave Astronomy: Molecular Chemistry & Physics in Space,* edited by W.F. Wall, Alberto Carramiñana, Luis Carrasco, P.F. Goldsmith
Hardbound, ISBN 0-7923-5581-4, May 1999

Volume 240: *Numerical Astrophysics,* edited by Shoken M. Miyama, Kohji Tomisaka,Tomoyuki Hanawa
Hardbound, ISBN 0-7923-5566-0, March 1999

Volume 239: *Motions in the Solar Atmosphere,* edited by Arnold Hanslmeier, Mauro Messerotti
Hardbound, ISBN 0-7923-5507-5, February 1999

Volume 238: *Substorms-4,* edited by S. Kokubun, Y. Kamide
Hardbound, ISBN 0-7923-5465-6, March 1999

For further information about this book series we refer you to the following web site: www.springeronline.com

To contact the Publishing Editor for new book proposals:
Dr. Harry (J.J.) Blom: harry.blom@springer-sbm.com